ADVANCES IN PLASMA PHYSICS

Volume 4

AN INTERSCIENCE SERIES

ADVANCES IN PLASMA PHYSICS

Advances in

PLASMA PHYSICS

Volume 4

Edited by

ALBERT SIMON

Department of Mechanical and Aerospace Sciences
and Department of Physics
University of Rochester
Rochester, New York

and

WILLIAM B. THOMPSON

Department of Physics
University of California at San Diego
La Jolla, California

INTERSCIENCE PUBLISHERS

A Division of John Wiley and Sons, Inc., New York · London · Sydney
Toronto

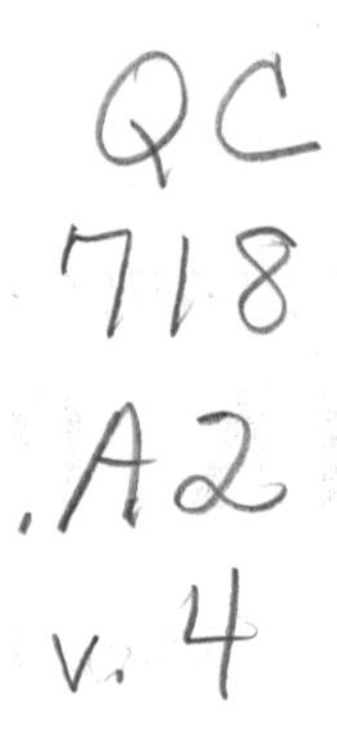

Library of Congress Catalogue Card Number: 67–29541

ISBN 0–471–79204–7

Printed in the United States of America.

10 9 8 7 6 5 4 3 2 1

To the memory of Harriet

Preface

In this volume, as in earlier ones, we have attempted to include a series of articles exhibiting something of the range of the subject. The role of plasma physics in interpreting natural phenomena is illustrated by a chapter on plasma resonances in the ionosphere; several chapters are concerned with the somewhat speculative area of fusion research, and one includes a discussion of as practical an object as the electrostatic precipitator. Two of our chapters, those of Briggs on the linear theory of the two-stream instability and of Bernstein on the adiabatic motion of charged particles, can perhaps be classed as last words on the subject, while that of Benford and Book on relativistic beam equilibria is a preliminary study. Of the two chapters stimulated directly by fusion research, that of Coppi is a thorough review of recent developments in the theory of geometry-dependent collective modes in the plasma, particularly in the study of the stability of multipole configurations; the other, by Spalding, examines the present state of research on a particular class of systems and considers its promise as a source of a future reactor. This is the chapter that mentions a 10-km-long 10,000-MW power plant. Perhaps the most unusual work is that of Sodha and Guha on the plasma produced by a colloidal suspension of metal in a hot gas, in which the degree of ionization and the electrical conductivity of such a system is discussed and its relevance to such practical matters as rocket exhausts indicated.

We naturally hope, and confidently expect, that this volume will appeal to a wide range of plasma physicists—theoretical, experimental, pure, and applied—and that its perusal will stimulate new manuscripts for later volumes.

ALBERT SIMON
WILLIAM B. THOMPSON

December 1970

Contents

ADVANCES IN PLASMA PHYSICS

Volume 4

AN INTERSCIENCE SERIES

The Interpretation of Plasma Resonances Observed by Ionospheric Topside Sounders

J. P. DOUGHERTY AND S. R. WATSON

Department of Applied Mathematics and Theoretical Physics, Cambridge University, England

Introduction

When an antenna is carried on a satellite through the earth's ionosphere, and is used both to transmit signals and receive their reflections, a new resonance effect is observed. The purpose of this article is to interpret these plasma resonances, but we would like to begin by sketching the historical background.

The most important parameter in a description of the ionosphere is the electron density, N, and much effort has been devoted to its determination. The results are usually presented in the form of the function $N(h)$, h being the height above the earth's surface; N is, of course, a function of geographical location and also time. Figure 1 is a typical example of an $N(h)$ curve. It is

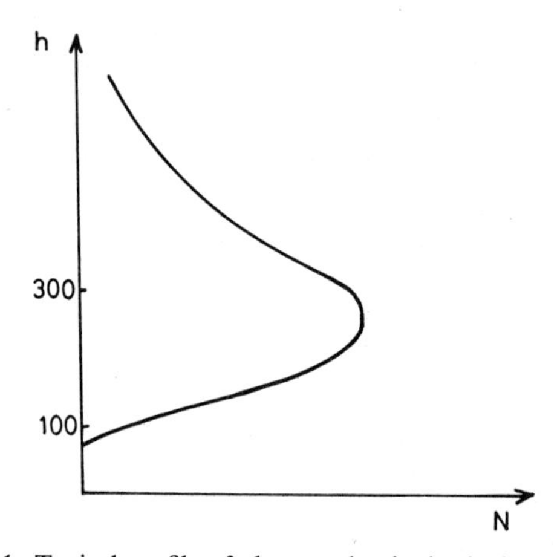

Fig. 1. Typical profile of electron density in the ionosphere.

usual to take h as the vertical ordinate for pictorial reasons. The maximum value of N is around 10^6 cm^{-3} by day but much smaller at night, and occurs at a height of about 300 km; this is known as the peak of the F region. A subsidiary maximum may occur at about 100 km, known as the E region.

The standard method for determining $N(h)$ is that of the reflection of radio waves, known as ionospheric sounding. It has been used for decades and was, in its simplest form, the means by which the ionosphere was originally discovered. According to magnetoionic theory, an electromagnetic wave is reflected by a plasma when the electron density reaches a sufficient value which depends on the frequency and on the strength of any steady magnetic field. The frequency of the wave that can just pass through the peak of an ionospheric layer is known as the penetration frequency. A wave incident on the ionosphere from below at a frequency lower than the penetration frequency is reflected at some height beneath the peak and suffers a group delay that depends on the profile $N(h)$ up to the reflection level. By measuring the group delay as a function of frequency, the profile itself can be determined. A further complication arises because a plasma in a magnetic field is birefringent, so that the above description applies separately to each polarization, the penetration frequency and group delay being different for the two characteristic waves.

Even when ionospheric sounding had been brought to a high degree of refinement, it had the inherent deficiency that it provided $N(h)$ only up to the peak of the layer. Anyone wishing to study the details of the formation and morphology was thus presented with a vast supply of data for the "lower side" of the ionosphere and virtually nothing for the "topside," though in fact there is no abrupt change in the physical properties at the peak and the behavior of the two regions needs to be considered together. There is, for example, diffusion of the electrons from one to the other, and the height of the peak itself changes with time.

In about 1958, two new methods for obtaining data for the topside began to look feasible. The recent launching of satellites led to the obvious suggestion that an ionospheric sounder could be carried in orbit at a suitable distance above the peak and sound the ionosphere from above. This idea became a reality with the launching on September 29, 1962 of the first topside sounder satellite, Alouette I. (Some short-lived rocket experiments in 1961 had helped to prepare the way for this. This was followed by Explorer XX (launched August 25, 1964), Alouette II (November 29, 1965) and ISIS-1 (January 30, 1969). Further topside sounders are also projected. A review of the data obtained has been given by Chapman and Warren (1).

It was as a by-product of topside-sounder work that the new plasma resonances were observed; to the plasma physicist they could even be said to be of greater interest than the original purpose of the project.

The second method for investigating the topside was that of incoherent scattering which uses very high-powered, ground-based radar. This has also engendered some interesting plasma physics, but it will not be discussed here; a review has been given by Gordon (2).

I. Description of Observational Results

In the standard form of ionospheric sounder a pulse is emitted in which the radio frequency, f, is swept continuously over a wide range (typically from 1 to 10 MHz). This pulse is in due course received after reflection from the ionosphere; the time delay, as a function of frequency, is obtained. Multiplying by $\frac{1}{2}c$, where c is the vacuum speed of light, gives the apparent range; for a ground-based sounder this is usually called the apparent height, h'. The raw data from one pulse thus consists of a graph of the function $h'(f)$. For a topside sounder, h' is the apparent depth and is conventionally plotted as increasing downward. The form that this graph could be expected to take follows from magnetoionic theory in very much the same way as in the ground-based sounding; we refer the reader to one of the standard texts (3–5) for the detailed theoretical development. The outcome of these considerations is shown in Figure 2.

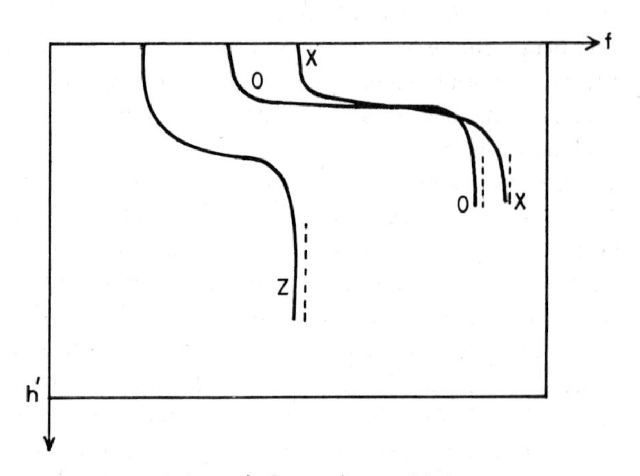

Fig. 2. Schematic form of a topside ionogram.

The curve marked O corresponds to the ordinary wave and those marked X and Z to branches of the extraordinary wave. The frequencies at which these curves meet the horizontal line ($h' = O$) are the "cutoff" frequencies, for the respective modes of propagation, in a plasma whose density has the value occurring at the transmitting antenna. They form the demarcation between propagation and evanescence. These have no analogue in the ground-based case, since there is then no plasma at the transmitter and the $h'(f)$ curve attains a limiting (nonzero) value of h' as f tends to zero. Following along one of these curves, by increasing f, one finds a finite value of h', corresponding to reflection at some level in the ionosphere, until f reaches the cutoff value appropriate to the plasma density at the peak, and thereafter there is no reflection; this is how the vertical asymptotes are formed.

Parenthetically, we may note that in the ground-based case, the wave that becomes the extraordinary mode in the ionosphere must be launched originally as a wave in vacuum, so that it can only be on the branch we have labeled X if linear ray theory is to be believed. This would mean that only the O and X curves should be present in ionograms, as if often the case. In practice, however, the third "Z-trace" does sometimes appear; to explain it, arguments involving coupling between modes have had to be invoked (see, for instance, Ref. 3, p. 424). In the case of topside sounders, the Z mode can be launched by the antenna itself, at suitable frequencies, on account of its being immersed in plasma, and no coupling need be assumed.

In Figures 3 and 4 we show some actual ionograms obtained by Alouette II. Below each is a simplified copy, on the same scale, with superfluous identification lines, and so on, removed. The O, Z and X curves described in Figure 2, though for technical reasons not complete, may be clearly identified. (The doubling of the Z trace to form the additional so-called Z' trace need not be discussed here; an explanation has been given by Calvert (6).)

A more striking feature of Figures 3 and 4, however, is the appearance of the heavy vertical marks centered on several particular frequencies; these are the plasma resonances. What they indicate is that signals are received at those frequencies continuously, rather than at a discrete time delay. Their reception can extend over as much as 10 msec, which is, of course, far longer than the original pulse. The first question arising here was that of the identification of the frequencies at which resonances occur. It was soon agreed that the frequencies were fixed by local properties of the plasma, rather than by properties at some distant point at which reflection might be taking place, for in the latter case there would clearly have to be a delay before the onset of the resonance. It was also agreed that the relevant (angular) frequencies are the local plasma frequency, ω_p; the local gyrofrequency of electrons, Ω, and harmonics thereof; and the hybrid frequency $\omega_m = (\omega_p{}^2 + \Omega^2)^{1/2}$. The second harmonic $2\omega_m$ also occurs commonly but does not fit naturally into the theoretical picture and is thought by some to be an instrumental effect. Lower frequencies, possibly subharmonics of Ω may also be observed and are as yet only tentatively explained.

The physical interpretation of the resonances is as follows. A plasma in a magnetic field is capable of sustaining a great variety of wave motions which have complicated dispersive and anisotropic behavior. Among these waves one can find cases in which the frequency and wave number are real (no damping) and the vector group velocity vanishes. This only happens for certain discrete

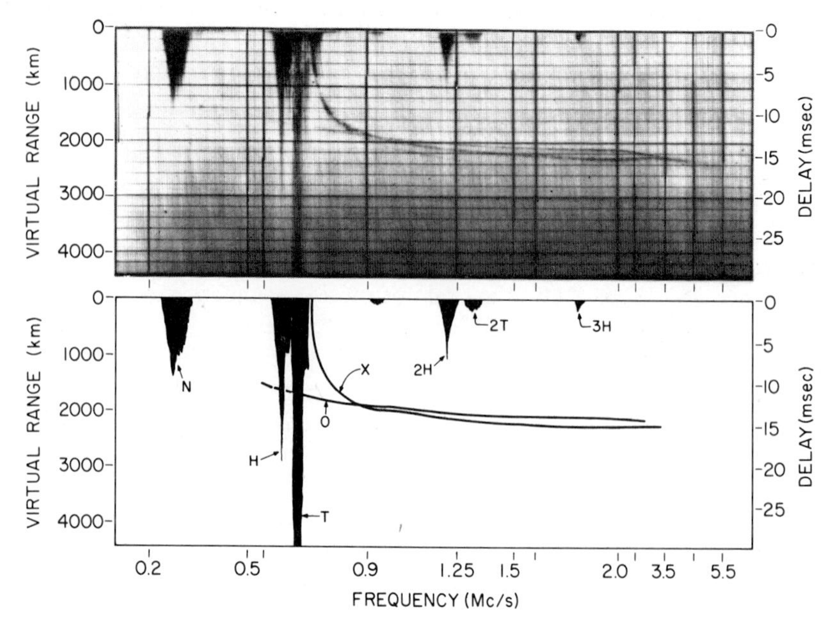

Fig. 3. An actual topside ionogram showing resonances in addition to the trace due to reflection; below is a simplified drawing to show the resonance more clearly. The symbols *N*, *H*, and *T* refer to the frequencies that in the text are denoted by ω_p, Ω, and ω_m, respectively. In this case $\omega_p < \Omega$ at the satallite. (Courtesy W. Calvert, Aeronomy Laboratory, ESSA, Boulder, Colo.).

frequencies. A packet of such waves would have the curious property that it remain at the same place. Now suppose that an antenna, immersed in the medium, excites a broad spectrum of waves, with a frequency band that includes one or more of these special frequencies, and that the same antenna is then used to detect waves in the medium. Most of the waves spread out in space, and will only be received again if reflected back, e.g., by a density gradient, but the wave packets with zero group velocity will continue to be present at the antenna and detected by it. It is these that cause the resonance effect, and their frequencies must correspond with those observed.

The detailed theoretical work thus involves the identification of the resonant frequencies by searching through the solutions of the dispersion relation for waves having zero group velocity; then follows the study of the efficiency with which packets of such waves can be excited, and detected, by an antenna of given form. It turns out that the most important waves for this

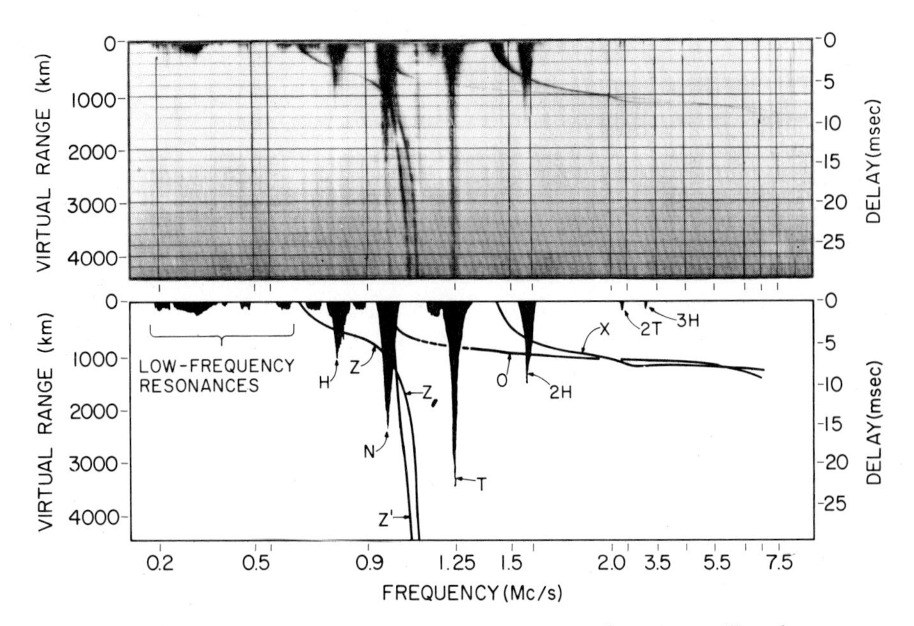

Fig. 4. A topside ionogram in the case in which $\omega_p > \Omega$ at the satellite; the presentation takes the same form as in Figure 3. (Courtesy W. Calvert, Aeronomy Laboratory, ESSA, Boulder, Colo.).

purpose are those for which the thermal distribution of electrons is important, though one can for many purposes use the electrostatic approximation (velocity of light infinite); in other words they are plasma oscillations and the so-called Bernstein modes. Electromagnetic waves described by the Appleton-Hartree equation, for which the thermal distribution may often be neglected, are relatively unimportant, though of course it remains correct to use electromagnetic waves to explain the reflection from the topside, and the $h'(f)$ curves which originally it was the sole aim of the topside sounder to produce. It is possible, however, that electromagnetic corrections to the Bernstein modes may be relevant.

The physical description above clearly needs some modification if the motion of the satellite is to be taken into account. One must search for waves whose group velocity is (vectorially) equal to that of the satellite, instead of zero. However, it emerges that this is a fairly small correction.

Plasma resonance phenomena have also been observed in laboratory experiments, e.g., by Landauer (7), and by Bekefi et al. (8). A review of experiments involving plasma waves and resonances has been given by Crawford (9).

II. Basic Theory

A. *Idealization of the Problem*

Since the resonances are a local effect, it is usual to neglect the spatial variation of plasma parameters in the ionosphere, thereby removing altogether the reflection process underlying the conventional $h'(f)$ curve (see, however, Section IV-F). With less clear-cut justification it is also usual to neglect the possibility that the antenna may seriously disturb the electron density close to it by the formation of a sheath. The situation is thus reduced to that of an antenna immersed in a uniform medium. Any applied magnetic field is also taken to be uniform.

Strictly speaking, one then has a boundary value problem in which the antenna has given geometry and voltages are applied at suitable terminals, but the distribution of current within the body of the antenna is one of the unknown variables. Some problems of this type have indeed been considered but they can only be made tractable by adopting a rather simple representation of the properties of the plasma. The plasma resonances, however, depend essentially on including rather subtle plasma behavior which has to be treated by using the Vlasov equations. Therefore, one has been obliged to adopt the further simplification of guessing the form of the current distribution within the antenna; indeed, as we shall see, it has been taken to be an infinitesimal dipole in many calculations. This procedure appears to be satisfactory if the antenna is short compared with the wavelength of whatever are the relevant waves. The problem is thus reduced to that of a given source radiating in a uniform (although complicated) medium. If we suppose, finally, that the disturbance in the medium is of sufficiently low amplitude that it may be treated in the linear approximation, standard transform techniques will be available. This last approximation must, however, be suspect close to the antenna.

A further simplification, the electrostatic approximation, is considered fully in the next sections.

B. *Sources in a Uniform Linear Medium*

We take a uniform unperturbed plasma, possibly with uniform external magnetic field $\mathbf{B}_0$. Wave fields may be considered in the first instance to be proportional to $e^{(\omega t - \mathbf{k}\cdot\mathbf{x})}$, and later superposed. We may divide the electric current density into an external part $\mathbf{J}$ representing the source, e.g., the current on the antenna, and the internal part $\mathbf{j}$ due to the motion of the charges that constitute the plasma. The fields within the plasma may be described by the oscillating magnetic field $\mathbf{B}$, together with a displacement vector $\mathbf{D} = \mathbf{E} + 4\pi\mathbf{j}/i\omega$. A linear treatment of the dynamics of the plasma must yield a relation of the type $j_p = \sigma_{pq}(\mathbf{k}, \omega)E_q$, where $\boldsymbol{\sigma}$ is a conductivity tensor. (In the cold case $\boldsymbol{\sigma}$ depends only on ω, not on $\mathbf{k}$). We can then introduce a dielectric tensor $\varepsilon_{pq}(\mathbf{k}, \omega)$ where

$$D_p = \varepsilon_{pq} E_q \tag{2-1}$$

and

$$\varepsilon_{pq} = \delta_{pq} + \frac{4\pi}{i\omega} \sigma_{pq} \tag{2-2}$$

Maxwell's equations then read

$$-ic\mathbf{k} \times \mathbf{E} = -i\omega\mathbf{B}, \qquad -ic\mathbf{k} \times \mathbf{B} = 4\pi\mathbf{J} + i\omega\mathbf{D} \tag{2-3}$$

and eliminating $\mathbf{B}$,

$$i\omega(\mathcal{M} - \boldsymbol{\varepsilon})\mathbf{E} = 4\pi\mathbf{J} \tag{2-4}$$

where the matrix

$$\mathcal{M}_{pq} = \frac{c^2}{\omega^2} (k\delta_{pq} - k_p k_q) \tag{2-5}$$

represents Maxwell's equations. Solving (2-4), we find that

$$\mathbf{E} = (4\pi/i\omega)(\mathcal{M} - \boldsymbol{\varepsilon})^{-1}\mathbf{J} \tag{2-6}$$

and $\mathbf{B}$ may be obtained at once from the first of eq. 2-3.

In some circumstances, eq. 2-6 may be replaced by a simpler expression obtained by using electrostatics, instead of the full set of Maxwell equations. This is equivalent to letting $c \to \infty$ in eq. 2-6, thus making the components of $\mathcal{M}$ all tend to infinity; a finite result is obtained as $\mathcal{M}$ is singular. The procedure is simpler to follow if we rcturn to the governing equations

$$-l\mathbf{k} \cdot \mathbf{D} = 4\pi\rho_{\text{ext}}, \qquad \mathbf{E} = -i\mathbf{k}\phi, \qquad \mathbf{B} = 0$$

where ρ_{ext} is the externally applied charge and ϕ is an electrostatic potential. By the equation of continuity $\rho_{\text{ext}} = \mathbf{k} \cdot \mathbf{J}/\omega$, and using eq. 2-1,

$$E_p = \frac{4\pi i k_p k_q}{\omega k^2 \varepsilon_{\text{long}}} J_q \tag{2-7}$$

Here

$$\varepsilon_{\text{long}} = k_p k_q \, \varepsilon_{pq}/k^2 \tag{2-8}$$

is the longitudinal dielectric constant, i.e., that appropriate for longitudinal electric fields ($\mathbf{E}$ parallel to $\mathbf{k}$, $\mathbf{D}$ measured parallel to $\mathbf{k}$). It is eq. 2-7 that is the desired approximation to eq. 2-6. It is difficult to give a precise general statement of the circumstances in which this is a close approximation but, roughly speaking, it is necessary either that the waves transmitting the disturbance should be primarily associated with space-charge effects (such as plasma oscillations), or that the field be observed at a point sufficiently close to the sources that retardation effects be negligible, as in the vacuum case (e.g., near the field of an oscillating dipole).

General fields may be treated by Fourier transforms of the type

$$\mathbf{E}(\mathbf{x}, t) = \int \mathbf{E}(\mathbf{k}, \omega) e^{i(\omega t - \mathbf{k} \cdot \mathbf{x}} \, d^3k \, d\omega \tag{2-9}$$

with similar expressions for **J** and **B**. The Fourier components of **E** and **J** are then related by eqs. 2-6 or 2-7 for each $\mathbf{k}$, ω. For a given source $\mathbf{J}(\mathbf{x}, t)$, one constructs $\mathbf{J}(\mathbf{k}, \omega)$, hence $\mathbf{E}(\mathbf{k}, \omega)$, and finally evaluates eq. 2-9; where necessary a causal limit must be applied, i.e., $\mathbf{E}(\mathbf{k}, \omega)$ is to be the analytic continuation of the function defined with ω in the lower half-plane (disturbance switched on in the past).

While straightforward in principle, this method encounters formidable difficulties in the evaluating eq. 2-9. This is because the expressions for $\boldsymbol{\varepsilon}$ are very complicated if all the relevant physics are to be included; e.g., they may involve integrations over velocity space arising from the solution of the Vlasov equation. The art in this problem is therefore to extract from eq. 2-9 results for identifying the resonances, and their properties, without evaluating the integrals completely.

C. *Particular Sources*

The following are particular model sources that have been considered in connection with antennas immersed in a plasma; we give also their Fourier transforms.

1. Infinitesimal Electric Dipole

For such a dipole, located at the origin and of moment $\mathbf{M}(t)$, we have

$$\mathbf{J}(\mathbf{x}, t) = \dot{\mathbf{M}}(t)\delta^3(\mathbf{x}) \tag{2-10}$$

with Fourier transform

$$\mathbf{J}(\mathbf{k}, \omega) = (2\pi)^{-3} i\omega \mathbf{M}(\omega) \tag{2-11}$$

where $\mathbf{M}(\omega)$ is the temporal Fourier transform of $\mathbf{M}(t)$. For a simple harmonic dipole with frequency ω_0, and amplitude **M**

$$\mathbf{M}(\omega) = \mathbf{M}\delta(\omega - \omega_0) \tag{2-12}$$

We may note that, for excitation of the type used in topside sounders, $\mathbf{M}(\omega)$

represents a fairly broad spectrum of frequencies, but, since $\mathbf{M}(t)$ vanishes for all sufficiently large positive and negative values of t, it follows that $\mathbf{M}(\omega)$ can have no singularities in the finite ω-plane.

2. Finite Current Element

A better representation of a dipole antenna would be to take an infinitesimally thin straight wire, lying (say) along the x-axis in the range $|x| < l$. In this case

$$\mathbf{J}(\mathbf{x}, t) = \dot{\mathbf{M}}(t) f(x)\delta(y)\delta(z) \tag{2-13}$$

where $\mathbf{M}$ is parallel to the x-axis and is again the dipole moment, while $f(x)$ describes the current distribution on the wire and is subject to

$$\int_{-l}^{l} f(x)\, dx = 1$$

A common choice is the triangular distribution

$$f(x) = l^{-1}[1 - |x|/l] \tag{2-14}$$

The Fourier transform eq. 2-11 is then modified by an additional factor:

$$\mathbf{J}(\mathbf{k}, \omega) = (2\pi)^{-3} i\omega\mathbf{M}(\omega) \frac{2(1 - \cos k_x l)}{k_x^2 l^2} \tag{2-15}$$

3. Cylindrical Antenna

This is similar to the preceding model, but the wire is assumed to be a cylinder of radius a with the current flowing on the surface. Then

$$\mathbf{J}(\mathbf{x}, t) = \dot{\mathbf{M}}(t) f(x) \frac{\delta(R - a)}{2\pi R} \tag{2-16}$$

where $R = (y^2 + z^2)^{1/2}$ and $f(x)$ is as before. This has the Fourier transform,

$$\mathbf{J}(\mathbf{k}, \omega) = (2\pi)^{-3} i\omega\mathbf{M}(\omega) \frac{2(1 - \cos k_x l)}{k_x^2 l^2} J_0(k_t a) \tag{2-17}$$

where $k_t = (k_y^2 + k_z^2)^{1/2}$ is the component to $\mathbf{k}$ transverse to the dipole.

D. *Dielectric Tensor for a Plasma*

Plasma physics enters our problem when we consider the form to be taken by the dielectric tensor $\varepsilon_{pq}(\mathbf{k}, \omega)$ in the relation (2-1). We may, however, quote the required results from earlier work; the details of the calculations are available in several texts (10–12).

1. Cold Plasma without Magnetic Field

In this case we have the well-known result

$$\varepsilon_{pq} = \varepsilon_0(\omega)\delta_{pq} \tag{2-18}$$

where $\varepsilon_0(\omega) = 1 - \omega_p{}^2/\omega^2$ and ω_p is the plasma frequency. Clearly, $\varepsilon_{\text{long}} = \varepsilon_0$ also.

2. Cold Plasma with Magnetic Field

Taking the field to be along the third axis, we obtain

$$\boldsymbol{\varepsilon} = \begin{pmatrix} \varepsilon_\perp & \varepsilon_{xy} & 0 \\ -\varepsilon_{xy} & \varepsilon_\perp & 0 \\ 0 & 0 & \varepsilon_0 \end{pmatrix} \tag{2-19}$$

where ε_0 is as before, while

$$\varepsilon_\perp = 1 - \frac{\omega^2}{\omega^2 - \Omega^2} \qquad \varepsilon_{xy} = \frac{i\omega_p{}^2\Omega}{\omega(\omega^2 - \Omega^2)} \tag{2-20}$$

and Ω is the electron gyrofrequency. Characteristic of the cold plasma is the fact that $\boldsymbol{\varepsilon}$ depends on ω but not $\mathbf{k}$. However, when we construct $\varepsilon_{\text{long}}$ as defined by eq. 2-8 the direction of $\mathbf{k}$ is involved; and if we let $\mathbf{k}$ be inclined to $\mathbf{B}_0$ at an angle θ, the result is

$$\varepsilon_{\text{long}} = \varepsilon_0 \cos^2\theta + \varepsilon_\perp \sin^2\theta \tag{2-21}$$

Collisions can if desired be included in eq. 2-19 by introducing a simple frictional term in the equation of motion of electrons, as in magnetoionic theory.

3. Hot Collisionless Plasma without Magnetic Field

The electrons are described by an unperturbed distribution function $f_0(\mathbf{v})$, where $\mathbf{v}$ denotes the thermal velocity. The general formula for $\boldsymbol{\varepsilon}$ is given in Ref. 12 (eq. 8.48 and 8.51), but as the electrostatic approximation appears to be adequate for practical purposes we simply quote the longitudinal component. For this purpose it is convenient to denote by $F(v)$ the one-dimensional distribution of velocities parallel to $\mathbf{k}$ after integrating over the velocity components perpendicular to $\mathbf{k}$. Then

$$\varepsilon_{\text{long}} = 1 + \frac{4\pi e^2}{mk}\int_C \frac{df_0/dv}{\omega - kv}\,dv \tag{2-22}$$

where C denotes the Landau contour. Here e and m are the charge and mass of the electron. When the electrons are in a Maxwellian distribution at temperature T,

$$\varepsilon_{\text{long}} = 1 + \frac{1}{2h^2k^2} G'(z) \tag{2-23}$$

where h is the Debye length, $z = (\omega/k)(m/2KT)^{1/2}$, $K =$ Boltzmann's constant and G is the plasma dispersion function.

4. Hot Collisionless Plasma with Magnetic Field

In the presence of a magnetic field, the distribution function is required to take the form $f_0(v_\parallel, v_\perp)$ where $v_\parallel$ and $v_\perp$ are the components of $\mathbf{v}$ parallel and perpendicular to B_0; we use a similar notation $k_\parallel, k_\perp$ for $\mathbf{k}$. Working again in the electrostatic approximation, we quote

$$\varepsilon_{\text{long}} = 1 + \frac{4\pi e^2}{mk^2} \int_{v_\parallel} \int_{v_\perp} \sum_{n=-\infty}^{\infty} \frac{J_n^2(\zeta)[(\partial f_0/\partial v_\parallel)k_\parallel + (n/\zeta)(\partial f_0/\partial v_\perp)k_\perp]}{\omega - k_\parallel v_\parallel - n\Omega} \times 2\pi v_\perp \, dv_\perp \, dv_\parallel \tag{2-24}$$

where $\zeta = k_\perp v_\perp/\Omega$. The formulas for the nine components of $\boldsymbol{\varepsilon}$ take a similar form; they may be found in Ref. 12 (eq. 9.105), together with the derivation. We have, however, given sufficient detail here to make clear the difficulty of handling eq. 2-6.

For reference a little later, we include here the formulas that apply when the electrons are in a Maxwellian distribution and $\mathbf{k}$ is perpendicular to $\mathbf{B}$, with $\mathbf{k}$ along the first axis and $\mathbf{B}$ along the third. They are

$$\varepsilon_{\text{long}} = 1 - \frac{\omega_p{}^2 e^{-\lambda}}{\Omega\lambda} \sum_{n=-\infty}^{\infty} \frac{nI_n(\lambda)}{\omega - n\Omega} \tag{2-25}$$

where $\lambda = kTk^2/m\Omega^2$ and I_n is the Bessel function of imaginary argument. Also in that case

$$\boldsymbol{\varepsilon} = \mathbf{1} - \frac{\omega_p{}^2 e^{-\lambda}}{\omega} \sum_{n=-\infty}^{\infty} \frac{1}{\omega - n\Omega} \begin{bmatrix} n^2 I_n/\lambda & -in(I_u' - I_n) & 0 \\ in(I_n' - I_n) & \dfrac{n^2}{\lambda} I_n + 2\lambda I_n - 2\lambda I_n' & 0 \\ 0 & 0 & I_n \end{bmatrix} \tag{2-26}$$

where $\mathbf{1}$ is the unit tensor.

E. *Waves in a Uniform Plasma*

When a Fourier transform such as eq. 2-9 is inverted, the integrand will contain the factor $(\mathbf{M} - \boldsymbol{\varepsilon})^{-1}$ which occurs in eq. 2-6, or $(\varepsilon_{\text{long}})^{-1}$ in the case of the electrostatic approximation. It follows that particular attention must be directed to the conditions

$$|\mathcal{M} - \boldsymbol{\varepsilon}| = 0 \tag{2-27}$$

or

$$\varepsilon_{\text{long}} = 0 \tag{2-28}$$

respectively. These are, of course, just the dispersion relations for plane waves; the contributions to the integrals that originate in these singularities simply correspond to the excitation of waves by the source. A necessary preliminary to the discussion of the integrals is therefore a knowledge of the solutions of eqs. 2-27 or 2-28. We may again refer to several texts (3,4,10,12) in which this subject is elaborated. Here we quote some particular results that will be useful.

1. Cold Plasma

If there is no magnetic field, we have the plasma oscillation $\omega = \omega_p$ for any $\mathbf{k}$, and the electromagnetic waves $\omega = ck(1 - \omega_p{}^2/\omega^2)^{-1/2}$ Figure 5*a* shows the dispersion curve as a plot of ω against k, a form we shall find particularly useful since, it exhibits at a glance the phase speed ω/k and the group speed $\partial\omega/\partial k$. When there is a magnetic field, the tensor eq. 2-19 leads to the Appleton–Hartree equation. The dispersion curves take a variety of forms, depending on θ and the parameter ω_p/Ω; at very low frequency the effect of the heavy ions should also be included. Figures 5*b* and 5*c* show plots of the same type as Figure 5*a*, for $\theta = 0$ and $\theta = \frac{1}{2}\pi$, respectively, in the case $\omega_p = 5^{1/2}\Omega$, which will be sufficiently illustrative for us. The cutoff frequencies ω_1, ω_2 are the positive roots of

$$\omega^2 \mp \Omega\omega - \omega_p{}^2 = 0 \tag{2-29}$$

They and ω_p are just the three cutoff frequencies referred to in our explanation of Figure 2; the existence of the three branches O (ordinary) and X and Z (extraordinary) in Figure 5*c*, which occur in a similar way at intermediate values of θ, also corresponds to our discussion at that point. The branch below $\omega = \Omega$ is the so-called whistler mode; its maximum frequency decreases as θ increases and it disappears when θ reaches $\frac{1}{2}\pi$ (or is limited to the lower hybrid frequency if one takes account of the positive ions).

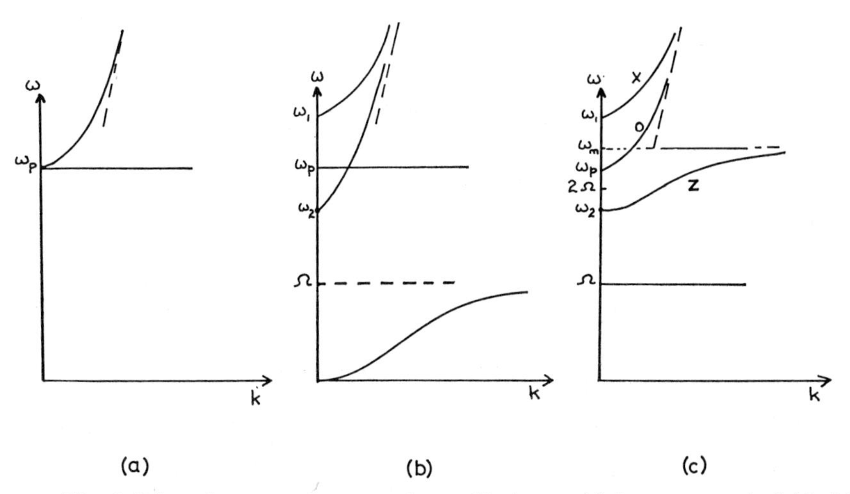

Fig. 5. Dispersion curves for waves in a cold plasma: (a) for no magnetic field; (b) for waves traveling parallel to an imposed magnetic field; (c) for waves traveling perpendicualr to an imposed field. Low-frequency branches made possible by ion motion are not shown.

2. Hot Plasma without Magnetic Field

The thermal effects are negligible for the electromagnetic modes, and it is well known that at long wavelength ($hk \ll 1$) eq. 2-22 leads to the correction

$$\omega_p = \omega_p(1 + \tfrac{3}{2}h^2k^2) \qquad (2\text{-}30)$$

For larger k, Landau damping enters, leading to complex ω. The modification to Figure 5*a* is shown in Figure 6. It is assumed that the thermal distribution does not contain features such as double streaming that could lead to instability.

3. Hot Magnetized Plasma

In the electrostatic approximation, the waves for $\theta = 0$ are unaffected by the field and are again given by Figure 6. At a general angle, and for long enough wavelengths (though not so long that the electrostatic approximation fails), there are waves given by $\varepsilon_{\text{long}} = 0$, where $\varepsilon_{\text{long}}$ is as in eq. 2-21, i.e., thermal corrections (and Landau damping) negligible. But as θ approaches $\frac{1}{2}\pi$, new modes (the Bernstein modes) appear close to each gyrofrequency harmonic; these are completely undamped when $\theta = \frac{1}{2}\pi$. An example of a dispersion curve, for a Maxwellian distribution, is shown in Figure 7; it corresponds to $\omega_p = 5^{1/2}\Omega$, as in Figure 5.

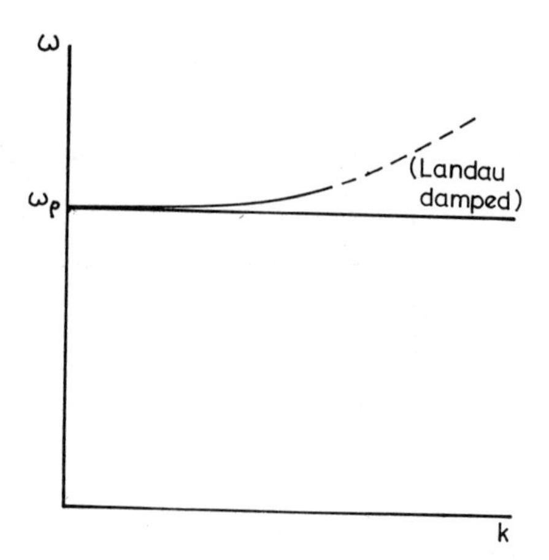

Fig. 6. Dispersion curve for plasma oscillations in a Maxwellian field-free plasma.

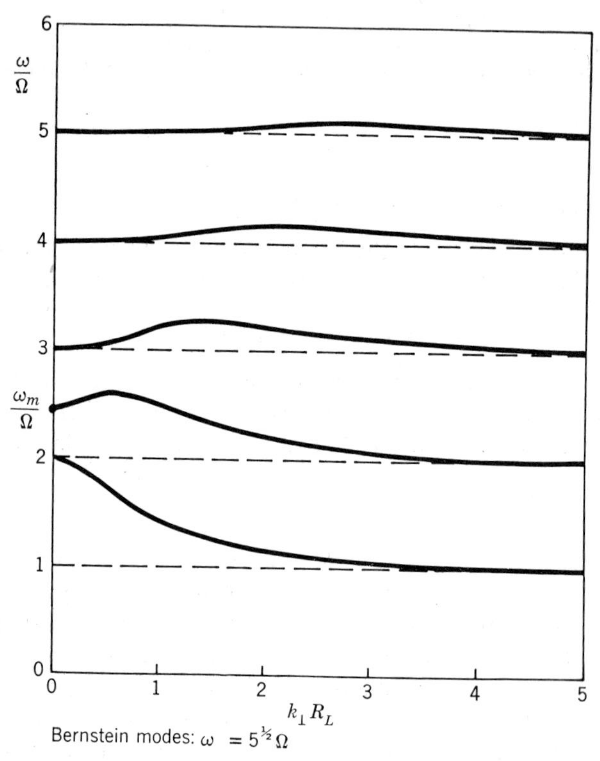

Fig. 7. Dispersion curve for electrostatic waves (Bernstein modes) in a Maxwellian plasma; propagation is perpendicular to the magnetic field.

When the full electromagnetic equations are used, the dispersion relation is extremely complicated. The only comments we can make here are the following. When $\theta = 0$, the plasma oscillation is still unaltered, and the high frequency branches of Figure 5*b* need not be corrected, since for them $\omega/k > c$. The low frequency (whistler) mode can be seriously affected, corresponding to Landau damping, or possibly excitation, of such waves, but this mode has not been thought to have any importance in the present context, nor have any possible new modes. When $\theta = \frac{1}{2}\pi$, the same remark applies to the *O* mode of Figure 5*c*, which remains purely electromagnetic. The two branches of the extraordinary mode must be examined closely, however, for it is they that may be identified with the electrostatic mode at sufficiently large k. Figure 7 is actually incorrect in a vertical wedge above the line $\omega \simeq ck$; on the scale appropriate to Figure 7, this wedge is very thin. Figure 8 shows how

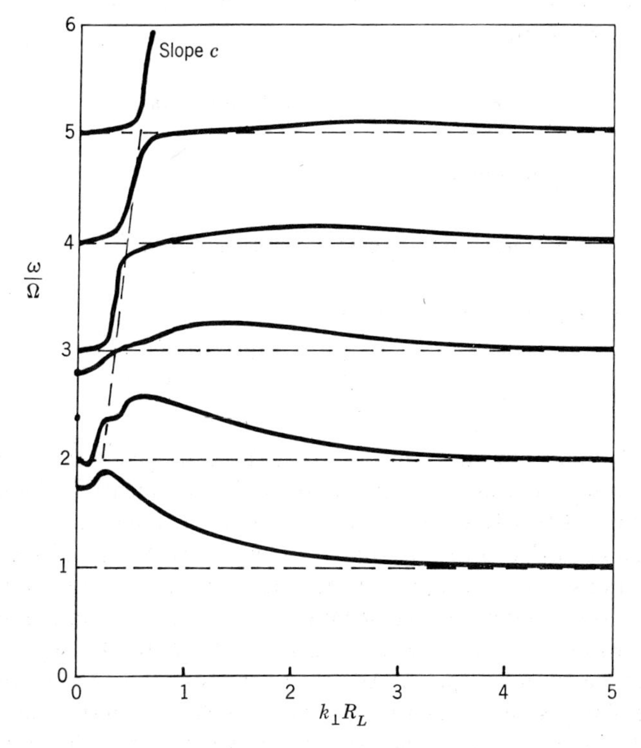

Fig. 8. Dispersion curve for waves traveling perpendicular to a magnetic field in a Maxwellian plasma, showing coupling between Bernstein modes and the Appleton-Hartree modes of Figure 5*c*.

Figures 5*c* and 7 might be reconciled when one has to include both finite temperature and finite speed of light. Unfortunately, detailed calculations of these dispersion curves do not appear to have been carried out.

III. Theory of Resonances

A. *Explicit Evaluation*

In two cases integral (2-9) can be made sufficiently simple for explicit evaluation; and though too much of the physics is omitted for the work to be useful in comparison with experiment, it gives a valuable demonstration of what is meant by a resonance.

In each case we shall suppose the source to be an infinitesimal dipole, as in eq. 2-11, without as yet specifying its temporal behavior $\mathbf{M}(\omega)$. Then eq. 2-9 may be written

$$\mathbf{E}(\mathbf{x}, t) = \int_{-\infty}^{\infty} \mathbf{R}(\mathbf{x}, \omega)\mathbf{M}(\omega)e^{i\omega t}\, d\omega \tag{3-1}$$

where the matrix $\mathbf{R}$ is given by

$$\mathbf{R}(\mathbf{x}, \omega) = \frac{1}{2\pi^2}\int (\mathcal{M} - \varepsilon)^{-1} e^{-i\mathbf{k}\cdot\mathbf{x}}\, d^3k \tag{3-2}$$

or, in the electrostatic approximation,

$$R_{pq}(\mathbf{x}, \omega) = -\frac{1}{2\pi^2}\int \frac{k_p k_q}{k^2 \varepsilon_{\text{long}}}\, e^{-i\mathbf{k}\cdot\mathbf{x}}\, d^3k \tag{3-3}$$

This matrix may be regarded as a Green's function, giving the response at $\mathbf{x}$ resulting from a unit dipole at the origin and of frequency ω. Now suppose for the moment that $\mathbf{R}(\mathbf{x}, \omega)$ has been evaluated; it remains to carry out the integration in eq. 3-1. If the medium is not inherently unstable, $\mathbf{R}$ must satisfy the conditions for causality (the Kramers-Kronig equations) which are equivalent to the statement that $\mathbf{R}$ must be analytic for ω in the lower half-plane. As $\mathbf{M}(\omega)$ is analytic (see Section II-C), it follows that eq. 3-1 must be analytic in the lower half-plane. Any singularities must lie on the real axis or above it; if $\mathbf{R}$ has any branch points, the branch cuts must be drawn in the upper half-plane. For t large and positive, one will therefore distort the path of integration to that shown in Figure 9, and the major contributions will be from residues at poles, or branch-cut integrals. The former will depend on

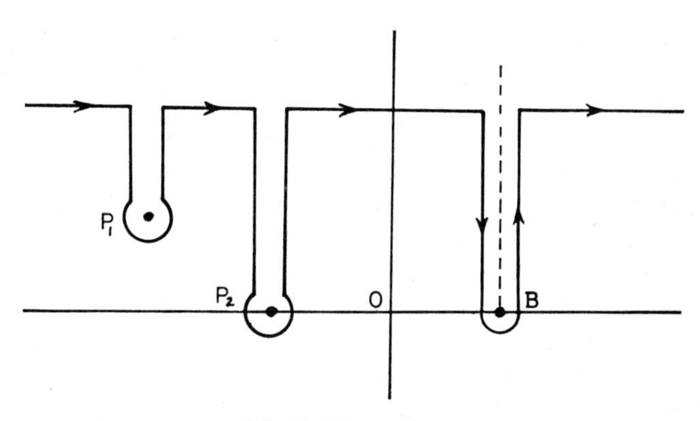

Fig. 9. The ω-plane.

time as $e^{i\omega_1 t}$ for a pole at $\omega = \omega_1$; the branch-cut integrals involve typically an additional factor that is a power of t. The singularity with the smallest imaginary part will dominate after a very long time.

The first tractable case is that of cold plasma with no magnetic field, for which $\boldsymbol{\varepsilon}$ is given by eq. 2-18. In a few steps one reaches

$$R_{pq} = \frac{1}{\varepsilon}\left[\frac{\varepsilon\omega^2}{c^2}\,\delta_{pq} + \frac{\partial^2}{\partial x_p\,\partial x_q}\right]\frac{\exp(-i\varepsilon^{1/2}\omega r/c)}{r} \tag{3-4}$$

where $r = |\mathbf{x}|$. This is just the standard result for a Hertz dipole in a medium of refractive index ε. The leading term close to the origin is

$$R_{pq} \simeq \frac{1}{\varepsilon}\frac{\partial^2}{\partial x_p\,\partial x_q}\left(\frac{1}{r}\right) \tag{3-5}$$

and this is also just what is obtained in the electrostatic approximation. The expression (3-5) has a pole, as a function of ω, where $\varepsilon = 0$, namely $\omega = \omega_p$. Provided that ω_p is included in the frequencies swept through by the transmitter (i.e., $\mathbf{M}(\omega_p) \neq 0$), there will be a contribution proportional to $r^{-3}e^{i\omega_p t}$ in the expression for $\mathbf{E}$. This electric field is what is detected by the antenna, acting as a receiver, long after the original excitation has been removed. (Strictly, one ought to solve a boundary value problem for the receiving antenna, but we are omitting this for reasons similar to the case of the transmitting antenna.) This result may be regarded as the prototype of how a resonance occurs mathematically. It also has an immediate physical interpretation: the oscillating dipole includes a spectral component at $\omega = \omega_p$, and this feeds energy into plasma oscillations. These persist without damping and, as is well known, do not propagate; so they may be continually observed by the receiving antenna.

As we shall see in the next section, it is the possibility of the excitation of waves with zero group velocity that leads to resonance. Figure 5*a* is the dispersion curve for waves in a cold field-free plasma; the horizontal line shows the plasma oscillations which are still correctly treated in the electrostatic approximation. All these waves have zero group velocity; hence the strong resonant effect. But the electromagnetic branch also achieves the condition of zero group velocity at $\omega = \omega_p$, $k = 0$. Now this branch was removed altogether by the electrostatic approximation, but we can reinstate it by using eq. 3-4 instead of eq. 2-5. Continuing the expansion, of which eq. 3-5 was the leading term, we find that the next term to be singular is proportional to $\varepsilon^{1/2}(\omega^3/c^3)\delta_{pq}$, with the branch point at $\omega = \omega_p$, giving a contribution to the electric field proportional to $(\omega_p{}^3/c^3)t^{-3/2}e^{i\omega_p t}$. This electromagnetic contribution is in practice much weaker, since it does not have the geometrical factor r^{-3} and has the time factor $t^{-3/2}$; also since its resonance is at the same frequency as the electrostatic one, it would be impossible to detect it.

The second case is that of the cold plasma with magnetic field but proceeding direct to the electrostatic approximation. For this we combine eqs. 3-3 and 2-21 and the integrals over $\mathbf{k}$ to give

$$R_{pq}(\mathbf{x}, \omega) = \pm \frac{\partial^2}{\partial x_p \partial x_q} [\varepsilon_\perp(\varepsilon_0 x^2 + \varepsilon_0 y^2 + \varepsilon_\perp z^2)]^{-1/2} \tag{3-6}$$

(the indeterminancy of sign can be settled but is of no physical significance). This is just the generalization of eq. 3-5. It introduces a new branch-point resonance at $\varepsilon_\perp = 0$, which is just $\omega = \omega_m$, the hybrid frequency. The condition $\varepsilon_0 = 0$ still gives a resonance at $\omega = \omega_p$ but only on the plane $z = 0$. The hybrid frequency resonance can be traced to the horizontal asymptote of Figure 5*c*, where there are again waves with very small group velocity. In fact in the electrostatic approximation this branch of the curve is just a horizontal straight line like that of Figure 5*a*.

To deal with the full electromagnetic expression when $\mathbf{B}_0 \neq 0$ is already analytically very elaborate; however, as could be anticipated from the case $\mathbf{B}_0 = 0$, it results in additional very weak resonances. These occur at the three frequencies because of the cutoff of the electromagnetic modes ($k = 0$) of Figures 5*b* and 5*c*. The frequencies ω_1 and ω_2, being different from any resonance listed so far, might have some chance of being observed; moreover, this part of the theory would not be subject to thermal corrections. These three frequencies are, of course, just the zero-range condition that we discussed at Figure 2. Their occurrence as resonances does not, however, appear to have been confirmed. Another possibility is of a resonance at Ω due to the horizontal asymptote of Figure 5*b* (upper limit of the "whistler" mode).

This informal connection between the condition of zero group velocity and the resonances will be given a basis in the next two sections. More details of calculations sketched in the present section may be found in Ref. 13.

B. *The Method of Pinching Poles*

Let us now revert to the possibility of a general source $\mathbf{J}(\mathbf{k}, \omega)$ and a medium with very general properties $\boldsymbol{\varepsilon}(\mathbf{k}, \omega)$ subject only to being linear and stable. The integral eq. 2-9, with $\mathbf{E}(\mathbf{k}, \omega)$ given by eqs. 2-6 or 2-7 is the formal solution to our problem, and we would like to be able to investigate possible resonances, even when the evaluation of the integral is impracticable. Separating the integrals over ω and $\mathbf{k}$, we may write

$$\mathbf{E}(\mathbf{x}, t) = \int \mathbf{F}(\mathbf{x}, \omega) e^{i\omega t}\, d\omega \tag{3-7}$$

where

$$\mathbf{F}(\mathbf{x}, \omega) = \int \mathbf{E}(\mathbf{k}, \omega) e^{-i\mathbf{k}\cdot\mathbf{x}}\, d^3k \tag{3-8}$$

If we imagine the k-integration to have been completed, $\mathbf{F}$ is known, and we would treat the final ω integration just as in the previous section, i.e., by means of the curve in Figure 9. The behavior of the field after a long time is governed by the singularities of $\mathbf{F}(\mathbf{x}, \omega)$; these occur at particular frequencies $\omega_1, \omega_2, \ldots,$ which are independent of $\mathbf{x}$, although their importance may depend on $\mathbf{x}$ (as we saw in the simple examples), and the behavior as a function of time will depend on the type of singularity. The resonant frequencies $\omega_1, \omega_2, \ldots$ could therefore be found merely by identifying the singularities of $\mathbf{F}$, and the detailed behavior of a resonance could be found from a discussion of $\mathbf{F}$ very close to the singularity (e.g., evaluation of the residue there in the case of a pole).

Some of this information can be extracted from eq. 3-8 without explicit evaluation of the integral over k. The singularities of $\mathbf{F}$ as a function of ω must be related to the singularities of $\mathbf{E}(\mathbf{k}, \omega)$, for where the latter is regular nothing drastic can happen to $\mathbf{F}$. But $\mathbf{E}$ contains the factors $\mathbf{J}(\mathbf{k}, \omega)$, describing the source, and $(\mathscr{M} - \boldsymbol{\varepsilon})^{-1}$ or $(\varepsilon_{\text{long}})^{-1}$ describing the medium. The former is quite well behaved for a realistic source (as appears in our examples, section II-C), while the singularities of the latter are just the solutions of the dispersion relation. To explain the role of these, suppose for the moment that space were one-dimensional, so that there is only a single component of k, and consider the performance of the integration eq. 3-8 along the real axis of the k-plane,

Figure 10. For a given value of ω, singularities such as s_1, s_2, ... of $E(k, \omega)$ occur in the k-plane but do not in general lie on curve C. In that case $F(\omega)$ is well behaved, provided that the integral converges. If we now let ω vary, the

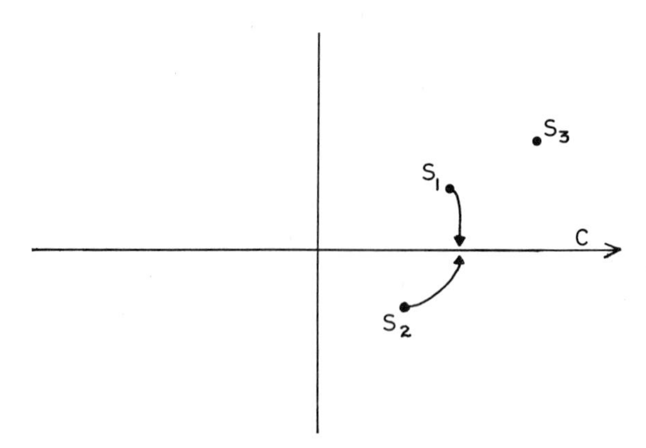

Fig. 10. The k-plane.

singularities move around smoothly, and if they meet the contour we may have to deform it to obtain the analytic continuation of $F(\omega)$. It is only when *two* singularities *merge* on the contour from either side that a singularity of $F(\omega)$ occurs; the contour may by then have been already deformed, although in the example shown on Figure 10 it has not. This criterion is known as the "pinching" of poles. A simple example to show this is

$$F(\omega) = \int_{-\infty}^{\infty} \frac{dk}{\omega^2 + k^2} \equiv \frac{2\pi}{\omega}$$

where the poles at $k = \pm i\omega$ pinch on $k = 0$ as $\omega \to 0$, thus locating correctly the singularity of $F(\omega)$. The point is that it may be possible to trace the pinching of poles in cases in which the integral is too cumbersome for explicit evaluation.

Now write $f(k, \omega) = 0$ as the dispersion relation. As we have seen, this gives the poles of $E(k, \omega)$. A necessary condition for poles to merge in the k plane is that $f(k, \omega) = 0$ should have a *double* root, so that $\partial f/\partial k = 0$ also. But if we solve the dispersion relation in the form $\omega(k)$, the group velocity is

$$\frac{d\omega}{dk} = -\frac{\partial f}{\partial k}\bigg/\frac{\partial f}{\partial \omega} \tag{3-9}$$

The merging of poles thus implies that the group velocity should vanish.

The above theory can be extended to multiple contour integrals, leading to the result that, if the dispersion relation is solved as $\omega(\mathbf{k})$, singularities of $\mathbf{F}(\mathbf{x}, \omega)$ occur at those frequencies for which

$$\frac{\partial \omega}{\partial k_p} = 0 \qquad p = 1, 2, 3 \tag{3-10}$$

that is, the vector group velocity vanishes.

A further extension of the theory is necessary if it is desired to classify a singularity of $F(\mathbf{x}, \omega)$ and to find such things as the residue at the singularity if it is a pole. Recent monographs by Pham (14) and by Hwa and Teplitz (15) go some way toward providing such material. Integrals of the type

$$I(\omega) = \int \frac{g(\omega, \mathbf{k})}{f(\omega, \mathbf{k})}\, d\tau \tag{3-11}$$

where $g(\omega,\mathbf{k})$ is regular, $f(\omega, \mathbf{k})$ has discrete zeros, and $d\tau$ is an n-dimensional differential form in the components of $\mathbf{k}$ are considered. The term $I(\omega)$ can only have singularities if ω allows $f = 0$, $\partial f/\partial k_i = 0$ to be solved simultaneously, say at $(\omega_0, \mathbf{k}_0)$. Provided that

$$\det\left(\frac{\partial^2 f}{\partial k_i \partial k_j}\right) \neq 0 \tag{3-12}$$

at (ω_0, k_0), leading terms in expressions for $I(\omega)$ in the neighborhood of ω_0 can be derived. Unfortunately, the situations arising in the plasma resonance case are so singular that eq. 3-12 may not be satisfied. This strange behavior is removed, however, if one takes account of the motion of the antenna relative to the plasma. Detailed use of results such as those of Refs. 14 and 15 does not, however, appear to have been made.

C. *The Method of Stationary Phase*

The idea in the previous section was to take the integrations over $\mathbf{k}$ first, then that over ω; here we modify eq. 2-9 by taking the integrations in the opposite order. So we write

$$\mathbf{E}(\mathbf{x}, t) = \int \mathbf{G}(\mathbf{k}, t) e^{-i\mathbf{k}\cdot\mathbf{x}}\, d^3k \tag{3-13}$$

where

$$\mathbf{G}(\mathbf{k}, t) = \int \mathbf{E}(\mathbf{k}, \omega) e^{i\omega t}\, d\omega \tag{3-14}$$

So, for fixed $\mathbf{k}$, we perform eq. 3-14 and the procedure is again that of Figure 9, but the singularities in the ω plane are just the solutions of the dispersion relation $f(\mathbf{k}, \omega) = 0$. (Causality will require the contour to be indented *below* any singularities on the real axis.) This (for t large and positive) gives an expression of the type

$$\mathbf{G}(\mathbf{k}, t) = \sum_r \mathbf{A}_r(\mathbf{k}) e^{i\omega_r(\mathbf{k})t} \tag{3-15}$$

where r labels the possible modes and the coefficients $\mathbf{A}_r(k)$ are their amplitudes, obtainable from the residues at the singularities; $\omega_r(\mathbf{k})$ are the solutions of the dispersion relation. Thus we have

$$\mathbf{E}(\mathbf{x}, t) = \sum_r \int \mathbf{A}_r(\mathbf{k}) e^{i\omega_r(\mathbf{k})t} e^{-i\mathbf{k}\cdot\mathbf{x}} \, d^3k \tag{3-16}$$

The previous method first found the resonant frequencies, and then estimated the behavior of the signal at large times by calculating the relevant function near the resonant frequencies. This is obviously what is required in evaluating the actual strengths of the fields long after the exciting pulse has been switched off, but while the resonance is still sounding. So the relevant treatment of eq. 3-16 is to derive an asymptotic series for large times. The expression immediately lends itself to the method of stationary phase; in three dimensions the expressions for the asymptotic series derived by this method are complicated, but a good account is given by Chako (16).

Once again it is found that the critical points, that is those about which the asymptotic series is formed, are determined by

$$\frac{\partial \omega}{\partial k_p} = 0 \qquad p = 1, 2, 3 \tag{3-17}$$

the same relationship as eq. 3-10. Again it is found that a condition of the form of eq. 3-12 should be satisfied for the series to be at all simple; but here, in contrast to the method of the previous section, it is possible to find the leading asymptotic term when this condition is not satisfied, as is the case for the simplest resonances. An account of this approach and its relation with the method of pinching poles has been given by Watson (17). Both methods give the same expression for the leading asymptotic term in the series, but we shall not demonstrate this here; instead, we shall simply quote the result.

For a general source whose Fourier transform is $\mathbf{J}(\mathbf{k}, \omega)$, the electric field at a point $\mathbf{x}$ and time t is

$$\mathbf{E}(\mathbf{x}, t) = \iint \frac{4\pi}{i\omega} (\mathcal{M} - \mathbf{e})^{-1} \mathbf{J}(\mathbf{k}, \omega) e^{i(\omega t - \mathbf{k}\cdot\mathbf{x})} \, d^3k \, d\omega \tag{3-18}$$

on combining eqs. 2-6 and 2-9. In the electrostatic approximation, this becomes

$$E_p(\mathbf{x}, t) = 4\pi i \iint \frac{k_p k_q}{\omega k^2 \varepsilon_{\text{long}}} J_q(\mathbf{k}, \omega) e^{i(\omega t - \mathbf{k}\cdot\mathbf{x})} \, d^3k \, d\omega \tag{3-19}$$

on using 2-7. Accepting eq. 3-19 and using either of the two methods just described, we obtain the asymptotic formula,

$$E_p(\mathbf{x}, t) = \sum_{\substack{\text{all} \\ \text{resonances}}} \frac{e^{i(\omega_0 t - \mathbf{k}_0 \cdot \mathbf{x})}}{t^{3/2}} \frac{k_{0p} k_{0q}}{\omega_0}$$

$$J_q(\mathbf{k}_0, \omega_0)(2\pi)^{5/2} e^{3i\pi/4} \left. \frac{[\partial D/\partial \omega]^{1/2}}{[\det(\partial^2 D/\partial k_i \, \partial k_j)]^{1/2}} \right|_{\omega_0, \mathbf{k}_0} \left(1 + 0\left(\frac{1}{\varepsilon}\right)\right) \tag{3-20}$$

where $D = k^2 \varepsilon_{\text{long}}$. This assumes that at each resonant point (k_0, ω_0) the dispersion relation $D = 0$ is nonsingular in the sense that $|\partial^2 D/\partial k_i \partial k_j| \neq 0$. When eq. 3-18 must be used, there is an analogous, yet more complicated formula to replace eq. 3-20.

IV. Application of the Theory

A. *Sturrock's Contribution*

A considerable effort has been given to the theoretical interpretation of plasma resonances, and we shall attempt in this chapter to indicate how the work of various authors fits into the framework above and what the inter-relations between them are. In the final chapter we shall compare the results with observations.

Ever since plasma resonances were first observed in the ionosphere (18–21), when tentative theoretical explanations (20) were put forward, the development of the theory has been continuous. Sturrock (22) was one of the first authors to attempt an accurate mathematical treatment of the problem, although, for reasons we have already indicated, he had to idealize the problem considerably. He assumed that the plasma and the magnetic field were uniform, and that the source was a stationary infinitesimal dipole, as in Section II-C-1 above. He first considered the plasma to be cold, and was able to calculate the strengths of the electric fields for the resonances at the plasma frequency and at the two frequencies ω_1, ω_2 given by eq. 2-29; his method uses an approximation that is equivalent to finding the first term only in the asymptotic series for the resonance fields. As we pointed out in Section III-A, this is fairly straightforward, even without the electrostatic approximation, provided that the plasma is cold. When thermal effects are included, however, as is necessary to explain resonances at the multiples of the gyro-frequency, this becomes far more difficult. Sturrock's analysis here necessarily becomes a little imprecise but he manages to obtain approximations to the behavior of the electric fields for resonance at the hybrid frequency and harmonics of the cyclotron frequency. He uses his results to obtain estimates for the times that resonances should be heard after pulsing; but because of the indeterminate nature of the approximations involved, these can be no more than rough indications. He does, however, demonstrate that the electric fields decay as some negative power of time, a result appearing often in subsequent literature.

Here we can note a more physical approach to the problem made at that time by Fejer and Calvert (23), although we shall discuss their work more fully below. They used approximate forms of the relevant electrostatic dispersion relations to estimate the time taken for a packet of near-resonant waves to disperse but admitted that no quantitative prediction could be made. They concluded that these resonances can be adequately explained by using the electrostatic approximation, although this contention has been disputed by other authors.

B. *The Work of Nuttall, Johnston, and Shkarofsky*

At the same time Nuttall and Johnston (24–26) were investigating this problem in a similar way. Nuttall (25) was the first to point out that the frequencies at which resonances occur can be found without working out the complete fields, by the method of pinching poles, as described in Section III-B. To find the principal parts of the singularities of the function $\mathbf{F}(x, \omega)$, defined by eq. 3-8, which is essentially a Green's function for the problem, Nuttall approximated the dielectric constant near the singularities by the first few terms in the Taylor series expansion about those points. By this method he obtained the "most singular" part of the Green's function at ω_p, the plasma frequency, and at $\omega_m = (\omega_p{}^2 + \Omega^2)^{1/2}$, the hybrid frequency. He went on to use these results in Ref. 26, to predict the absolute strength of a resonant signal, its asymptotic behavior, and its dependence on the orientation of the satellite antenna. No attempt was made here, however, to consider resonances occurring at or near cyclotron harmonic frequencies, nor to take into account effects of nonuniformity or of satellite motion.

The theory was pursued, however, to try to take into account these effects by Nuttall's co-workers in Montreal. Shkarofsky and Johnston (27,28) approached the discussion of resonances near the cyclotron frequencies by more physical considerations. Because of the complicated nature of the dispersion relations for propagation across the field at small wave number, where the Appleton-Hartree cold electromagnetic modes must be combined with the Bernstein hot electrostatic modes (cf. Figure 8), these authors stressed the importance of including the satellite velocity in the mathematical model. In this case the critical points, giving the frequency and wavelength of the resonant waves, are determined by the relation

$$\frac{\partial \omega}{\partial k_p} = V_p \qquad p = 1, 2, 3 \tag{4-1}$$

where **V** is the satellite velocity.

This condition emerges naturally from the mathematics, whether we consider work in the coordinate frame of the antenna (17) or of the plasma (29). It is also the obvious condition, namely, that the resonant waves are those whose energy stays in the vicinity of the antenna, so that their group velocity must be equal to the satellite velocity. Shkarofsky and Johnston attempted to find points on the dispersion curves where the slope is equal to the satellite velocity, and then estimated the strength and duration of a resonant signal at these points. Reference 28 is a good study of resonances at harmonics of the cyclotron frequency, but unfortunately it lays great stress on the need to use relativistic dispersion relations (30,31) a view that Shkarofsky has since modified (32), since, in practice, field and plasma non-uniformities are likely to have a more serious effect on the results. Because of the exceedingly complex structure of the dispersion relations, it is not possible to explain simply the complicated results that Shkarofsky and Johnston obtain, but the reader is recommended to study Ref. 28, in which these are presented. We may summarize their work by pointing out that they have made a realistic study of resonances at very long wavelength near harmonics of the gyrofrequency, taking into account the relevant physical effects but necessarily at the expense of mathematical rigor. Their expressions for the time decay of resonances can therefore be no more than an indication of the order of magnitude expected.

C. *The Contributions of Dougherty and Co-workers*

Meanwhile, Dougherty and Monaghan (13) made a detailed analysis of the problem to predict the frequencies at which resonances could be expected but did not go on to estimate the strength and ringing times of resonant signals. The plasma and magnetic field were assumed uniform, and satellite motion neglected. They first used the electrostatic approximation and the theory of pinching poles, as described in Section III-B, to predict the frequencies of resonance when the longitudinal dielectric constant was described by each of eqs. 2-18, 2-21, 2-23, and 2-25. Selecting the last of these as the most general, Dougherty and Monaghan found resonances at the plasma frequency ω_p, the hybrid frequency ω_m, harmonics of the cyclotron frequency and at frequencies slightly shifted from these harmonics, provided that $\omega > \omega_m$. Of these, ω_p alone arises from waves traveling parallel to **B** and results from the very low group velocity in the limit $k \to 0$ (see Figure 6). Figure 7 shows how ω_m and $n\Omega$ ($n \geqslant 2$) arise from waves with $\mathbf{k}$ perpendicular to $\mathbf{B}$, in the limit $k \to 0$. The last-mentioned series of resonances is a result of the waves with finite k that are at the "humps" of the curves in Figure 7; their wavelength is roughly the thermal Larmor radius for electrons and their frequencies are given by the approximate formula

$$\omega = \Omega \left[n + \frac{C\omega_p{}^2}{n^2\, \Omega^2} \right] \tag{4-2}$$

where $C = (27/2\pi e^3)^{1/2} = 0.464$. It might be thought that a still further series could arise from waves with $k \to \infty$ in Figure 7 but these are probably unimportant in practice, owing to finite antenna size.

The authors went on to consider the full electromagnetic dispersion relations which had to be approximated before any answer could be produced; they discovered resonances at, or nearly at, the frequencies listed above, but it is not clear that every resonance was deduced, particularly the coupling resonances considered by Shkarofsky and Johnston (27). However, they do show that a resonance could be expected at the cyclotron frequency Ω for small wave number, which is not predicted in the electrostatic approximation. The nature of the source is irrelevant for the prediction of resonant frequencies as pointed out in Section III-B; but to go on to calculate the actual fields of a resonant signal, we do need to know the actual form of the radiating antenna. Watson (17,33) has continued the work of Dougherty and Monaghan. He assumes an antenna with a current distribution of the forms of eqs. 2-15 and 2-17, and attempts to find the asymptotic form of the fields at large time near the satellite. The methods of Sections III-B and III-C are both used, and leading terms in the asymptotic series obtained for all the resonances given by Dougherty and Monaghan (13) in the electrostatic case. Because the dispersion relations become singular (in the sense that eq. 3-12 is not satisfied) at zero-wave number, Watson (33) considers the problem in a coordinate frame fixed relative to a slowly moving satellite. The resonant points now move slightly away from the origin in wave-number space, so that the methods of stationary phase and pinching poles may be used. As do most other authors he tries to compare his results with experiment; this will be discussed in Section V.

D. *The work of Fejer et al.*

We have already mentioned the earlier work of Fejer and Calvert (23) to predict the resonant frequencies. Fejer went on to make a more exact analysis of this problem, this time with Deering (29). Their model was of an infinitesimal dipole with an impulsive source current. They obtained expressions for the electric fields due to this source as in eq. 3-1 and then evaluated the integrals at the different resonances by various approximations. They considered first the field-free case, and at the plasma frequency they used an expansion of the plasma dispersion function for small wave number to reduce the denominator of the integrand to a simple quadratic expression. Even with this simplification, complete evaluation of the resonant fields at the plasma frequency is not possible, but explicit expressions are found in two limiting situations. For the longitudinal (i.e., electrostatic) part of the asymptotic radiated field, one must divide space into regions $0 < r \ll a(t)$, $a(t) \ll r \ll v_{\text{th}} t$ and $v_{\text{th}} t < r$, where v_{th} is the thermal speed of the electrons and the radius $a(t)$ is given approximately by $v_{\text{th}} t^{1/2}/\omega_p{}^{1/2}$. Within the first region, the electric field is approximately uniform, parallel to the dipole, and proportional to $t^{-3/2}e^{i\omega_p t}$. Within the second region it is radial, proportional to $\cos\theta$ (θ being the angle between $\mathbf{r}$ and the dipole), and has a similar time dependence to that of the first region but with an additional term in the phase proportional to r^2/t. In the third region the field is presumed to be small. The eletromagnetic part of the field is also approximated in a similar way but turns out to be much smaller, just as we saw in our complete evaluation for the cold case (3-1).

In the presence of a magnetic field they use the electrostatic approximation, and by similar methods derive expressions for the fields at the plasma frequency, the hybrid frequency, and multiples of the cyclotron frequency. They are the only authors to consider the resonance at the lower hybrid frequency which depends on the presence of ions; every other author assumes that only the electron motion is significant in determining the plasma characteristics, largely because the frequencies involved in ion resonances are so low. Their expressions for the fields at harmonics of the cyclotron frequency are somewhat tentative because the approximations necessary for an analytic solution are so vague. They go on to discuss the range of validity of their expressions but do not attempt to compare them with experiment. More recently, Fejer and Yu (34) have given a more systematic formulation of the problem, once again using an infinitesimal impulsive source in a uniform plasma. They use the method of stationary phase, as given in Section III-C to find the fields at a point moving with arbitrary velocity from the initial position of the source. Thus fields can be found not only near the dipole but anywhere in space. The results they obtain are similar to those of Watson (17); expressions for the fields at the plasma frequency both in the presence and absence of a magnetic field are given. At frequencies near multiples of the cyclotron frequency the calculations are more difficult because of the nature of the dispersion relations. Fejer and Yu suggest that the asymptotic approximation they obtain for the resonant fields at these frequencies are only likely to be good for times that are much longer than the ringing observed in topside sounding experiments. Fejer and Yu (34) do not consider here the effect of nonuniformities or finite antenna size on their results but their approach has the advantage of giving the fields at all points in space for a sufficiently large time. A more recent work of Fejer and Yu (44) will be discussed in Section IV-F.

E. *The Work of Crawford and His Co-Workers*

The other group of authors that has contributed to plasma resonance theory has been led by F. W. Crawford of Stanford University, Stanford, California. Initially, their interest was practical, stimulated by experiments on the propagation of cyclotron harmonic waves in the laboratory (35–37). The method used by Tataronis (38) and Tataronis and Crawford (39) is similar to that used by Nuttall (25,26). The frequencies of resonance are found by the pinching poles technique, but instead of following the rigorous method of Pham (14) to find the most singular part of the resulting integral, these authors merely expand the integrand around the resonant frequencies. The electrostatic approximation is used, but care is taken to include the resonances at finite k; see eq. 4-2 and the preceding discussion. The source is taken to be a stationary infinitesimal dipole, and so the results must be treated with a certain circumspection. But Tataronis and Crawford make a significant contribution to the theory in stressing that the quantity actually measured is the current in the antenna, rather than the electric field strength in the plasma. They give numerical values when the source is taken to be a pair of parallel grids, and the results are shown to compare favorably with experiment.

F. *Oblique Echoes in a Stratified Medium*: *McAfee's Theory*

Early in the paper we indicated that the resonance phenomenon should be interpreted in terms of plasma waves in a uniform medium (though ionogram traces would involve reflection from distant gradients), and the theory up to now has been developed on that basis. Intuitively, one might first expect that nonuniformity of the plasma parameters would only modify the resonance theory if the wavelength of the relevant waves is extremely long, that is, where they correspond to near-zero **k**. By "long," we mean comparable with the scale of nonuniformity. If this were so, the "zero-range" resonances (2-29) and the electromagnetic corrections to the ω_p resonance would be the only cases affected (see Figure 5*c*). The physical effect here is straightforward: the satellite is situated on the boundary between propagation and evanescence for waves at these frequencies, and this manifests itself in the vertical tangent to the O, X, and Z traces at $h' = 0$ in Figure 2—hence the name. The Bernstein modes with $k = 0$ have also figured in our theory (Figure 7), but as we have explained the corrections in the "relativistic wedge" (Figure 8), possibly combined with the effect of the satellite velocity, are enough to shift the resonance to a finite wavelength. In practice this wavelength is already much smaller than the typical scale for variation of plasma parameters.

Credit goes to McAfee (40–43) for realizing the importance of another, more subtle, effect of nonuniformity. This has been investigated mainly for the resonance $\omega = \omega_p$, although McAfee (42) has also considered the upper hybrid resonance. Close to the solution ($\omega = \omega_p$, $\mathbf{k} = 0$) of the electrostatic dispersion relation for a hot plasma, there are solutions with very small group velocity. This is evident from eq. 2-30 for **k** parallel to **B**; if **k** is given a small inclination θ to **B**, the group velocity remains small in magnitude but its direction varies very rapidly with θ. The frequencies of these waves lie only slightly above ω_p. Moreover, if such waves are launched in a slightly nonuniform medium, the direction of the group velocity is a very sensitive function of ω_p. This results in rather sharply curved ray paths, along which a wave packet proceeds comparatively slowly.

We may now consider a satellite moving slowly through a horizontally stratified medium (with ω_p increasing downward), emitting a spectrum of waves including frequencies just above the local value of ω_p. We refer to Figure 11. If the satellite is at S_0 at the moment of emission, it may be possible

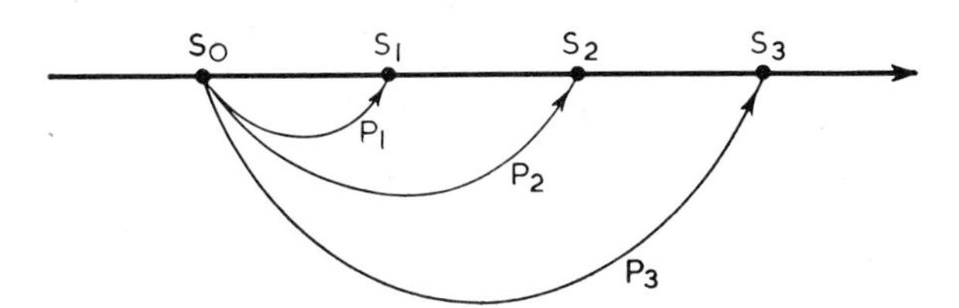

Fig. 11. Illustrating the idea of McAfee's theory of resonances.

to find ray paths P_1, P_2, P_3, ... at frequencies ω_1, ω_2, ω_3, ..., so that packets return to the satellite at subsequent positions S_1, S_2, S_3, ... with the correct time delay. The result will be that similar to a plasma resonance, provided that the frequencies ω_1, ω_2, ω_3, ... are so close to ω_p that the shift is unimportant. McAfee has computed ray paths, using a simple thermal correction to the cold dispersion relation, which show that this idea can work in practice.

McAfee has also shown that there are sometimes two possible paths such as P_1 reaching a given point S_1 with the correct delay. These two relate to slightly different frequencies which beat to form an interference pattern. Such patterns have been observed with the fixed frequency sounder of Explorer

XX (45). Two distinct cases arise, depending on the ratio ω_p/Ω. If the plasma frequency is the greater (41), there are generally two paths which lie below the satellite, i.e., in denser plasma. When the plasma frequency is less than the cyclotron frequency (43), there is only one path and it lies above the path of the satellite. There is also a maximum delay time, above which no paths are possible, so that the resonance would be limited to a short delay time. This is somewhat analogous to the loss of resonances in the uniform case if the satellite speed exceeds the maximum available group speed. In the case of the hybrid resonance (42), a similar change of properties occurs when the hybrid frequency crosses the *second harmonic* of the gyrofrequency. Even in the uniform case (13), some features of the resonances change at that critical value.

McAfee's work (40–43) was carried out throughout by consideration of rays in a slightly inhomogeneous plasma, and his concern lay in computing the paths. A more detailed treatment would require that one calculate the flux of signals propagating along such paths, and consider too the validity of a ray theory. Some work in this direction has recently been published by Fejer and Yu (44). They solve, in a particular case, for the electric fields by a WKB approximation. Several features of McAfee's heuristic explanation are confirmed. In particular, it is shown that the inhomogeneity substantially enhances the amplitude of the resonance as compared with the uniform case.

V. State of the Problem

A. *Resonant Frequencies*

In this final section, our aim is to summarize the theoretical work on this problem and to see how the predictions of the mathematical model compare with experiments. First, we shall consider identification of the frequencies of resonance and then go on to consider the strengths of resonance; in both cases we shall try to show whether the results from topside sounding satellites are adequately explained.

Table 1 lists all the resonant frequencies that have been observed or predicted. The principal ones were correctly identified in the very early observations (18–20), apart from some temporary confusion of the upper hybrid frequency with a zero range condition. The table makes no allowance for the slight shifts necessary if the plasma is not stationary, as is the case for satellite applications. In fact, these corrections are generally second order in the ratio V/v_{th}, where V is the satellite speed and v_{th} the electron thermal speed, and $V \ll v_{\text{th}}$.

The resonance at the *plasma frequency* was the most easy one to identify and is the only one that can exist in the absence of a magnetic field. Calvert

TABLE I
Resonant Frequencies

Frequency of resonance	Wavenumber and direction of propagation	References in which theory was developed	References in which observation was reported	Comments
ω_p, plasma frequency	$k = 0$, $\parallel \mathbf{B}$	13, 22, 23, 26	21	
ω_m, the upper hybrid frequency $(=(\omega_p^2 + \Omega^2)^{1/2})$	$k = 0$, $\perp \mathbf{B}$	13, 22, 23, 26	21	Wavenumber slightly shifted if c is finite (see Ref. 13).
$2\omega_m$	?	—	21	Not understood.
Ω, the cyclotron frequency	$k = 0$, $\perp \mathbf{B}$ $k = \infty$, $\perp \mathbf{B}$	13, 13	21	Only appears when c is finite.
$n\Omega$, harmonics of the cyclotron frequency $(n \geqslant 2)$	$k = 0$, $\perp \mathbf{B}$	13, 39	20, 21	Electrostatic approximation used.
$n\Omega$, harmonics of the cyclotron frequency	k small, $\perp \mathbf{B}$	27, 28	20, 21	Electromagnietc equations used.
$\approx n\Omega$, $\omega > \omega_m$ (eq. 4-2)	k finite and nonzero $\perp \mathbf{B}$	13, 39	50	
ω_1, ω_2, the zero range frequencies (eq. 2-29)	$k = 0$, $\perp \mathbf{B}$	13, 22	—	Very weak, probably indistinguishable from zero range limits of X and Z ionogram traces.
ω_L, the lower hybrid frequency	$k = 0$, $\perp \mathbf{B}$	29	—	Involves dynamics of the positive ions.

and Goe (21) checked that one of the resonances observed on the Alouette ionograms was at the plasma frequency, by comparison with the other traces on the ionogram and of sequences of ionograms. The comparison was excellent. As we have seen above, this resonance is predicted theoretically even if we use the simple cold electrostatic model for the plasma, so it is not surprising that it is always observed.

Calvert and Goe (21) also observed that a resonance was nearly always excited at the *upper hybrid frequency* $\omega_m = (\omega_p{}^2 + \Omega^2)^{1/2}$. As we have noticed in previous chapters, resonance at this frequency is to be expected. It corresponds to the condition $\varepsilon_\perp = 0$, where $\varepsilon_\perp$ is given by eq. 2-20.

Resonance at the *fundamental of the cyclotron frequency* is not invariably observed, although it was reported by Calvert and Goe (21). It is very often not observed if $\omega_p > \Omega$, and this is reflected by the relevant dispersion relation (Figures 7 and 8), remembering that resonances occur where the slope vanishes. Fejer and Calvert (23) endeavor to give an explanation of this resonance for propagation along the field but their analysis is questionable.

In the literature on resonances, probably most attention has been paid to resonances at *harmonics of the cyclotron frequency*. With the use of using the electrostatic approximation, the dispersion relation for waves propagating across the magnetic field is shown in Figure 7. From this, Sturrock (22), Dougherty and Monaghan (13), and Tataronis and Crawford (39) showed that resonances can be expected to occur at zero wave number, exactly at the harmonics of the cyclotron frequency. Resonances occurring at a series of harmonics were identified from the first Alouette I results (21) but their identification with harmonics of the cyclotron frequency depended on the correct evaluation of the magnetic field at the satellite (46). Benson (47,48) has made a thorough analysis of some results from Alouette I and II and has shown that the relations between the series of resonant frequencies differs from the true harmonic relationship by less than 0.2±. This improved an earlier result of Barrington and Herzberg (49).

There has, however, been serious criticism of the use of the electrostatic approximation in describing cyclotron harmonic resonances. Shkarofsky and Johnston (27,28) point out that, as k tends to zero, it is incorrect to assume that the velocity of light can be taken to be infinite; the dispersion diagram of Figure 7 must be replaced by a more accurate one such as Figure 8, resulting from the full electromagnetic equations, as we mentioned in Section II. Because of the complicated nature of the dispersion relation, however, these authors are not able to give exact results for the wave number of resonant waves; this is all the more difficult in the relativistic framework that they use. Now, the points on dispersion curves corresponding to resonance for a moving satellite are not the flat parts of the curves, but points at which the slope is equal to the satellite velocity. From Figure 8 we see that, no matter what the satellite velocity is, there are points on the electromagnetic dispersion curves that equal it, whereas in the electrostatic approximation this is not so; at higher harmonics the curves become too flat to accommodate even a small slope. In fact, for ionospheric parameters the electrostatic model only gives resonance points for cyclotron harmonics up to the hybrid frequency. It thus seems that higher resonances can only be explained using the full electromagnetic theory, but for the lower resonances the electrostatic model is adequate.

Next we must consider the "finite k" resonances (13, 39) occurring just above those cyclotron harmonics that exceed ω_m. For some time it was held that these were not observed, but it seems nearer the truth to say that they were not recognized. They do, in fact, quite commonly occur in the ionograms, though they are admittedly weaker than the set that are at the harmonics. Their identification (in 1968) was due to Warren and Hagg (50), who called them "Q-resonances." They do not occur at harmonics higher than about the fourth: the explanation of this may be that satellite motion alters the resonance condition so that it cannot be satisfied, an idea we described above in connection with the $k = 0$ resonances. Figure 12 is an ionogram which shows examples of the Q-resonances. To test the identification, Warren and Hagg constructed Figure 13. The dispersion relation for the Bernstein modes (eqs. 2-25 and 2-28) may be rewritten in the form

$$\frac{\omega}{\Omega} = F\left(\lambda, \frac{{\omega_p}^2}{\Omega^2}\right) \tag{5-1}$$

where, as before, $\lambda = KTk^2/m\Omega^2$ is a normalized value of k^2, and we normalize frequencies to the gyrofrequency. Whereas in Figure 7 we plotted ω against k for fixed ω_p/Ω, Figure 13 consists of a plot of ω/Ω against ${\omega_p}^2/\Omega^2$. The solid curves are theoretical values and are labeled by λ as a parameter. In this form of presentation the zero group velocity waves lie on the envelope of the sets of curves in each band $n < \omega/\Omega < n + 1$; these envelopes are shown dashed and would be straight if eq. 4-2 were exact. The dots are observational points. The conclusion that they appear to agree well with theory does of course involve an assumption that the correction for satellite motion is negligible.

The zero range resonances have not so far been observed experimentally, and we have already commented on their nature in Section IV-F.

So far, the resonance phenomena have been explained only in terms of electron dynamics. Deering and Fejer (29) pointed out that resonance should also occur at the lower hybrid frequency

$$\omega_L = \left[\frac{\Omega\Omega_i(\Omega\Omega_i + {\omega_p}^2)}{\Omega^2 + {\omega_p}^2}\right]^{1/2}$$

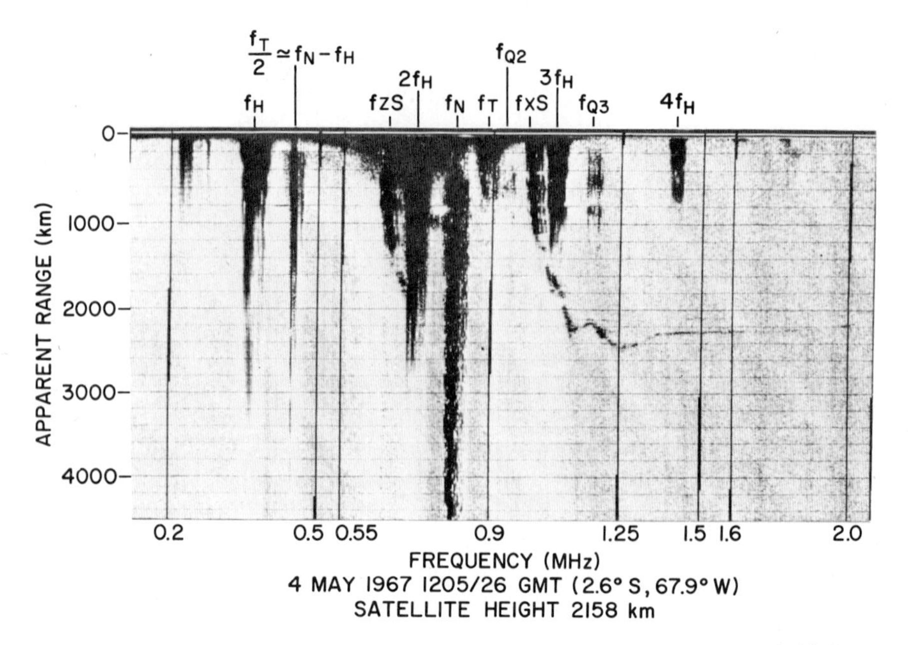

Fig. 12. A topside ionogram showing "*Q*-resonances" at the second and third cyclotron harmonic. There is also a "floating resonance." (Courtesy E. S. Warren and E. L. Hagg, Communications Research Centre, Dept. of Communications, Shirley Bay, Canada).

where Ω_i is the ion gyrofrequency. The existence of this resonance on satellite records does not appear to have been convincingly demonstrated, since it is generally at very low frequency.

One other resonance, which we have not so far mentioned, is the "floating" resonance. This is observed only rarely and occurs when, simultaneously, $\omega = \frac{1}{2}\omega_m = \omega_p - \Omega$. If this is correct, its rarity is a natural consequence of the fact that the second equality here does not generally hold, as it requires $\omega_p \approx 2.22\Omega$. It has the further very distinctive feature that there is a small time delay before it is received, so that instead of being attached to the top line $h = 0$ on an ionogram, it appears to float a little way below. Figure 12 was selected to show this feature, in addition to the *Q*-resonances. It should be emphasized that its appearance is far more rare than that of the latter.

B. *Strength and Duration of Resonances*

We now come to the more difficult problem of what we can say about the strength of the resonances observed. How closely can we predict their amplitude and duration, and how easily can we test our predictions? Unfortunately,

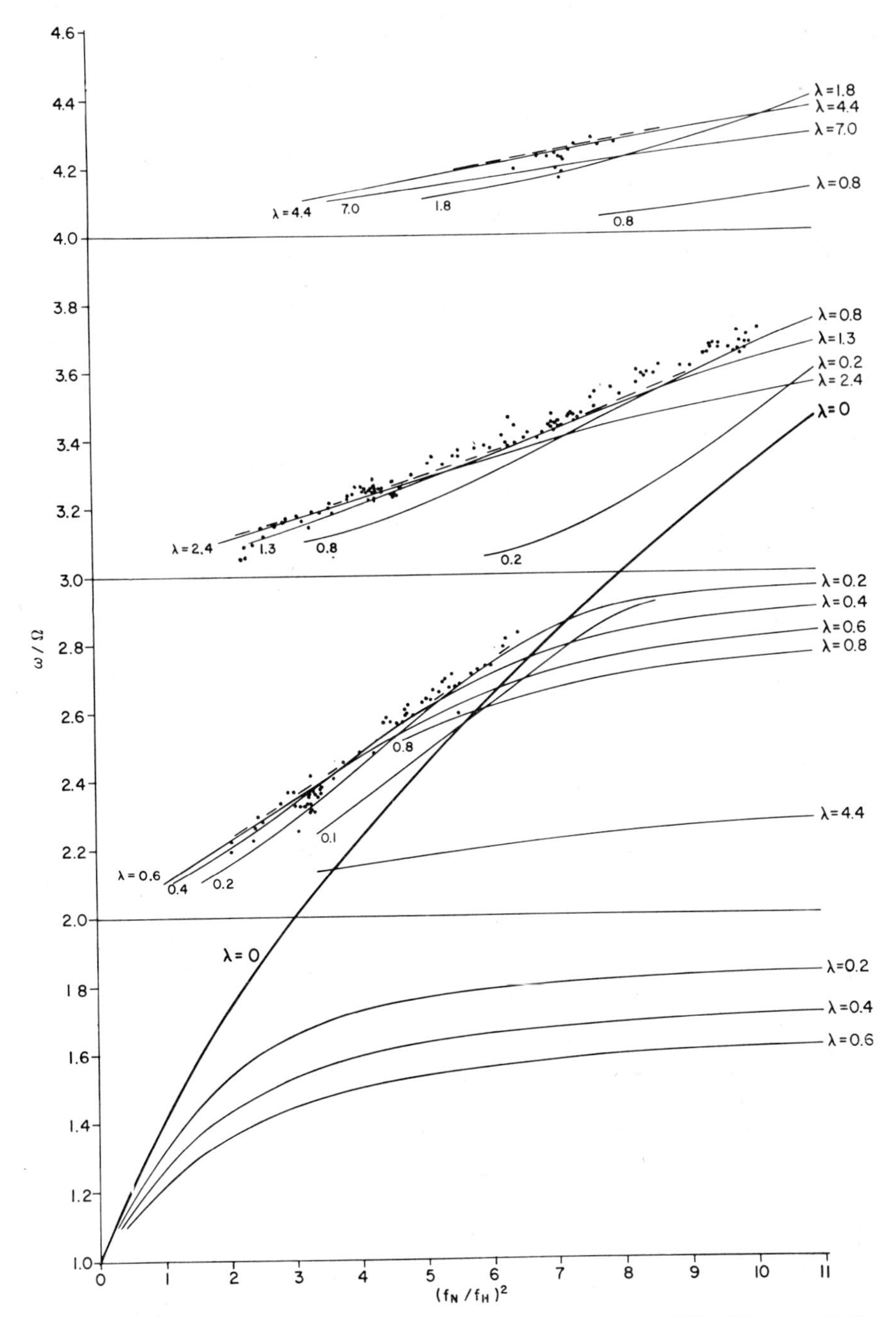

Fig. 13. Analysis of the observed frequencies of the *Q*-resonances, (50). (Courtesy E. S. Warren and E. L. Hagg.)

we can only answer these questions rather poorly, in both cases because of the complexity of the physical situation. Before exploring these questions, we should describe what the experimental arrangement was on the topside sounding satellites.

In each case the satellite was equipped with a pair of antennas between 20 and 40 meters long. The antenna was fed from the center with a pulse lasting 100 μsec; then it was switched off for the same period; and then it was converted into a receiver for 15 msec, while the operator listened for any possible echo. If the original pulse contained a contribution at a resonance frequency, the echo would sound as soon as the receiver was switched on.

All the theoretical models we have considered so far are obviously a rather poor representation of this physical situation. First, we ought to solve the self-consistent problem of determining the electric fields around the antenna from the given voltage applied at its ends, using the boundary condition on the surfaces; this should be more confused by accounting for the plasma sheath which appears around any conductor placed in a plasma, but which we have studiously avoided. Second, the actual readings taken by the receiver are of incoming antenna current, not of the field strength; sometimes the two are related only in a rather complicated way because of the automatic gain control used on the satellites.

Despite all this, attempts have been made to compare theory with experiment, although only one approach seems possible. Sturrock (22) was the first to notice that one aspect of the resonance observations which could be simply related to the observed electric field strengths was the time for which the resonant signal could be distinguished from background galactic noise. These times are easily picked out from the ionograms (see Figure 3) and are nearly always between 0.5 and 20 msec. Sturrock (22) evaluated these times from his mathematical model and produced rather longer times than these; but on correcting a mistaken assumption by him about galactic noise level, noticed by Shkarofsky and Johnston (28), they become reasonable. However, the same argument applied by these latter authors to their own mathematical model yields ringing times of only about 20 μsec which would die away before the receiver is switched on. Incidentally, independent measurement of the galactic noise level has been given by Hartz (51).

Provided that one can avoid singular points, the basic formula for this problem is eq. 3-20. Actually, most of the authors we have cited set up their theory in such a way that singular points arise; but we have seen that they can be avoided, with a closer approach to the physics, by introducing a satellite velocity or (equivalently) using a reference frame fixed with the antenna, so that the plasma streams at the negative of the satellite velocity (17). It is then relatively straightforward to insert typical experimental parameters into the first term of the series (3-20) to work out the time taken for the electric field strengths of the resonances to decay to the level of background galactic noise. Considering the many assumptions and approximations made in this theory, we find it gratifying that the ringing times obtained by this method are close to those observed for several of the resonances (17,28).

We can, obviously, use expressions of the form (3-20) for further predictions about the resonances. Dependence on orientation of the antenna with respect to the magnetic field, and on the velocity, can be estimated for example, but the asymptotic nature of the expression stops us from attaching much significance to this result. The calculation of ringing times is probably the most reliable fact we can draw from theoretical calculations of field strength, and even this is dubious because of our lack of knowledge of the range of validity of the asymptotic expression in eq. 3-20. Only if we had some measure of the time at which the expression becomes a valid approximation would we be justified in making further deductions from our model, and this seems to be a very difficult task because of the nature of the expresssion (3-20).

C. *Conclusion*

This concludes our review of the theoretical work that has been stimulated by the discovery of the plasma resonances by ionospheric topside sounders. A considerable level of understanding has been reached, e.g., in the identification of the resonant frequencies. The work is, to a certain extent, a satisfying illustration in detail of some of the complex properties of plasmas: the propagation of Bernstein modes and the role of the vector group velocity. However, rather extensive approximations and assumptions have been made; moreover, there are several properties such as the fringe patterns and the floating resonance, as yet inadequately treated.

The way forward in this field would seem to be on several fronts. A detailed analysis should be conducted on how resonances and their strengths are changed when the directions of antenna orientation, magnetic field, and satellite velocity vary with respect to each other. The general problem of the characteristics of an antenna, both in transmitting and receiving, should be analyzed more closely when it is immersed in a plasma surrounded by a sheath. But perhaps the most interesting problem that would profit from further study is the subtle behavior of the dispersion relations when the plasma parameters move through certain critical values, and when the directions of propagation departs from being along or across the magnetic field. In perhaps no other branch of plasma physics are the observations so crucially dependent on the balance of all the factors involved.

References

1. J. H. Chapman, and E. S. Warren, *Space Sciences Reviews*, **8**, 846 (1968).
2. W. E. Gordon, *Revs. Geophys.*, **5**, 191 (1967).
3. K. G. Budden, *Radio Waves in the Ionosphere*, Cambridge University Press, London, 1961.
4. J. A. Ratcliffe, *Magneto-ionic Theory*, Cambridge University Press, London, 1959.
5. K. Davies, "*Ionospheric Radio Propagation*," U. S. Department of Commerce, National Bureau of Standards Monograph 80, 1965.
6. W. J. Calvert, *Geophys. Res.*, **71**, 5579 (1966).
7. G. Landauer, *J. Nucl. Energy*, Part C, **4** 395 (1962); *Phys. Letters*, **25A** 562 (1967).
8. G. Bekefi, J. D. Coccoli, E. B. Hooper, and S. J. Buchsbaum, *Phys. Rev. Letters*, **9**, 6 (1962).
9. F. W. Crawford, Proc. NATO Advanced Study Institute, Røros, Norway; Edinburgh University Press, Edinburgh, Scotland, 1969.
10. T. H. Stix, *The Theory of Plasma Waves*, McGraw-Hill, New York, 1962.
11. D. C. Montgomery, and D. A. Tidman, *Plasma Kinetic Theory*, McGraw-Hill, New York, 1964.
12. P. C. Clemmow, and J. P. Dougherty, *The Electrodynamics of Particles and Plasmas*, Addison Wesley, Reading, Mass. 1969.
13. J. P. Dougherty, and J. J. Monaghan *Proc. Roy. Soc.*, *Ser.* **289**, 214 (1965).
14. F. Pham, *Introduction a l'Étude Topologique des Singularitées de Landau*, Gauthier-Villars, Paris, 1967.
15. R. C. Hwa, and U. L. Teplitz, *Homology and Feynmann Integrals*, Benjamin, New York, 1966.
16. N. Chako, *J. Inst. Math. and Appl.*, **1**, 373 (1965).
17. S. R. Watson, Proc. NATO Advanced Study Institute, Røros, Norway; Edinburgh University Press, Edinburgh, Scotland, 1969.
18. R. W., Knecht, T. E. Van Zandt, and S. Russell, *J. Geophys. Res.*, **66**, 3078 (1961).
19. R. W. Knecht, and S. Russell, *J. Geophys. Res.*, **67**, 1178 (1962).
20. G. E. K. Lockwood, *Can. J. Phys.*, **41**, 190 (1963).
21. W. Calvert, and G. B. Goe, *J. Geophys. Res.*, **68**, 6113 (1963).
22. P. A. Sturrock, *Phys. Fluids*, **8**, 88 (1965).
23. J. A. Fejer, and W. Calvert, *J. Geophys. Res.*, **69**, 5049 (1964).
24. T. W. Johnston, and J. Nuttall, *J. Geophys. Res.*, **69**, 2305 (1964).
25. J. Nuttall, *Phys. Fluids*, **8**, 286 (1965).
26. J. Nuttall, *J. Geophys. Res.*, **70**, 1119 (1965).
27. I. P. Shkarofsky, and T. W. Johnston, *Phys. Rev. Letters*, **15**, 51 (1965).
28. I. P. Shkarofsky, and T. W. Johnston, RCA Victor Research Report 7-801-35, 1965.
29. W. D. Deering, and J. A. Fejer, *Phys. Fluids*, **8**, 2066 (1965).
30. I. P. Shkarofsky, *Phys. Fluids*, **9**, 561 (1966).
31. I. P. Shkarofsky, *Phys. Fluids*, **9**, 570 (1966).
32. I. P. Shkarofsky, *J. Geophys. Res.*, **15**, 4859 (1968).
33. S. R. Watson, Ph.D. Thesis, University of Cambridge, Cambridge, England, 1968.
34. J. A. Fejer, and Wai-Mao Yu, *J. Plasma Phys.*, **2**, 227 (1968).
35. T. D. Mantei, Stanford University Institute for Plasma Research Report No. 194, 1967.
36. F. W. Crawford, R. S. Harp, and T. D. Mantei, *Phys. Rev. Letters*, **17**, 626 (1966).
37. F. W. Crawford, R. S. Harp, and T. D. Mantei, *J. Geophys. Res.*, **72**, 57 (1967).

38. J. A. Tataronis, Stanford University Institute for Plasma Research Report No. 205, 1967.
39. J. A. Tataronis, and F. W. Crawford, Stanford University Institute for Plasma Research Report No. 234, 1968.
40. J. R. McAfee, *J. Geophys. Res.*, **73**, 5577 (1968).
41. J. R. McAfee, *J. Geophys. Res.*, **74**, 802 (1969).
42. J. R. McAfee, *J. Geophys. Res.*, **74**, 6403 (1969).
43. J. R. McAfee, *J. Geophys. Res.*, **75**, 4287 (1970).
44. J. A. Fejer, and Wai-Mao Yu, *J. Geophys. Res.*, **75**, 1919 (1970).
45. W. Calvert, and T. E. Van Zandt, *J. Geophys. Res.*, **71**, 1799 (1966).
46. E. L. Hagg, *Physics in Canada*, **19**, 3 (1963).
47. F. R. Benson, Proc. NATO Advanced Study Institute, Røros, Norway; Edinburgh University Press, Edinburgh, Scotland, 1969.
48. R. F. Benson, *Proc. IEEE*, **57**, 1139 (1969).
49. R. E. Barrington, and L. Herzberg, *Can. J. Phys.*, **44**, 987 (1966).
50. E. S. Warren, and E. L. Hagg, *Nature*, **220**, 466 (1968).
51. T. R. Hartz, *Nature*, **203**, 173 (1964).

Two-Stream Instabilities*

RICHARD J. BRIGGS

Department of Electrical Engineering,
Research Laboratory of Electronics,
Massachusetts Institute of Technology, Cambridge, Massachusetts

Introduction

The two-stream instability is perhaps the most extensively investigated microinstability in plasma physics. Since the first positive experimental demonstration of beam-plasma amplification (1), there have been a large number of experimental investigations of beam-plasma interactions. Many theoretical treatments of this problem have included effects that are not often faced in plasma theory such as the finite geometry of the system. At this time,

* This work was supported by the National Science Foundation Grant GK-10472.

however, the detailed correlation between experiment and theory, even in the simplest configurations and even well within the linear regime, is still not wholly satisfactory (2–4). Certainly, at least part of the blame must be laid on the simplifications that are still inherent in a tractable theory.

There have already been many review articles and books that treat the subject of two-stream instability. (5–10) These reviews have provided good bibliographies of the work prior to about 1964; hence the references in this article will be largely limited to those that are particularly basic or that have appeared in more recent years. The main goal of this article is to present a fairly coherent story of linear two-stream instability theory in a way that emphasizes some of the methodology that is employed in the analysis and description of microinstabilities. For example, we shall use the concept of negative energy waves to interpret a number of the results. With this approach, it is hoped that the present article will be of use to intermediate-level graduate students as a "case study in plasma microinstabilities," as well as to those readers interested in the details of two-stream instabilities. We have not included any descriptions of published experimental results and their comparison with theory, partly in the interests of brevity and partly because of the large gaps still remaining in the overall picture.

Our approach will be to analyze the simplest possible limiting cases first, in order to provide a framework in which one can understand and estimate the influence of various physical effects left out of the zero-order theory. In this way, we shall be able to show how the "hydrodynamic" description of monoenergetic beams connects with the velocity distribution description. For clarity and to limit the scope of the investigation, we shall concentrate on those modes that are not dependent upon the magnetic field

In the first section, the equations describing a single monoenergetic stream in an infinite uniform system are developed. The two-stream instability in its simplest form is then illustrated by superimposing two such monoenergetic streams. We also show how a superposition of a large number of streams can describe the modes in a plasma system with a continuous velocity distribution. In the second section, we study the interaction of a low-density beam with a stationary plasma, again in the limit of infinite uniform media. The case of a low-density beam in a stationary plasma is of particular importance, since this is very often the situation that is realized in practice. By taking the ratio of beam density to plasma density as a small expansion parameter, we are able to develop an analytic theory for cases of nonzero plasma and beam temperature, and nonzero collision frequencies. In the general case, the results are rather complicated; therefore one of the main purposes of our analysis in this section is to show how one can determine criteria for deciding when a given physical effect is important. We can then show under what circumstances one has the classical "bump in tail" configuration and when the monoenergetic beam description is the appropriate one.

In the third section, we begin the analysis of radially finite systems. The basic assumption in Section III is that a very strong axial magnetic field confines the particle motion to the axial direction. In this case, one has a set of discrete eigenmodes, and rather general density profiles can be handled in a straightforward way. The major new phenomena that occurs is that the system can be stable for all wavelengths if the diameter is small enough, even with monoenergetic streams.

In Section IV, we treat the case of finite inhomogeneous systems in the absence of a magnetic field. It is perhaps surprising that this case raises a number of far more subtle and complicated questions than the infinite magnetic field situation. For a cold beam, there are no discrete eigenmodes when the density profiles are identical. It turns out in this case that there is a continuous spectrum of unstable eigenfrequencies. Almost all previous analyses of finite systems in finite or zero magnetic fields have assumed sharp boundaries for the beam and plasma density profiles, and in this case one does deal with discrete eigenmodes. The analysis in the last section brings us to the limit of research as of 1968; it indicates that a number of important physical effects may arise when smooth gradients in the densities are included in the theory.

I. The Multistream Model

In this section the simplest possible situation is analyzed; namely, an infinite uniform system in the absence of an applied magnetic field. A single monoenergetic stream is analyzed first, and its small-signal energy aspects are discussed. Then, in Section I-B, elementary situations with two monoenergetic streams are presented to illustrate the basic two-stream mechanism. In Section I-C, the dispersion relation for an arbitrary velocity distribution is developed as the limit of a large number of monoenergetic streams and general Nyquist methods for obtaining the stability boundaries are discussed.

A. *Properties of a Single Monoenergetic Stream*

In order to understand many of the features of the two-stream instability, it is useful to consider first the wave properties of a single directed stream of charged particles. The basic equations used to describe the longitudinal space-charge oscillations of a zero-temperature particle stream are the force law

$$\frac{dv}{dt} = \frac{\partial v}{\partial t} + v\frac{\partial v}{\partial z} = \frac{q}{m}E \tag{1-1}$$

charge conservation,

$$\frac{\partial J}{\partial z} = -\frac{\partial \rho}{\partial t} \tag{1-2}$$

and Gauss's law,

$$\frac{\partial E}{\partial z} = \frac{\rho}{\varepsilon_0} \tag{1-3}$$

where the velocity, v, the current, J, and the electric field, E, are z-directed and ρ is the charge density in the stream. (Note that we use mks units in the present paper.) In order to discuss the small amplitude waves that vary as

$$\exp i(kz - \omega t),$$

we linearize the above equations about the unperturbed state of constant drift. For example,

$$v = v_0 + v_1 \tag{1-4}$$

$$\rho = \rho_0 + \rho_1 \tag{1-5}$$

and

$$J = \rho_0 v_0 + \rho_1 v_0 + \rho_0 v_1 = J_0 + J_1 \tag{1-6}$$

where we use a subscript zero to indicate unperturbed quantities and a subscript one to indicate perturbation quantities. From the linearized form of eqs. 1-1 and 1-2 and the assumed wave dependence, we can derive the following form of the perturbation current:

$$J_1 = i\omega\varepsilon_0 \frac{\omega_p{}^2}{(\omega - kv_0)^2} E_1 \tag{1-7}$$

where we have introduced the plasma frequency,

$$\omega_p{}^2 = \frac{q\rho_0}{\varepsilon_0 m} \tag{1-8}$$

If we add the displacement current to the conduction current, we can formally define a dielectric constant as

$$J_1 - i\omega\varepsilon_0 E_1 = -i\omega\varepsilon_0 \varepsilon E_1 \tag{1-9}$$

with

$$\varepsilon = \left(1 - \frac{\omega_p{}^2}{(\omega - kv_0)^2}\right) \tag{1-10}$$

The dispersion relation for our system can be shown from the linearized form of eqs. 1-2 and 1-3 to be given by $\varepsilon = 0$, or

$$\omega = \begin{cases} kv_0 + \omega_p & \text{(the "fast space-charge wave")} \\ kv_0 - \omega_p & \text{(the "slow space-charge wave")} \end{cases} \tag{1-11}$$

These space charge waves are, of course, simply the well-known space-charge oscillations of a stationary plasma doppler-shifted by the beam velocity, v_0.

One of the interesting aspects of the space-charge waves is the fact that the directed energy of the stream can lead to a negative small-signal energy. To show this, we recall that the time-average, small-signal energy density of these electrostatic oscillations is given by (11)

$$\langle W_{ek} \rangle = \frac{\varepsilon_0}{4} \frac{\partial}{\partial \omega} (\omega \varepsilon) |E_1|^2 \tag{1-12}$$

If we evaluate the small-signal energy using the dielectric constant given in eq. (1-10), we find that

$$\langle W_{ek} \rangle = \pm \frac{\omega}{\omega_p} \frac{1}{2} \varepsilon_0 |E_1|^2 \tag{1-13}$$

where the upper sign applies to the fast space-charge wave and the lower sign applies to the slow space-charge wave. The slow space-charge wave carries negative small-signal energy (for $\omega > 0$, or $kv_0 > \omega_p$, clearly).

We can attach a physical meaning to this negative energy by calculating the relative phase of the perturbation velocity and the perturbation density. It is easily shown that the perturbation velocity for the fast space-charge wave is in phase with the density perturbation, whereas the velocity perturbation for the slow space-charge wave is 180° out of phase with the density perturbation. The total density and velocity of the stream would therefore appear as shown in Figure 1. For the fast space-charge wave, in regions in which the density is increased by the perturbation, the particles are traveling faster than the average velocity; conversely, in regions of decreased density, the particles are traveling slower than the average velocity. Therefore, the net kinetic energy carried by the stream is *greater* than the unperturbed energy when the fast space-charge wave is excited. Conversely, for the slow space-charge wave the electrons travel slower than the average in regions of increased density, and faster than the average in regions of decreased density; hence, the net kinetic energy carried by the stream is *less* in the presence of the perturbation

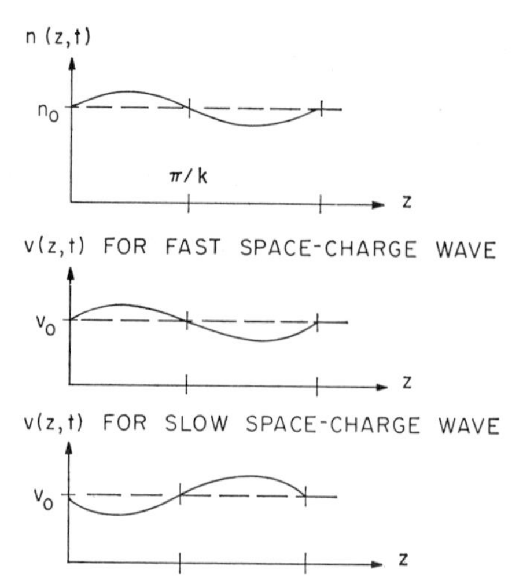

Fig. 1. View of density and velocity in a single stream for the physical interpretation of the negative-energy slow wave.

than it is for the unperturbed system. This is the origin of the negative perturbation energy associated with the slow space-charge wave. The physical significance of a negative perturbation energy is that one must *remove* energy from the wave in order to *increase* its amplitude.

The knowledge of these energy aspects of the space-charge waves allows us to understand situations that otherwise might seem somewhat puzzling. As an example, consider a small perturbation in the preceding situation by imagining the electron stream to permeate a medium with a very small conductivity σ. Since the conducting medium leads to a positive dissipation ($\frac{1}{2}\sigma|E_1|^2$), we should expect on the basis of small-signal energetics that the fast space-charge wave would be damped by the conducting medium, whereas the slow space-charge wave would become unstable. We can, of course, derive this result directly by adding the conduction current, σE_1 to the particle and displacement currents in eq. 1-9. If we do this and solve the resulting dispersion equation, we obtain (for $\sigma \ll \omega\varepsilon_0$)

$$\omega - kv_0 = \pm\omega_p\left(1 - i\frac{\sigma}{2\omega\varepsilon_0}\right) \tag{1-14}$$

As expected, the fast space-charge wave (the upper sign in eq. 1-14) is exponentially damped in time, whereas the slow space-charge wave (the lower sign in eq. 1-14) increases exponentially in time.

An important consequence of these ideas for the two-stream problem is that losses are not necessarily a stabilizing influence, as we shall see in the following sections. In the case of an electron stream passing through a stationary plasma, collisional losses or losses due to Landau damping in the plasma will, under some circumstances, act to reduce the growth rate of an instability; however, under other circumstances these losses may actually provoke instability in regions where the "lossless" model predicted stable waves.

B. *Basic Two-Stream Interaction*

One of the simplest limiting cases that we can analyze is that of a number of one-dimensional monoenergetic streams. The dispersion relation for this system follows immediately from our previous analysis of the single monoenergetic beam. The current contributed by all of the streams must now be included in eq. 1-9. The perturbation current of each stream is given by eq. 1-7, and therefore the dispersion relation is

$$\varepsilon = 1 - \sum \frac{{\omega_{pj}}^2}{(\omega - kv_{0j})^2} = 0 \tag{1-15}$$

where ω_{pj} and v_{0j} are the plasma frequencies and drift velocities of the jth stream. For two streams, we have the dispersion relation

$$1 = \frac{{\omega_{p1}}^2}{(\omega - kv_1)^2} + \frac{{\omega_{p2}}^2}{(\omega - kv_2)^2} \tag{1-16}$$

In Figure 2, we illustrate the real values of ω and k obtained from eq. 1-16 for the case in which the drift velocities of the stream are in the same direction (case *a*), in opposite directions (case *b*), and for the case in which one of the streams is stationary (case *c*). In all of these examples, there is a band of real k for which complex conjugate values of ω are obtained. The limits of this band occur at the points $k = \pm k_c$, where two of the real roots of ω disappear; therefore, in all cases there is a *minimum* wavelength $(2\pi/k_c)$ for instability. As an example, for equal density streams $(\omega_{p1} = \omega_{p2} = \omega_p)$, one can show from eq. 1-16 that the critical wave number is

$$k_c = \frac{2\sqrt{2}\,\omega_p}{|v_1 - v_2|} \tag{1-17}$$

and that the maximum *temporal* growth rate (max Im ω for real k) is

$$(\omega_i)_{\max} = \tfrac{1}{2}\omega_p \tag{1-18}$$

In any laboratory investigation of the two-stream instability, the apparatus *length* is limited and the description of the unstable wave in terms of the *temporal* growth rate (Im ω for real k) is often not the most useful one. The majority of such experiments involves an excitation of the system at a given

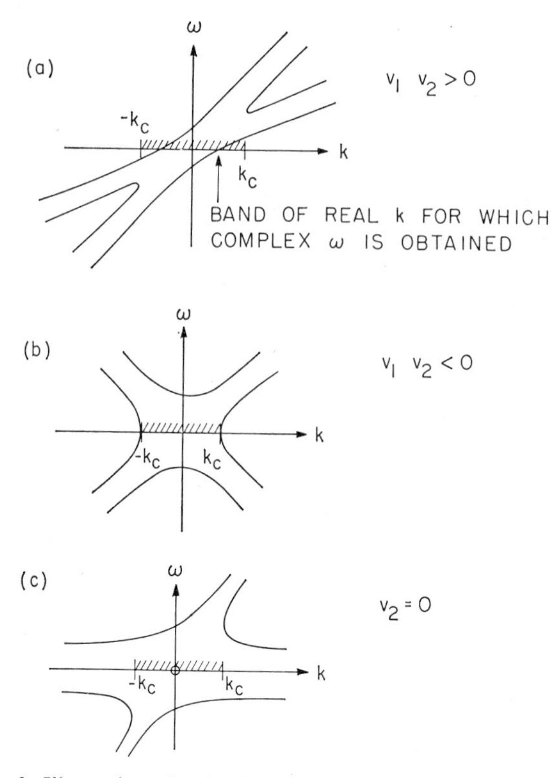

Fig. 2. Illustration of ω-k relation for two monoenergetic streams.

frequency and an observation of the *spatial* amplification of the signal. A detailed discussion of the connection between the descriptions in terms of spatial and temporal growth, with applications to the two-stream instability problem, have been given elsewhere (8). Briefly stated, the essential point is the distinction as to whether an initially localized disturbance (on an infinite system) grows exponentially with time locally (absolute instability) or ultimately decays because of the propagation of the growing disturbance away from its point of origin (convective instability). For a *convective* instability the description in terms of a spatial amplification of a sinusoidal oscillation is generally more useful, since it gives directly the maximum growth of the disturbance in a given length. This spatial amplification rate is, of course, obtained by solving the dispersion relation for Im k with ω real; we note in Figure 2 that in all cases there is a band of real ω for which complex conjugate k roots are obtained.

The mathematical technique for determining which roots of k correspond to spatial amplification (and not decay) consists of following the Im k for a particular root as Im ω goes from 0 to $+\infty$; if Im k changes sign in the process, that root represents spatial growth for real ω. Absolute instability, on the other hand, obtains whenever there is a double root of k forming for some ω (with $\omega_i > 0$), where the merging roots have different signs of Im k for $\omega_i \to +\infty$. Using this technique, one can show that the counterflowing streams (case b in Figure 2) result in absolute instability, in agreement with our physical intuition (there is internal feedback). A solution of the dispersion relation for complex k and real ω is meaningless in this case. For the coflowing streams (case a), on the other hand, it can be shown easily that Im $kv_{1,2} \to +\infty$ as $\omega_i \to +\infty$ for all roots, and hence the instability is necessarily convective. The roots of complex k with Im $kv_{1,2} < 0$ for real ω represent spatial amplification (in the direction of the stream motion). The "marginal" case of one stream stationary (case c) is a bit special; since previous discussions of this case have not been strictly correct, it is discussed in Section V in more detail.

In the following sections, we shall generally not discuss in detail the distinctions between absolute and convective instabilities, and the calculations of spatial as well as temporal growth rates for the examples in question. This omission is not intended to imply that the distinction is not important for a complete theory, but the existence of temporal growth is certainly a *necessary* condition for any form of instability or amplification, and the emphasis in this review is on the description of the influence of various physical factors (temperature, geometry, etc.) on the instability mechanisms.

In order to discuss the physical meaning of the two-stream instability, let us consider the case of a stream permeating a stationary plasma. This particular case is of considerable interest in many experimental configurations, and we might also note that in the context of the infinite medium problem, we could always transform into a reference frame for which one of the streams appears stationary. The dispersion relation for this situation is

$$\frac{\omega_{pb}{}^2}{(\omega - kv_0)^2} = \varepsilon_p = 1 - \frac{\omega_p{}^2}{\omega^2} \tag{1-19}$$

where we shall always use ω_{pb} and v_0 to indicate quantities appropriate to the beam in the stream-plasma situation.

Let us now imagine that we have a *frozen perturbation* of charge density on the beam of the form exp ikz moving through the plasma and inquire about the forces acting on the beam particles. In the laboratory frame, the plasma will see an alternating field of frequency kv_0, and if $kv_0 < \omega_p$, the plasma is equivalent to a dielectric with a negative dielectric constant, as far as the phase of electric field relative to the *beam*-charge perturbation is concerned. (See eq. 1-19). That is, in the presence of a dielectric with a negative dielectric constant, the electric field relative to the beam-charge density "bunches" would be *reversed* from its free-space value. Therefore the forces on the beam bunches are such as to enhance the bunching for $kv_0 < \omega_p$; the perturbation is unstable.

This argument, reasoning from an imagined "frozen" beam-charge perturbation, is strictly valid only in the limit $\omega_{pb} \to 0$. To determine the precise real k values for which complex ω is obtained, we derive the real k and real ω values for which $\partial k/\partial \omega = 0$ from eq. 1-19. The result is that the wave is unstable for real k in the range

$$kv_0 \leq \omega_p \left(1 + \frac{\omega_{pb}^{2/3}}{\omega_p^{2/3}}\right)^{3/2} \tag{1-20}$$

which is roughly $kv_0 < \omega_p$ for $n_b \ll n_p$.

C. *Dispersion Relations for Velocity Distributions*

The essential physical features of the two-stream instability are most clearly illustrated by the limit of monoenergetic streams; these results are also directly applicable to many situations involving injected beams in plasmas. In other applications, and particularly in situations in which the two-stream instability is an undesirable effect, a study of the stabilizing influence of a spread in the unperturbed velocities ("temperature" in the stream or plasma) is of paramount importance.

We can easily generalize our previous results to describe a general distribution of unperturbed velocities within a stream. To do this, we proceed from eq. 1-15 and pass to the limit of an infinite number of streams (12). For a continuous unperturbed distribution of velocities, $f_0(v)$, the plasma frequency or density of the jth stream is given by

$$\omega_{pj}{}^2 = \omega_p{}^2 f_0(v)\Delta v \tag{1-21}$$

where the unperturbed velocity of the jth stream lies between v and $v + \Delta v$. The sum in eq. 1-15 becomes an integral, and we obtain the dispersion relation

$$0 = 1 - \omega_p{}^2 \int_{-\infty}^{+\infty} \frac{f_0(v)\, dv}{(\omega - kv)^2} \tag{1-22}$$

where we assume for simplicity that all streams are comprised of a single type of particle (say electrons). To include different types, we merely sum over the various species. An alternate form of the above dispersion relation can be obtained by integrating by parts over velocity to give

$$0 = 1 - \frac{\omega_p{}^2}{k^2} \int_{-\infty}^{+\infty} \frac{f_0'(v)\, dv}{(v - \omega/k)} \tag{1-23}$$

The dispersion relation that we have obtained is identical to the one that is obtained by working with the Vlasov equation (13). This is not surprising, since in our formulation of the "multistream model," we have implicity assumed that the electric field in eq. 1-1 is identifiable with the macroscopic electric field. This is precisely the same assumption that one makes in the collisionless Vlasov equation. To proceed from the Vlasov equation, we use the linearized form

$$\frac{\partial f_1}{\partial t} + v \frac{\partial}{\partial z} f_1 + \frac{qE_1}{m} n_0 \frac{\partial f_0}{\partial v} = 0 \tag{1-24}$$

where f_1 is the perturbation in the distribution function f which is the number of particles contained in a volume element in phase space, $\Delta v \, \Delta z$. The unperturbed velocity distribution function f_0 is defined such that $n_0 f_0$ is the unperturbed density of particles in phase space. In the distribution function analysis, the macroscopic charge density perturbation is $\int_{-\infty}^{+\infty} f_1 \, dv$, so that Poisson's equation becomes

$$\varepsilon_0 \frac{\partial E_1}{\partial z} = q \int_{-\infty}^{+\infty} f_1 \, dv \tag{1-25}$$

If we assume the wave dependence exp $i(kz - \omega t)$, we obtain the dispersion relation eq. 1-23.

In deriving the dispersion relation for the finite temperature stream, we have ignored one important aspect of eq. 1-23. That is, the function

$$F\left(\frac{\omega}{K}\right) = \int_{-\infty}^{+\infty} \frac{f_0'(v) \, dv}{(v - \omega/k)} \tag{1-26}$$

is discontinuous across the line $\mathrm{Im}(\omega/k) = 0$ for all real values of $\omega/k = u$ for which $f_0'(u) \neq 0$. This phenomenon is related to the existence of a continuous spectrum of modes in the limit of a continuous distribution of velocities; extensive discussions of this aspect of the problem are given in many plasma texts and articles (10,14–16). Since we are interested in *unstable* modes in the present article, we can deal with the function $F(\omega/k)$ as defined for Im $\omega \geq 0^+$ and ignore the continuous mode spectrum, as far as stability calculations are concerned.

For smooth distribution functions, one can formulate rather general conditions on $f_0(v)$ for the existence of unstable modes (10,16,17). In terms of the function $F(u)$, where we use u as the phase velocity ω/k, the dispersion relation is

$$F(u) = k^2/\omega_p{}^2 \tag{1-27}$$

For an unstable mode, there must exist a value of ω with a positive imaginary part that satisfies eq. 1-1 for some real value of k. Since k is a free parameter in this analysis, a necessary and sufficient condition for instability that

$$\operatorname{Re} F > 0, \qquad \operatorname{Im} F = 0 \tag{1-28}$$

for some $u = u_r + iu_i$ with $u_i > 0$. We discuss explicitly only the case of k positive; the case of negative k carries through in exactly the same way and is equivalent to letting $f_0(v) \to f_0(-v)$ (17).

To apply eq. 1-28, we consider the mapping of the upper half u-plane into the complex F-plane, shown in Figure 3. Moving along the real u axis from $-\infty$ to $+\infty$, we traverse a closed curve in the F-plane. (The return path at

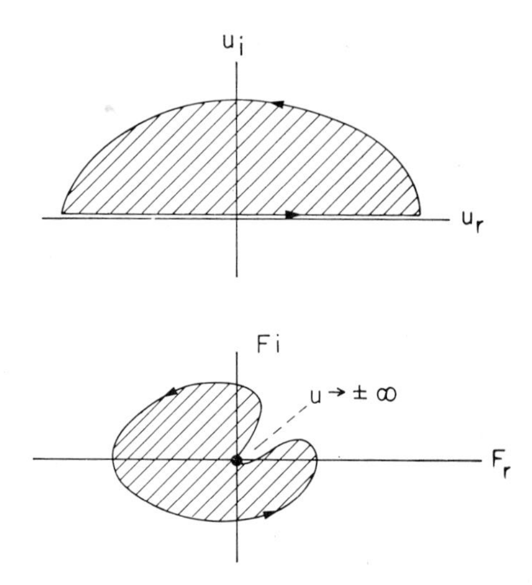

Fig. 3. Conformal mapping of u-plane into F-plane.

$u = \infty$ corresponds to the origin in the F-plane.) Since the region inside the closed contour in the u-plane (shaded region in Figure 3) maps into the shaded region in the F-plane, we can conclude that the necessary and sufficient conditions for an unstable mode are that eq. 1-28 be satisfied for a *real* value of u, $u = u_r + i0$. Physically, this result implies that the existence of an unstable mode occurs if and only if there also exists a marginally stable (or neutrally stable) mode for some real value of k.

For real values of u, in the limit as u_i tends to zero from positive values, the function $F(u_r)$ becomes

$$F(u_r + j0) = P \int_{-\infty}^{+\infty} \frac{f_0'(v)\, dv}{(v - u_r)} + i\pi f_0'(u_r) \tag{1-29}$$

where P denotes the principal value of the integration which is real. It is clear, therefore, that a *necessary* condition for instability is that

$$f_0'(u_r) = 0 \tag{1-30}$$

The point of zero slope in $f_0(v)$ must also be a local minimum of "sufficient depth," the precise criteria follows from the additional requirement that the real part of $F(u_r)$ be positive in eq. 1-29. Integrating by parts (cf eq. 1-22), we obtain

$$\int_{-\infty}^{+\infty} \frac{f_0(v) - f_0(u_r)}{(v - u_r)^2}\, dv \geq 0 \tag{1-31}$$

where the principal value prescription is unnecessary, since $f_0'(u_r) = 0$ by assumption. Equations 1-30 and 1-31 constitute necessary *and* sufficient conditions on the distribution function, $f_0(v)$, for the existence of unstable modes. They were first derived by Penrose and are sometimes referred to as the Penrose criterion (17).

II. Interaction of a Low-Density Beam with a Stationary Plasma

The most common way of realizing the two-stream environment in practice is by injecting an electron beam into a stationary plasma. In such cases the density of the beam is typically much less than the density of the stationary plasma, i.e., $n_b \ll n_p$. In this section we examine this situation in some detail and calculate the growth rates for various ranges of the parameters.

We shall investigate first the limit of a zero-temperature beam and plasma, and we will then consider the temperature effects of each component separately. One of the main objectives of the analysis in this section is to develop techniques and intuition for understanding the transitions between these various limiting cases.

A. *The Cold Beam and Cold Plasma Limit*

The dispersion equation for electrostatic waves in a zero-temperature plasma permeated by a monoenergetic beam is (cf eq. (1.19)):

$$\varepsilon(\omega, k) = 1 - \frac{{\omega_p}^2}{\omega^2} - \frac{{\omega_{pb}}^2}{(\omega - kv_0)^2} = 0 \tag{2-1}$$

In this section, the plasma ions will generally be ignored, except for a few closing remarks about possible ion interactions in a hot-electron plasma.

It has often been assumed that the instability described by eq. 2-1 is convective; a solution for the complex k values for real ω then shows that the steady-state amplification rate is

$$\operatorname{Im} k = \frac{\omega_{pb}/v_0}{[(\omega_p^2/\omega^2) - 1]^{1/2}} \tag{2-2}$$

which tends to *infinity* as $\omega \to \omega_p$. It is shown in Section V that this result is not strictly correct, since the transient solution for *any* excitation will, in fact, tend to infinity at $t \to \infty$. This result is not particularly surprising, since infinite steady-state amplification does imply that a *finite* amplitude will never obtain for $z > 0$. Finite collisions or plasma temperature do limit the amplification rate to finite, though generally rather large, values.

We turn now to the problem of describing the instability in terms of its temporal growth rate, i.e., in terms of complex ω for real k. This description is particularly useful when we come to estimate the stabilizing effects of temperature and collisions. Assuming $n_b \ll n_p$, the unstable modes have

$$\delta = \omega - kv_0 \tag{2-3}$$

much less than $|\omega|$. Therefore, the dispersion relation in this regime can be approximated by

$$\frac{\omega_{pb}^2}{\delta^2} = (\varepsilon_p)_{\omega = kv_0} + \left(\frac{\partial \varepsilon_p}{\partial \omega}\right)_{\omega = kv_0} \delta + \cdots \tag{2-4}$$

When the frequency ($\omega_r \simeq kv_0$) is well removed from ω_p, we can ignore the second term on the right-hand side in eq. 2-4. We then find that the growth rate of the instability for $\omega < \omega_p$ is linearly proportional to ω_{pb}. For frequencies above ω_p, ε is positive and the system is stable. We can easily show from eq. 2-4 that the maximum growth rate of the instability occurs *at* $\omega = \omega_p$, and that this maximum growth rate is

$$(\omega_i)_{\max} = \max \operatorname{Im} \delta = \frac{\sqrt{3}}{2}\left(\frac{\omega_{pb}^2\, \omega_p}{2}\right)^{1/3} \equiv \omega_{g0} \tag{2-5}$$

The qualitative behavior of the growth rate as a function of wave number (or frequency) is shown in Figure 4. Note from eq. 2-5 that the maximum instability growth rate is the order of 0.1 ω_p, even for beam densities as low as $n_b = 10^{-3} n_p$. The instability of a monoenergetic beam is a rather violent one.

The effect of collisions in the plasma on these results is easily determined. If we use a simple relaxation form of the collision frequency, ν_c, the plasma dielectric constant becomes (10, 11).

$$\varepsilon_p = 1 - \frac{\omega_p^2}{\omega(\omega + i\nu_c)} \tag{2-6}$$

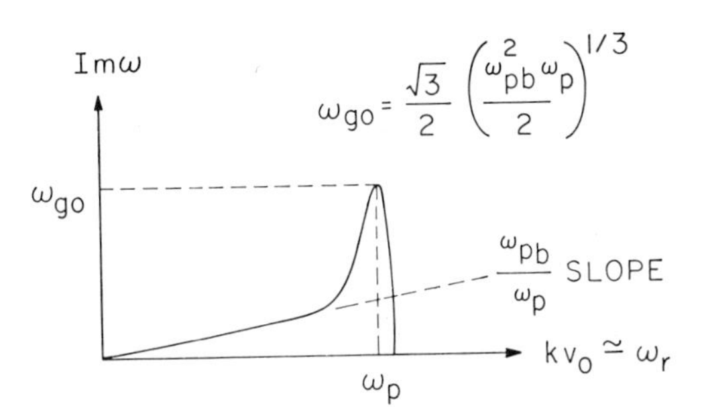

Fig. 4. Temporal growth rate vs. k for a weak beam in a plasma.

From eq. 2-4, if we ignore the second term on the right-hand side, we find that the maximum growth rate is

$$(\omega_i)_{\max} = \omega_{pb}\left(\frac{\omega_p}{2\nu_c}\right)^{1/2} \tag{2-7}$$

The condition for ignoring the second term in eq. 2-4 is basically the same as the condition that the growth rate from eq. 2-7 be much *less* than the growth rate calculated on the basis of $\nu_c = 0$ (eq. 2-5). Therefore, the collision frequency required for a reduction in the maximum growth rate of the streaming instability is

$$\frac{\nu_c}{\omega_p} > \left(\frac{n_b}{n_p}\right)^{1/3} \tag{2-8}$$

Even for very small values of the ratio of beam to plasma densities, we see that this requires a relatively high value of collision frequency; hence the collisionless approximation is generally quite adequate in describing the temporal growth. We should also note in passing that, whereas in the absence of collisions the system is stable for $kv_0 > \omega_p$, in the presence of collisions *this is no longer the case*, as can easily be seen from eq. 2-4. The small growth rate at frequencies well above ω_p is, of course, just a manifestation of the resistive instability that we discussed in Section I.

B. *Beam Temperature Effects*

If the beam velocity, v_0, is much larger than the thermal velocity of the plasma electrons, the plasma is still adequately described by the cold approximation, since the unstable modes must have $\omega/k \simeq v_0$ (see Figure 5). On the other hand, if the phase velocity of the wave falls within the velocity distribution of the beam (in the sense of a displacement from v_0 in the complex plane),

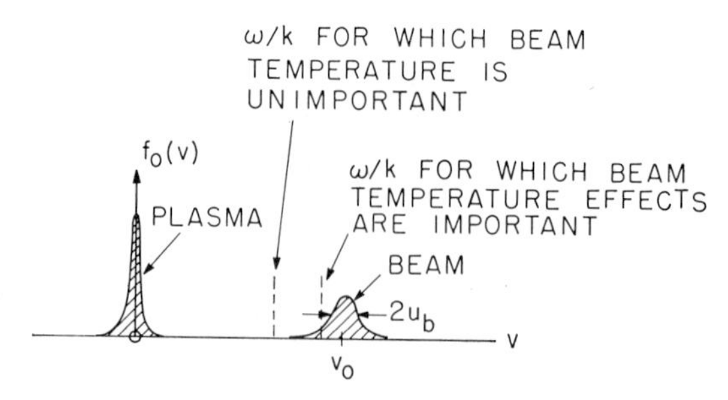

Fig. 5. Beam and plasma distribution functions, illustrating the phase velocity regions, where beam temperature is important.

then the beam can no longer be adequately described by a delta function in velocity space. The condition for the monoenergetic beam theory to hold is therefore

$$\left| \frac{\omega}{k} - v_0 \right| \gg u_b \tag{2-9}$$

where u_b is a measure of the thermal spread of the beam electrons, as, e.g., in the shifted Maxwell-Boltzmann distribution

$$f_{0b}(v) = \frac{1}{\sqrt{2\pi u_b}} \exp\left[-(v - v_0)^2/2u_b{}^2\right] \tag{2-10}$$

An alternate argument which is often given is that, if the instability e folds several times before the particles can spread over one wavelength, then the instability is well described by monoenergetic theory. This condition is essentially $\text{Im}\,\omega \gg ku_b$, which is equivalent to eq. 2-9 if $\text{Im}\,\omega \sim |\omega - kv_0|$. In many situations however, this is not the case, as with the resistive modes discussed in Section I (see Ref. 20 for another important example in particle accelerators). The previous argument based on the *magnitude* of the shift in v_{ph} away from v_0 (in the sense of a shift in the complex plane) is more reliable in general.

We can estimate the left-hand side of inequality 2-9 by using the value of $|\omega/k - v_0|$ deduced from the monoenergetic theory (eq. 2-5). We obtain

$$\frac{u_b}{v_0} \ll \left(\frac{n_b}{n_p}\right)^{1/3} \tag{2-11}$$

as the condition on beam velocity spread for which the beam is adequately described by a delta function in velocity space (as far as the *temporal* growth rate of the two-stream instability is concerned) (18, 19). Note that for density ratios as low as $n_b/n_p \sim 10^{-3}$, the delta-function approximation is valid as long as $u_b \gtrsim 0.1\, v_0$. This is an extremely high velocity spread, as far as beams injected into plasmas are concerned; we are lead to conclude that most injected beams are well represented by the monoenergetic beam approximation (in the initial development of the instability, anyway).

It is not correct to assume, however, that when the beam velocity spread is such that inequality 2-11 is reversed, that the two-stream instability is stabilized. For a (lossless) zero-temperature plasma, the instability persists even for large velocity spreads in the beam.

If we assume that the imaginary part of the frequency is small enough, then the dispersion relation for a hot beam in a cold plasma can be written in the form (c.f. eq. 1-29):

$$1 - \frac{\omega_p{}^2}{\omega^2} - \frac{\omega_{pb}{}^2}{k^2} P \int \frac{f'_{0b}(v)}{(v - \omega_r/k)} - i\pi \frac{\omega_{pb}}{k^2} f'_{0b}\left(\frac{\omega_r}{k}\right) = 0 \qquad \text{(2-12)}$$

The real part of eq. 2-12 can be satisfied, for $\omega_{pb} \ll \omega_p$, *only by choosing a frequency close to the plasma frequency* ($\omega_r \simeq \omega_p$). The balancing of the imaginary part of this equation requires, to the lowest order in n_b/n_p,

$$\frac{2\omega_i}{\omega_p} = \pi \frac{\omega_{pb}{}^2}{k^2} f'_0\left(\frac{\omega}{k}\right) \simeq \pi \frac{\omega_{pb}{}^2}{\omega_p{}^2} v_0{}^2 f'_0\left(\frac{\omega}{k}\right) \qquad \text{(2-13)}$$

For the case considered here, in which we assume both that the beam density is very much less than the plasma density and that the velocity distributions are well separated (see Figure 5), the wave is unstable for any phase velocity in which the distribution function of the beam has a positive slope. The present limiting situation is closely related to the so-called bump in tail configuration, since the overall distribution function must clearly have a very small diffuse bump, representing the beam, for the analysis to apply. For higher plasma temperatures, the tail of the main distribution will begin overlapping the beam distribution—this limit is discussed in the following section.

The hot-beam limit of the two-stream instability is much weaker than the delta-function type of instability. The physical reason for this is that, in the former case, the wave interacts strongly only with a small percentage of the beam electrons; namely, those with velocities very close to the phase velocity. The physical description of the instability in this limit is also a bit different: the electrons moving slightly less than $v_{ph} = \omega/k$ are (on the average) accelerated by the wave and gain energy, while those moving faster are decelerated; if $f'_0(v_{ph})$ is positive, net energy is given to the wave by the synchronous electrons. Note also that growth is obtained *only* for $\omega \simeq \omega_p$ with a hot beam, in contrast to the cold beam case (see Figure 4).

The maximum growth rate of the instability in this limit obtains at the phase velocity where the slope of the distribution function is greatest. For the Maxwellian distribution in eq. 2-10 this maximum growth rate occurs at $\omega/k - v_0 = -u_b$, and is

$$(\omega_i)_{\max} = \frac{1}{2}\sqrt{\frac{\pi}{2e}}\,\frac{{\omega_{pb}}^2}{\omega_p}\,\frac{{v_0}^2}{{u_b}^2} \tag{2-14}$$

The derivation of this growth rate from eq. 2-12 is only valid when

$$\omega_i \ll ku_b \simeq \frac{\omega_p u_b}{v_0} \tag{2-15}$$

which requires

$$\left(\frac{n_b}{n_p}\right)^{1/3} \ll \frac{u_b}{v_0} \tag{2-16}$$

i.e., just the *opposite* of inequality 2-11.

The growth rate in eq. 2-14 can be written in the form (outside of numerical factors the order of unity) as

$$(\omega_i)_{\max} \sim \left[\frac{v_0}{u_b}\left(\frac{n_b}{n_p}\right)^{1/3}\right]^2 \omega_{g0} \tag{2-17}$$

where ω_{g0} is the monoenergetic beam growth rate, eq. 2-5. The hot-beam growth rate is, therefore, reduced from ω_{g0} by a factor less than unity (eq. 2-16). In Figure 6, the effect of beam temperature on the maximum growth rate is schematically illustrated by a sketch of $(\omega_i)_{\max}$ vs. the beam temperature (u_b/v_0) for fixed densities.

As we have shown, the instability persists for a lossless plasma, even for very large beam temperature; however, if we include a finite collision frequency in the plasma, the system is stable for (18)

$$\frac{\nu_c}{\omega_p} \geq \sqrt{\frac{\pi}{2e}}\,\frac{{\omega_{pb}}^2}{{\omega_p}^2}\,\frac{{v_0}^2}{{u_b}^2} \tag{2-18}$$

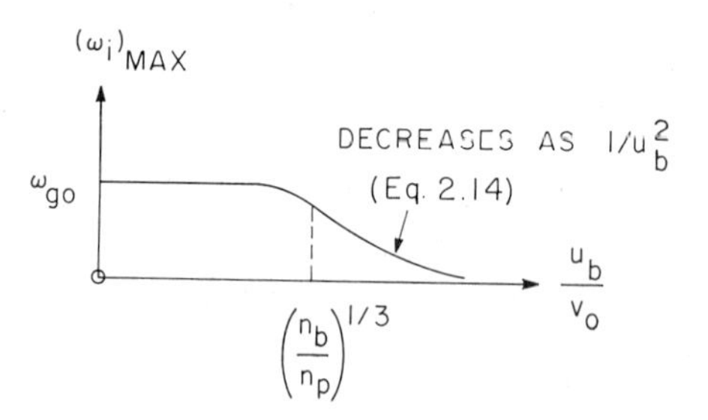

Fig. 6. The maximum growth rate vs. the beam temperature for fixed n_b/n_p.

as is easily derived from eq. 2-12 when the finite collision frequency is introduced into the plasma dielectric constant (e.g., eq. 2-6).

C. *Plasma Temperature Effects.*

As we mentioned in the last section, the effects of plasma temperature are generally not of great importance as far as temporal growth rates are concerned, unless the thermal velocity of the plasma electrons (v_T) is comparable to v_0 (say $v_T \gtrsim 0.3\, v_0$). For a weak beam the unstable modes have $\omega_i \ll \omega_p$; we can therefore often evaluate the equivalent plasma dielectric constant by taking ω/k as essentially real, so that

$$\varepsilon_p = 1 - \frac{\omega_p{}^2}{k^2} P \int \frac{f_0'(v)\, dv}{v - \omega/k} - i\pi \frac{\omega_p{}^2}{k^2} f_0'\left(\frac{\omega}{k}\right) \tag{2-19}$$

The case of $\omega/k \simeq v_0 \gg v_T$ would be only a slight perturbation on the cold plasma results. For v_T comparable to v_0, the "Landau damping" term (the Im ε_p in eq. 2-19) affects the growth rate in much the same way as collisional damping. We shall discuss the stabilization of the two-stream instability by Landau damping in a moment. First, however, let us note that for a *hot* plasma, where $\omega/k \simeq v_0 \ll v_T$, the plasma dielectric constant becomes

$$\varepsilon_p \simeq 1 + \frac{\omega_p{}^2}{k^2 v_T{}^2}\left(1 + i\sqrt{\frac{\pi}{2}}\frac{\omega}{k v_T}\right) \tag{2-20}$$

for the Maxwellian distribution

$$f_0(v) = \frac{1}{\sqrt{2\pi}\, v_T}\, e^{-v^2/2v_T{}^2} \tag{2-21}$$

In this high-temperature regime, Re $\varepsilon_p > 0$ for all real k and the negative dielectric constant mechanism (reactive medium effects) clearly cannot be operative. On the other hand, it is apparent from the approximate dispersion relation, eq. 2-4, that a nonzero Im ε_p does lead to instability for a monoenergetic beam for all real k. The mechanism in this case is basically a resistive medium effect, where the plasma loss comes about because of Landau damping. In Figure 7, the growth rate of this mode as a function of wave number (frequency) is presented for several values of v_T/v_0 in the range $v_T \gg v_0$. The resistive mechanism is a relatively weak one, particularly for large v_T/v_0.

In the hot plasma regime, there is also the possibility of a significant direct interaction of the electron beam with the ions in the plasma. If we

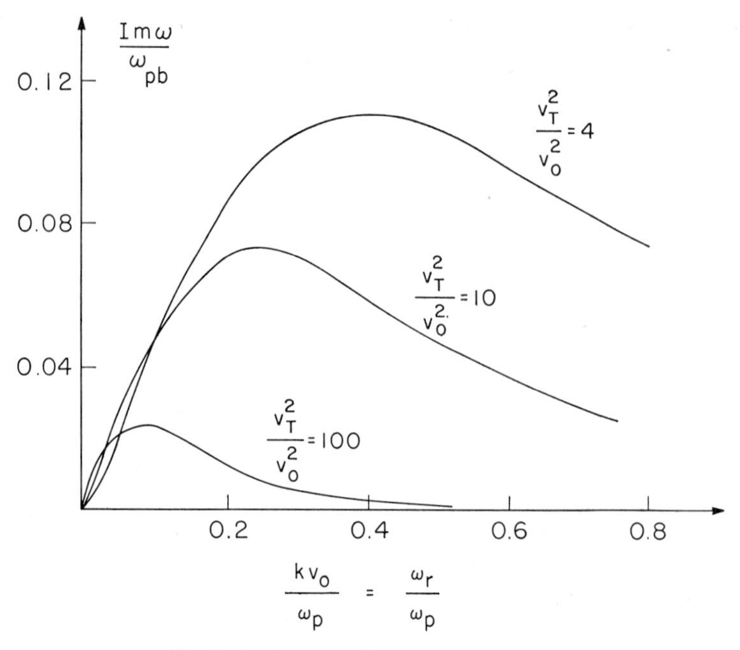

Fig. 7. Resistive-media growth in a hot plasma (Ref. 8).

include the contribution of cold ions, the plasma dielectric constant for $\omega/k \ll v_T$ is approximately

$$\varepsilon_p \simeq 1 - \frac{\omega_{pi}^2}{\omega^2} + \frac{\omega_{pe}^2}{k^2 v_T^2} \tag{2-22}$$

where ω_{pi} is the ion plasma frequency. When $\omega/k < (\omega_{pi}/\omega_{pe})v_T$, the dielectric constant can be negative for $\omega < \omega_{pi}$, and we have the possibility of a low-frequency instability that involves the ion dynamics. This problem has been considered at some length elsewhere (8); the condition for instability for $\omega < \omega_{pi}$, using the approximate ε_p given in eq. 2-22 (which neglects the Im ε_p), is

$$\frac{\omega_{pb}^2}{\omega_{pe}^2} \frac{v_T^2}{v_0^2} \left[1 + \left(\frac{\omega_{pi}}{\omega_{pb}}\right)^{2/3}\right]^3 > 1 \tag{2-23}$$

(The neglect of Landau damping rules out the weak resistive growth discussed above and concentrates on only the strong low-frequency interactions.)

For a very weak beam ($\omega_{pb} \ll \omega_{pi}$), the above condition becomes just $v_T/v_0 > \omega_{pe}/\omega_{pi} = \sqrt{M/m}$, as would be expected, since this is identical to the condition $\varepsilon_p \leq 0$ for $\omega/k \simeq v_0$.

Let us now return to the moderate temperature regime ($v_0 \gtrsim v_T$) and the high-frequency interactions. If the beam is assumed monoenergetic, then there is always an instability, as we have already noted. Calculations of the growth rate for a weak beam would proceed by inserting the expression for ε_p in eq. 2-19 into eq. 2-4. The qualitative nature of the reduction in growth rates due to a small plasma temperature would be similar to the effects of collisions in a cold plasma, as discussed in Section II-A. With a finite velocity spread in the beam, the Landau damping can result in a complete stabilization just as with collisional effects. Note that, with finite beam and plasma temperatures, we have approached the classical bump in tail configuration. The marginal condition for stability is obtained by equating the maximum imaginary part of the beam term (eqs. 2-12 to 2-14) with the plasma Landau damping term (eq. 2-19), which results in the following condition for stability.

$$\frac{n_b}{n_p}\frac{v_0^2}{u_b^2} < \left(\frac{v_0}{v_T}\right)^3 \exp\left(v_0^2/2v_T^2\right) \tag{2-24}$$

In deriving the above, we have assumed Maxwellian velocity distributions for both the beam and the plasma, with $u_b \ll v_0$, so that $\omega/k \simeq v_0$ in the unstable domain. For reasonable densities, the Landau damping by the plasma is an effective stabilizing influence only when $v_T/v_0 \lesssim 0.1$ to 0.2.

III. Finite Systems—Infinite Magnetic Field

The previous sections have all been concerned with the theory of two-stream instabilities in infinite homogeneous systems. For laboratory experiments, this model is a rather crude approximation, although it does presumably illustrate many of the basic physical mechanisms involved. One model of a finite system that is quite tractable theoretically is a beam-plasma system immersed in an infinite axial magnetic field, and this model has some applicability to physical situations in which the plasma frequency is much less than the cyclotron frequency.

A. *General Dispersion Relations*

In the presence of an infinite magnetic field in the z-direction, the particle velocity and current are necessarily z-directed, and hence the perturbation current of a single monoenergetic beam is still given by eq. 1-7, even when the transverse variation of density is arbitrary ($\omega_p^2 = \omega_p^2(r)$).

An exact analysis could be carried through with very little additional effort, but for simplicity we shall restrict ourselves to a quasistatic approximation ($\omega/k \ll c$), where

$$\bar{E}_1 \simeq -\nabla\phi$$

From eq. 1-7, the perturbation charge density of a single monoenergetic beam is given by

$$\begin{aligned}\rho_1 &= ik\varepsilon_0 \frac{\omega_p{}^2(r)}{(\omega - kv_0)^2} E_{z1} \\ &= k^2\varepsilon_0 \frac{\omega_p{}^2(r)}{(\omega - kv_0)^2} \phi\end{aligned} \tag{3-1}$$

Note that k is the wave number in the z-direction and that the system must be uniform in z to allow the dependence exp $i(kz - \omega t)$. Since

$$\nabla^2\phi \equiv \nabla_T{}^2\phi - k^2\phi = -\frac{\rho_1}{\varepsilon_0} \tag{3-2}$$

we have the following governing equation for $\phi(r)$:

$$\nabla_T{}^2\phi - k^2\left(1 - \sum \frac{\omega_{pj}{}^2(r)}{(\omega - kv_{0j})^2}\right)\phi = 0 \tag{3-3}$$

In eq. 3-3, we have included the possibility of a multistream system by summing the charge density of the various streams; a continuous velocity distribution can be handled by passing over to an integral, in exactly the same way as was done in Section I for the infinite system.

The formulation of the basic equations for a *finite temperature* multistream system in an infinite magnetic field is therefore quite straightforward, and this is probably the main reason for the extensive use of this appoximation in the literature. (21–24) It is perhaps somewhat surprising that the study of inhomogeneous systems in the absence of a magnetic field raises a number of far more subtle issues, as is discussed in the following section.

We now take, as a simple and somewhat representative example, (23,24), a cylindrical system in which *all zero-order densities vary in the same manner with radius*, so that

$$\omega_{pj}{}^2(r) = g(r)\omega_{p0j}^2 \tag{3-4}$$

with $g(r)$ the density shape function ($g \leq 1$) and ω_{p0j} the maximum plasma frequency. The differential equation for $\varphi(r, \theta) = \varphi_l(r)e^{il\theta}$ is then

$$\frac{1}{r}\frac{\partial}{\partial r} r \frac{\partial \varphi_l}{\partial r} - \frac{l^2}{r^2}\varphi_l + k^2(\lambda g(r) - 1)\varphi_l = 0 \tag{3-5}$$

with

$$\lambda \equiv \sum_j \omega_{p0j}^2 \int \frac{f_{0j}(v)\, dv}{(\omega - kv)^2} \tag{3-6}$$

(In eq. 3-6 we explicitly allow for a number of finite temperature streams.) With a perfectly conducting wall at $r = a$, we have $\varphi_l = 0$ ar $r = a$; in the absence of any wall we demand $\varphi \to 0$ at infinity. For either boundary condition, eq. 3-5 poses a standard eigenvalue problem for λ, where the eigenvalues (λ_n) depend *only on* k. That is, imposing the boundary condition on φ_l to the solutions of eq. 3-5, we obtain an (infinite) set of eigenvalues, $\lambda = \lambda_n(k)$. The dispersion equation for any one of these radial modes is obtained by inserting $\lambda_n(k)$ into eq. 3-6.

We can easily show that all $\lambda_n(k)$ are real, positive, and greater than unity for all real k. Multiply eq. 3-5 by φ^* and integrate over the entire cross section of the system to obtain, after an integration by parts of the first term,

$$\lambda = \frac{\int |\nabla_T \varphi|^2\, da + k^2 \int |\varphi|^2\, da}{k^2 \int g(r) |\varphi|^2\, da} \tag{3-7}$$

where $|\nabla_T \varphi|$ is the magnitude of the transverse electric field. Since $g(r) \leq 1$ for all r by definition, it is clear that $\lambda \geq 1$ for all real k. We should also point out that eq. 3-7 is a variational expression for λ; therefore, approximations to the eigenvalues can be found by inserting various trial functions, $\varphi(r)$, into this expression (23,24).

The main general point that we wish to make is that the dispersion equation for the various modes (in a system with identical density profiles for all species) is of exactly the same *form* as for a one-dimensional system, with the replacement of all plasma frequencies by $\omega_q \equiv \omega_{p0}/\sqrt{\lambda(k)}$. That is, all of the one-dimensional results discussed in Sections I and II can be carried over to the present problem, when the plasma frequency is replaced by the above *k-dependent* reduced plasma frequency. This reduction of the effective plasma frequency in finite systems is well known in the theory of space-charge waves on electron beams; it occurs because the "fringing" of the perturbation electric field reduces the axial restoring force on the charge bunches. Since larger densities in the one-dimensional theory lead to larger growth rates, the lowest radial mode (with the smallest $\lambda_n(k)$) is clearly the most important for stability studies.

The k dependence of the effective (reduced) plasma frequency does bring some qualitatively new features. For monoenergetic streams the infinite system is always unstable for long enough wavelengths; while, for a finite system, it is shown in Section III-B that the system can be stabilized for *all* wavelengths if the dimensions are small enough. The procedure for calculating thermal effects is a straightforward extension of the one-dimensional results, and therefore in the next sections we shall concentrate on the monoenergetic limit for clarity.

We should emphasize again that the results derived in this section were for the special case in which the density profiles of all streams are identical. In many practical situations, the mean beam radius is often considerably smaller than the plasma radius, and some of the new features that can arise in this case are discussed in Section III-C.

B. *Cold Beam Plasma Interaction in a "Filled Pipe"*.

As a simple example, consider a cold beam-plasma system with *uniform* densities of both the beam (ω_{pb}) and plasma (ω_p) inside a perfectly conducting cylinder of radius a. The eigenfunctions in this case are, from eq. 3-5

$$\varphi_l(r) = AJ_l(p_n r) \tag{3-8}$$

with

$$p_n^2 \equiv k^2[\lambda_n - 1] \tag{3-9}$$

From the boundary conditions at $r = a$,

$$p_n a = \varepsilon_n \tag{3-10}$$

is required, where ε_n denotes the nth root of $J_l(\varepsilon_n) = 0$ (the lowest such value being 2.405). For any particular mode, therefore, p_n is a *fixed* number that scales inversely with the system radius, a. From eq. 3-6 and 3-9, specializing to the case of a cold beam plasma system, we can write the dispersion relation as

$$D(\omega, k) \equiv (\omega - kv_0)^2\left(1 - \frac{\omega_p^2}{\omega^2} + \frac{p_n^2}{k^2}\right) - \omega_{pb}^2 = 0 \tag{3-11}$$

First consider the dispersion of plasma waves in the absence of the beam, as illustrated in Figure 8. In contrast to the infinite medium that has $\omega = \omega_p$ for all k, the finite column exhibits propagating waves with a nonzero group velocity for $\omega \le \omega_p$.

In the presence of a very low density beam, the unstable waves must have $\omega \simeq kv_0 + \delta$ with $\delta \ll \omega$, and an approximate solution for δ from eq. 3-11 shows that the modes are *stable for all* k if $v_0 < v_A = \omega_p/p_n$ (see Figure 8). For $v_0 > v_A$, the system exhibits instability for all $k \le k_0$ ($\omega \le \omega_0$), as is clear by inspection of eq. 3-11. The fastest growth rate is at $\omega = \omega_0$, and for low beam density it is

$$(\text{Im}\ \omega)_{\max} = \frac{\sqrt{3}}{2}\left(\frac{\omega_{pb}^2\omega_p}{2}\right)^{1/3}\frac{\omega_0}{\omega_p} \tag{3-12}$$

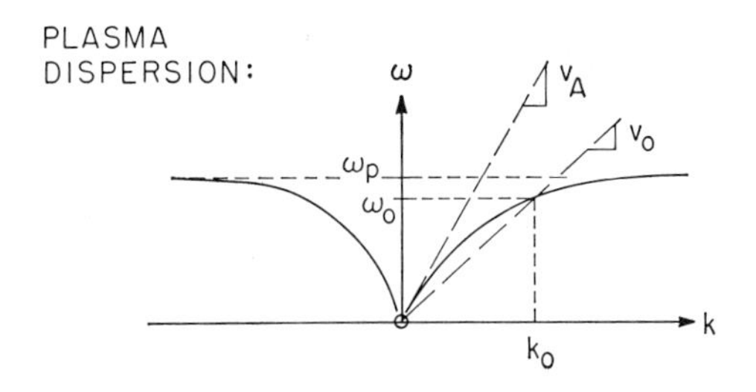

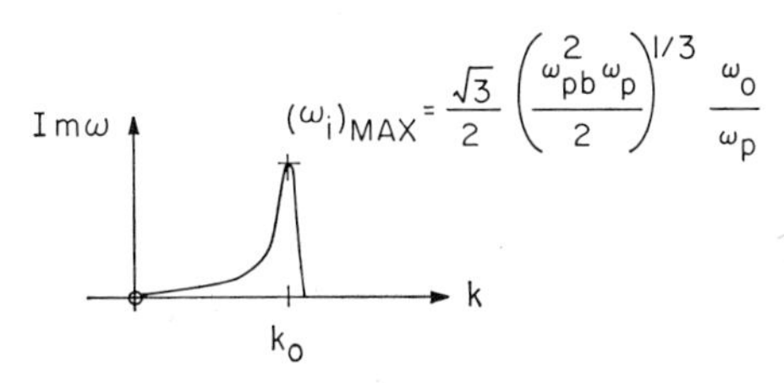

Fig. 8. Dispersion of waves on a plasma column in an infinite magnetic field. In an exact analysis (nonquasistatic), $v_A = (\omega_p/p_n)(1 + \omega_n^2/p_n^2c^2)^{-1/2}$ which is approximately ω_p/p_n if $v_A \ll c$.

(cf. eq. 2.5). That is, this peak growth obtains at the point of synchronism of the beam velocity with the phase velocity of a propagating plasma wave, much as in an ordinary traveling wave tube (see Figure 8).

The possibility of a complete stabilization of the two-stream instability in a radially finite system deserves some comment. In the infinite media we discussed the instability mechanism in terms of a negative plasma dielectric constant. In the finite system the transverse dielectric constant is always equal to its free space value because of the infinite magnetic field, and the *effective dielectric constant* that must be negative in the present case is a weighed average of the longitudinal and transverse dielectric constants, namely, $k^2(1 - \omega_p^2/\omega^2) + p^2(1)$. For large enough p^2 (small enough radii), the positive transverse dielectric constant dominates and the system is stable.

The precise stability condition, for arbitrary beam density, is easily obtained from eq. 3-11. Since all coefficients are real (for real k), complex ω must occur in conjugate pairs; hence the onset of complex ω is signaled by a double-order *real* ω root from eq. 3-11. The condition for such a double-order root is obtained by solving $\partial D/\partial\omega = 0$ and $D = 0$; this yields the following stability condition for the nth mode.

$$v_0^2 \geq \frac{\omega_p^2}{p_n^2}\left[1 + \left(\frac{\omega_{pb}}{\omega_p}\right)^{2/3}\right]^3 \tag{3-13}$$

Before closing this section we might remark that the above stability condition is also valid for inhomogeneous systems when the density profile of both the beam and the plasma are the same. In the inhomogeneous case it can be shown that the appropriate value of p_n^2 to use in eq. 3-13 is determined from the limiting value of the eigenvalue λ_n as $k \to 0$; namely,

$$p_{n0} = \lim_{k \to 0} (\lambda_n k^2) \tag{3-14}$$

From eq. 3-7, it is clear that the above limit is finite; in fact, various trial functions could be used in eq. 3-7 to obtain approximations to the appropriate p_n to use for an inhomogeneous system in stability condition 3-13.

C. *Thin Beam in a Uniform Plasma*

The previous sections have dealt exclusively with systems in which the density profiles of the beam and plasma are identical. In this section we illustrate the new features that arise when these profiles differ, as is the case in many practical situations. The specific example treated is an inhomogeneous monoenergetic beam ($\omega_{pb}^2(r)$) in a uniform density cylindrical plasma of radius a, bounded by a conducting wall of the same radius (Figure 9). The differential equation for φ is, from eq. 3-3,

$$\nabla_T^2 \varphi - k^2\left(1 - \frac{\omega_p^2}{\omega^2}\right)\varphi + k^2 \frac{\omega_{pb}^2(r)}{(\omega - kv_0)^2}\varphi = 0 \tag{3-15}$$

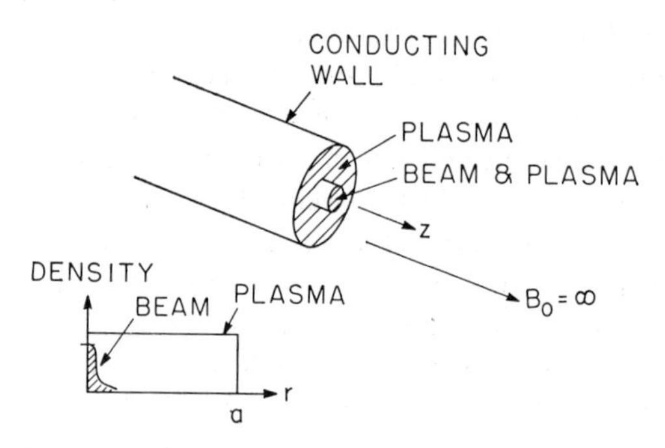

Fig. 9. Illustration of system treated in Section III-C.

For simple profiles such as a uniform density beam of radius $b < a$, we could set up a transcendental dispersion relation (8). However, it is perhaps more instructive to "solve" eq. 3-15 by a mode expansion technique, since this shows directly that the nature of the results is not dependent on the beam profile (or, in fact, the plasma profile, although we shall not consider this generalization here). Let $\varphi_n(r, \theta)$ represent the complete set of orthogonal functions satisfying

$$\nabla_T{}^2\varphi_n + p_n{}^2\varphi_n = 0 \tag{3-16}$$

with the boundary condition $\varphi_n(r = a) = 0$, and normalized so that

$$\int |\varphi_n|^2 \, da = \int_0^a \int_0^{2\pi} |\varphi_n|^2 r \, d\theta \, dr = 1 \tag{3-17}$$

For the cylindrical system we can show that the appropriate φ_n are

$$\varphi_n = [-\pi a^2 J_{l-1}(p_n a)J_{l+1}(p_n a)]^{-1/2} J_l(p_n r)e^{il\theta} \tag{3-18}$$

where $J_l(p_n a) = 0$ determines the eigenvalues p_n, which we should probably write as p_{ml}, since there are actually an infinite set for each l value. In the following, we consider each azimuthal mode (l value) separately, and suppress additional notation. Note that all $p_n{}^2$ are real and positive, a general property of the eigenvalues of eq. 3-16.

We now expand φ in terms of this complete set, as

$$\varphi = \sum_n a_n \varphi_n \tag{3-19}$$

Inserting this expansion into eq. 3-15, and using the orthogonality property of the φ_n's ($\int \varphi_n \varphi_p^* \, da = 0$ for $n \neq p$), we obtain the following expression for the amplitudes, a_n:

$$a_n\left[p_n{}^2 + k^2\left(1 - \frac{\omega_p{}^2}{\omega^2}\right)\right] = \frac{k^2}{(\omega - kv_0)^2} \sum_p a_p \int \omega_{pb}{}^2(r)\varphi_n^* \varphi_p \, da \tag{3-20}$$

For a particular beam profile, described by $\omega_{pb}{}^2(r) = \omega_{b0}{}^2 s(r)$ with $s(r) \leq 1$, the solution of the infinite set of eq. 3-20 leads to the consideration of an infinite-order determinant. The special case for which a great simplification occurs is when the beam is also uniform in $r \leq a$, since eq. 3-20 then leads to uncoupled equations for each a_n because of the orthogonality of the φ_n. We can now appreciate the reason for the great simplification introduced by treating systems in which the beam and plasma profiles are the same: the modes of the composite system in that case match those of each subsystem and hence modes of the beam (say) join only *one* of the modes of the plasma, since they are orthogonal to all other modes.

We shall now confine our attention to the study of a very low-density beam ($\omega_{b0}{}^2 \ll \omega_p{}^2$) and to the most interesting regions of $\omega - k$ space, namely, the regions in which the beam is close to synchronism with one of the propagating plasma modes. That is, we assume that ω and k are close to the point where

$$p_n{}^2 + k^2(1 - \omega_p{}^2/\omega^2) \simeq 0 \tag{3-21}$$

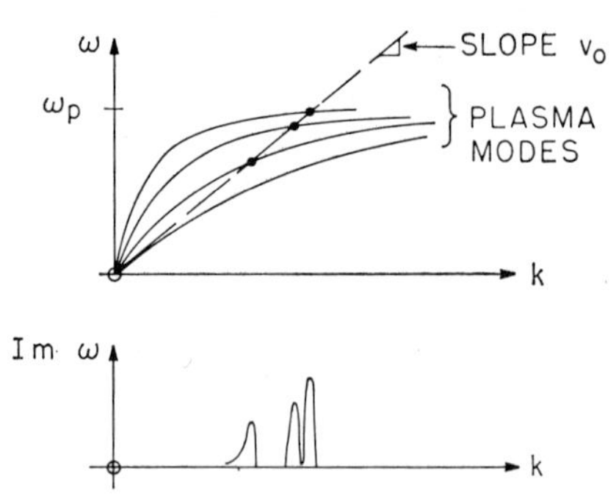

Fig. 10. Coupling of a thin beam to several plasma modes in a plasma-filled waveguide. The top figure indicates the condition for synchronism of the beam with the propagating plasma modes; the bottom figure indicates the qualitative nature of the growth rate vs. k

for $\omega \simeq kv_0$ (see Figure 10). In this case, $|a_n| \gg |a_p|$ for $p \neq n$; hence an approximate dispersion relation near this region can be obtained by ignoring all terms with $p \neq n$ on the right-hand side of eq. 3-20,

$$(\omega - kv_0)^2\left(1 - \frac{\omega_p^{\,2}}{\omega^2} + \frac{p_n^{\,2}}{k^2}\right) = \omega_{b0}^{\,2} \int s(r)\,|\,\varphi_n\,|^2\,da \qquad (3\text{-}22)$$

For any given n, this is in the same form as eq. 3-11, except that the beam-plasma frequency is reduced from its peak value by the factor R_n, where

$$R_n^{\,2} \equiv \int s(r)\,|\varphi_n|^2\,da < 1 \qquad (3\text{-}23)$$

Near synchronism of the beam with the nth plasma mode, therefore, the results of the weak beam analysis in a "filled pipe" carry over to the present case with the replacement $\omega_{pb} \to R_n\,\omega_{b0}$. The main qualitative difference in the present case is that the growth rate can exhibit *several* peaks as a function of frequency, rather than just one (Figure 10).

In the filled-pipe case, we found that a given mode was stable if the beam velocity were always larger than the phase velocity of the plasma wave i.e., if $v_{ph} \leq \omega_p/p_n < v_0$ (see Figure 8). In the present case we can show that, if $v_0 > \omega_p/p_n$ for *all* p_n, then a low-density beam is stable. To show this note that in matrix notation eq. 3-20 is a standard characteristic value problem of the form

$$\mathbf{La} = \lambda\mathbf{ba} \qquad (3\text{-}24)$$

where

$$L_{nm} = \int \varphi_n^*\varphi_m\,da,\; b_{nm} \equiv (p_n^{\,2}/k^2 + 1 - \omega_p^{\,2}/\omega^2)\delta_{nm}$$

and the eigenvalue $\lambda \equiv (\omega - kv_0)^2/\omega_{b0}^{\,2}$. For a weak beam, we can evaluate b_{nm} at $\omega = kv_0$ to check for unstable roots, and since $\mathbf{L}$ is Hermitian, it follows that all the eigenvalues λ are positive and real if $b_{nm} > 0$ for all n. That is, the "thin" beam system is also stable if $v_0 > \omega_p/p_n$ for the lowest p_n (which is $2.405/a$ for the cylindrical system).

IV. Finite Systems—Zero Magnetic Field

In this section we shall study the two-stream instability in a radially inhomogeneous (cold) beam plasma system in the *absence* of a magnetic field. For smooth and identical density profiles, we shall show that there *are* no discrete eigenmodes (25). The mode spectrum is entirely continuous; to exhibit this spectrum of unstable frequencies, we shall find it necessary to examine the formal solution to the initial-value problem in much greater detail than we have done previously.

A. *Discrete Eigenmodes*

For no magnetic field, it is easily shown that the perturbation current in a cold stream is given by

$$\bar{J}_1 = i\omega\varepsilon_0 \frac{\omega_p^2(\bar{r}_T)}{(\omega - kv_0)^2} \bar{E}_1 \tag{4-1}$$

where the unperturbed velocity is $\bar{v}_0 = \bar{i}_z v_0$ and the wave dependence $\exp i(kz - \omega t)$ has been assumed (cf. eq. 1-7). For nonrelativistic streams, the quasistatic approximation is valid and the differential equation for the potential ($\bar{E}_1 = -\nabla\phi$) is

$$\nabla_T\, \varepsilon(\bar{r}_T)\nabla_T\, \phi - k^2\, \varepsilon(\bar{r}_T)\phi = 0 \tag{4-2}$$

where once again the "dielectric constant" is

$$\varepsilon(\bar{r}_T) = 1 - \sum \frac{\omega_p^2(\bar{r}_T)}{(\omega - kv_0)^2} \tag{4-3}$$

In a cylindrical system and with no θ variation, the governing equation for $\phi(r)$ becomes

$$\frac{1}{r}\frac{d}{dr}\left[r\varepsilon(r)\frac{d\phi}{dr}\right] - k^2\varepsilon(r)\phi = 0 \tag{4-4}$$

An important aspect of this differential equation is the existence of singular points at any radius $r_s(\omega, k)$, where $\varepsilon(r_s) = 0$. For the case of identical density profiles of all species, where

$$\varepsilon(r) = 1 - g(r) \sum \frac{{\omega_{p0}}^2}{(\omega - kv_0)^2} \equiv 1 - \lambda(\omega, k)\, g(r) \tag{4-5}$$

with $g(r) \leq 1$, we shall show that there are no eigenmodes with regular $\phi(r)$. (A similar proof can be given for an inhomogeneous beam in a homogeneous plasma (25).) Multiply eq. 4-4 by $r\phi^*$ and integrate to obtain

$$\int_0^\infty \varepsilon(r)\left[r\left|\frac{d\phi}{dr}\right|^2 + k^2 r|\phi|^2\right] dr = 0 \tag{4-6}$$

or

$$\int_0^\infty Q(r)\,dr = \lambda(\omega, k)\int_0^\infty g(r)Q(r)\,dr \tag{4-7}$$

with

$$Q(r) \equiv r\left|\frac{d\phi}{dr}\right|^2 + k^2 r|\phi|^2 \tag{4-8}$$

Since $Q(r)$ is real and positive for real k, we must have $\lambda(\omega, k)$ real, and it is also clear from eq. 4-7 that

$$\lambda(\omega, k) \geq 1 \tag{4-9}$$

since $g(r) \leq 1$ by definition. Therefore, $\varepsilon(r)$ must be real, and we must have $\varepsilon = 0$ somewhere.

As we mentioned before, the potential $\phi(r)$ is singular at any radius $r = r_s(\omega, k)$, where $\varepsilon(r_s)$ vanishes. (If we used only the solution of eq. 4-4 that is regular at r_s, $\phi(r)$ then could not be regular at both $r = 0$ and ∞ (25)). The conclusion in this case is that there *are no* discrete eigenmodes with well-behaved ϕ. To understand this somewhat surprising result and properly determine the role of the singularities in $\phi(r)$, we must examine the solution to the initial value problem in more detail.

B. *Continuous Spectra*

In stability calculations we are concerned with normal modes, primarily because of their role in determining the response of the system to some assumed initial perturbation. The present problem is one in which the spectrum of eigenvalues is at least partly continuous, and the clearest way to exhibit this continuous spectrum is to write down a formal solution to the initial value problem.

We shall apply a Fourier transform in z and concern ourselves only with the behavior of a single wave number (all quantities vary as exp ikz). If we apply a Laplace transform in time, i.e.,

$$\phi_k(r, t) = \int_L \tilde{\phi}_k(r, \omega)e^{-i\omega t}\frac{d\omega}{2\pi}$$

$$\tilde{\phi}_k(r, \omega) = \int_0^\infty \phi_k(r, t)e^{+i\omega t}\,dt \tag{4-10}$$

and assume, for simplicity, that only the perturbed densities are nonzero at $t = 0$, we obtain the following differential equation for $\tilde{\phi}_k(r, \omega)$.

$$\frac{1}{r}\frac{d}{dr} r\varepsilon(\omega, k, r)\frac{d\tilde{\phi}_k}{dr} - k^2\varepsilon(\omega, k, r)\tilde{\phi}_k = \sum \frac{\rho_1(t=0, r)}{i(\omega - kv_0)} \equiv S(r) \quad \text{(4-11)}$$

The initial values enter as "source" terms on the right-hand side of eq. 4-11; with $S = 0$, we recover the homogeneous differential equation discussed before in connection with the normal mode problem (eq. 4-4). Note also that we have considered only circularly symmetric perturbations; the case of a perturbation with an exp $(il\,\theta)$ dependence carries through in exactly the same manner.

By a Green's function technique we can write down the following formal solution to eq. 4-11:

$$\tilde{\phi}_k(r, \omega) = \frac{1}{D(\omega, k)}\left\{\phi_2 \int_0^r \frac{r'\phi_1(r')S(r')\,dr'}{\varepsilon(r')} + \phi_1 \int_r^\infty \frac{r'\phi_2(r')S(r')\,dr'}{\varepsilon(r')}\right\} \quad \text{(4-12)}$$

where ϕ_1 and ϕ_2 are the solutions to the *homogeneous* equation,

$$\frac{1}{r}\frac{d}{dr} r \frac{d\phi_{1,2}}{dr} + \frac{[\partial\varepsilon(r)/\partial r]}{\varepsilon(r)}\frac{d\phi_{1,2}}{dr} - k^2\phi_{1,2} = 0 \quad \text{(4-13)}$$

with ϕ_1 regular at $r = 0$, and $\phi_2 \to 0$ at $r = \infty$ (or at the radius of a conducting wall surrounding the beam-plasma system). The function $D(\omega, k)$ is defined by the Wronskian of ϕ_1 and ϕ_2,

$$\phi_1\phi_2' - \phi_2\phi_1' = \frac{D(\omega, k)}{r} \quad \text{(4-14)}$$

The discrete eigenvalues correspond to those $\omega(k)$ that satisfy $D(\omega, k) = 0$, since the vanishing of the Wronskian of ϕ_1 and ϕ_2 implies that they are the same function outside of a constant multiplier. For these values of $\omega(k)$, $\phi_k = \phi_1 = c\phi_2$ satisfies the homogeneous equation and the boundary conditions at both $r = 0$ and $r = \infty$.

A continuous spectrum of eigenvalues, on the other hand, is reflected by branch cuts in $\tilde{\phi}_k(r, \omega)$ in the complex ω plane. In the present case we find that there is a cut in $\tilde{\phi}_k(r, \omega)$ along the line *in the* ω*-plane* defined by

$$\varepsilon(r, \omega) = 0, \quad \text{all real } r \quad \text{(4-15)}$$

This branch cut arises both because of the branch points in ϕ_1 and ϕ_2 at places where $\varepsilon = 0$, and because of the $1/\varepsilon$ terms in the integrals in eq. 4-12. As an example, consider again the case of identical density profiles (eq. 4-5), where

$$g(r) \sim 1 - r^2/L^2 \quad \text{(4-16)}$$

near $r = 0$, and where the maximum densities occur at the origin. Since ϕ_1 is regular at $r = 0$, the first integral in eq. 4-12 yields

$$\tilde{\phi}_k(\omega, r) \sim \int_0^r \frac{d(r')^2}{\varepsilon(r')} \sim \ln(\lambda - 1) \tag{4-17}$$

That is, there is a branch point of $\tilde{\phi}(\omega, r)$ *in the ω-plane* at the value of $\omega(k)$, where $\lambda = 1$, or

$$\varepsilon_0(\omega, k) = 1 - \sum \frac{\omega_{p0}{}^2}{(\omega - kv_0)^2} = 0 \tag{4-18}$$

The quantity we are ultimately interested in is the perturbed potential $\phi_k(t, r)$ which is obtained from the inverse Laplace transform, eq. 4-10. In performing the inverse transform, the dominant contribution to $\phi_k(t, r)$ for large time will come from the integral around the branch point at $\omega_0 = \omega_{0r} + i\omega_{0i}$, where ω_0 is the root of eq. 4-18 with the largest Im ω (Figure 11). This yields the asymptotic behavior

$$\phi_k(t, r) \sim \frac{1}{t} \exp(-i\omega_{0r} t + ikz + \omega_{0i} t) \tag{4-19}$$

The general result, then, is that the response of the finite inhomogeneous system grows exponentially at essentially the *infinite medium* growth rate corresponding to the *maximum* densities. A more detailed analysis of eq. 4-12 makes it clear that this result applies in general, and not just to cases in which the density profiles are identical. To see this, note that the branch cuts of $\tilde{\phi}_k(r, \omega)$ in the complex ω-plane lie along the lines defined by the solutions of

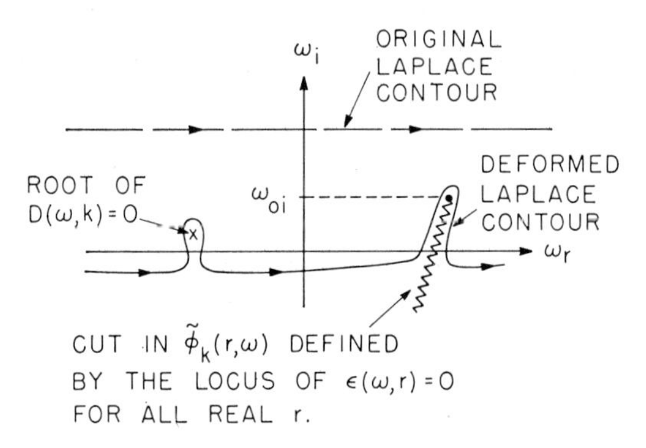

Fig. 11. Illustration of the deformation of the Laplace contour, and the contribution to $\phi_k(r, t)$ from the continuous spectra.

eq. 4-15; therefore the maximum Im ω along this line dominates the asymptotic response of the system. Solving eq, 4-15, of course, amounts to solving the infinite medium dispersion relation using the "local" densities. It is also very important to note that $\phi_k(t, r)$ does *not* (necessarily) grow at a rate determined by the dispersion relation appropriate to that radial position $r(\varepsilon(r) = 0)$. The potential at all radial positions grows at the *same* rate asymptotically.

Another point of considerable importance to note is the fact that singularities in the *transform* function $\tilde{\phi}_k(\omega, r)$, and also in ϕ_1 and ϕ_2, did *not* imply singularities in the physically meaningful quantity, $\phi(t, r)$. The proper interpretation of the logarithmic singularity in $\tilde{\phi}$ followed directly from the initial value treatment.

We have thus far ignored the contribution from the discrete modes, as contained in the poles of $\tilde{\phi}(\omega, r)$ (the zeros of $D(\omega, k)$). For identical density profiles (or for an inhomogeneous beam in an infinite plasma), we showed in the previous section that there are no such discrete modes. For more general profiles, there could perhaps be discrete modes, but we fully expect these growth rates to be less than the infinite medium growth rate corresponding to the maximum densities. The continuous spectrum, interestingly enough, is more important in this system.

C. *Discussion*

The great majority of the theoretical work on finite beam-plasma interactions has dealt with models in which the density profiles are stepwise constant with radius (see, e.g., Refs. 8,9,26,27). This type of model yields discrete spectra only, and the usual type of normal mode analysis applies. In the preceding section, we showed that, for a cold beam-plasma system with a smooth density profile (and in the absence of a magnetic field), continuous spectra dominate the scene. One can presumably trace through the transition from a continuous to a discrete spectrum as the profile is made sharper and sharper; however, in the limit of a sharp edge the linearization assumption for the edge electrons is in doubt. (28,29). For this reason, the sharp-edge limit is of doubtful validity as a physical model in the case of finite (or zero) magnetic fields.

A remaining question, for either model, is the effect of finite temperature. A velocity spread in the *axial* direction could easily be included in the preceding analysis by passing to the limit of a continuous distribution of "cold streams" in eq. 4-3. The effects of random velocity in the *perpendicular* direction, however, are not so easily included. The closely related problem of the transverse resonances of a finite temperature inhomogeneous column has received considerable attention in recent years (30–32). Recent work (33) has indicated that many phenomena can be interpreted on the basis of the cold plasma equations, but the detailed connections between the cold plasma results and the limiting forms of the finite temperature theory (34) are rather involved and are still a subject of current research. Therefore, at the present time we are not able to give a condition analogous to inequality 2-11 (which delineates the limits of validity of the cold beam-plasma approximation) for the case of an inhomogeneous system.

We have discussed only the case of zero applied magnetic field here, but it is easy to show that, with a finite magnetic field, the singularities and continuous spectra are governed by the zeros of $\varepsilon_\perp$ where

$$\varepsilon_\perp = 1 - \sum \frac{\omega_{pj}{}^2(\bar{r}_T)}{(\omega - kv_{0j})^2 - \omega_{cj}{}^2} \tag{4-20}$$

with ω_{cj} the cyclotron frequency of the jth specie (35). The major features in the analysis of the unmagnetized system apply also to the modes with $\omega^2 \sim \omega_p{}^2 + \omega_c{}^2$ in the finite magnetic field case. Modes with $\omega \sim \omega_p \ll \omega_c$ are, of course, well approximated by the infinite magnetic field limit.

V. Appendix

In this appendix, we consider the distinction between absolute and convective instability for the simple infinite uniform media, zero-temperature, beam-plasma interaction. From the pole-pinch criteria (8), one would be tempted to classify this instability as convective, since as Im $\omega \to +\infty$, the two roots of k from eq. 2-1 are in the upper-half k-plane (that is, both waves propagate in the $+z$ direction). The spatial amplification rate, however, is *infinite* at $\omega = \omega_p$ in the absence of plasma temperature or collisions, and this leaves the question of the ultimate limit of the transient response somewhat in doubt. From a theoretical standpoint, it is interesting to sort out this limiting case, even though the practical effects of finite geometries, temperature, etc., may considerably alter the results (36).

The question of convective instability (amplification) or absolute instability (spontaneous oscillation) is settled by a consideration of the impulse response of the system. For instance, if we have a pair of grids very close together, we can excite the system with a circuit electric field of the form

$$E_c(z, t) = E_0\, \delta(z)\delta(t) \tag{5-1}$$

Using Fourier-Laplace transforms, we find that the response is given by

$$E(z, t) = E_0 \int_L \int_F \frac{e^{i(kz - \omega t)}}{\varepsilon(\omega, k)} \frac{dk\, d\omega}{(2\pi)^2} \tag{5-2}$$

where F is the Fourier contour in the k-plane (Im $k = 0$), L is the Laplace contour in the ω-plane (Im $\omega \geq \sigma > 0$), and $\varepsilon(\omega, k)$ is given by eq. 2-1. The k-plane integration can easily be performed by closing the contour in the upper-half k-plane for $(z > 0)$ or lower-half k-plane for $(z < 0)$, and noting that both poles of the integrand (roots of $\varepsilon = 0$) are in the upper-half k-plane for large positive Im ω. We thus obtain

$$E(z, t) = \int_L \frac{ik_b E_0}{2\varepsilon_p^{3/2}} e^{-i\omega\tau}\{e^{ik_b z/\varepsilon_p{}^{1/2}} - e^{-ik_b z/\varepsilon_p{}^{1/2}}\} \frac{d\omega}{2\pi}; \qquad z > 0 \quad (5\text{-}3)$$

$$= 0; \qquad z < 0$$

where $k_b = \omega_{pb}/v_0$, $\varepsilon_p = 1 - \omega_p{}^2/\omega^2$, and $\tau = t - z/v_0$. To evaluate the asymptotic form of $E(z, t)$ as $t \to \infty$, we look for the saddle points of

$$\phi(\omega) = \omega\tau - \frac{k_b z}{\varepsilon_p^{1/2}} \quad (5\text{-}4)$$

The saddle point frequencies in the upper-half plane (roots of $\partial\phi/\partial\omega = 0$) are, for $\omega_p \tau \gg k_b z$,

$$\omega \simeq \omega_p[1 + \tfrac{1}{2}(k_b z/\omega_p \tau)^{2/3} e^{j2\pi/3}] \quad (5\text{-}5)$$

Using these frequencies in eq. 5-4, we find that the asymptotic dependence of $E(z, t)$ is

$$E(z, t) \simeq \exp(i\omega_p \tau) \exp \frac{3\sqrt{3}}{4} (k_b z)^{2/3} (\omega_p \tau)^{1/3}, \qquad z < 0 \quad (5\text{-}6)$$

$$\equiv 0, \qquad z < 0$$

This result is considerably different from the usual form of absolute instability, in which the response tends to infinity as $\exp(\alpha t)$, where α does not depend on position. This case is an example of how the pole-pinch criteria must be modified when the pinch occurs at *infinity* for *real* values of ω. It has been shown elsewhere (37) that the inclusion of finite temperature or collisions *does* make the instability convective, and the spatial amplification rate is then bounded.

References

1. G. D. Boyd, L. M. Field, and R. W. Gould, *Phys. Rev.*, **109**, 1393 (1958).
2. A. B. Cannara and F. W. Crawford, *J. Appl. Phys.*, **38**, 583 (1967).
3. D. M. Mills, E. E., Abraham, and F. W. Crawford, *J. Appl. Phys.*, **38**, 4767–4779 (1967).
4. J. R. Apel and A. M. Stone *Proc. Seventh International Conference on Phenomena in Ionized Gases*, 1966 **2**, 405–409 (1966); J. R. Apel, Tech. Memo TG-821, Applied Physics Laboratory, John Hopkins University, June 1966).

5. G. D. Boyd, R. W. Gould, and L. M. Field, *Proc. IRE*, **49**, 1906 (1961).
6. F. W. Crawford and G. S. Kino, *Proc. IRE*, **49**, 1767 (1961).
7. A. W. Trivelpiece, "Slow Wave Propagation in Plasma Waveguides," San Francisco, 1967. (Based on a thesis and articles published around 1958.)
8. R. J. Briggs, *Electron-Stream Interaction with Plasmas*, M.I.T. Press, Cambridge, Mass. (1964).
9. V. Y. Kislov, E. V. Bogdanov, and Z. S. Chernov *Advances in Electronics and Electron Physics*, Vol. 25, Academic Press, New York, 1965.
10. T. H. Stix, *The Theory of Plasma Waves*, McGraw-Hill, New York, 1962.
11. W. P. Allis, S. J. Buchsbaum, and A. Bers, *Waves in Anisotropic Plasmas*, M.I.T. Press, Cambridge, Mass., 1963.
12. J. M. Dawson, *Phys. Rev.*, **118**, 381 (1960).
13. O. Buneman, *Radiation and Waves in Plasmas*, M. Mitchner, ed., Stanford University Press, Stanford, California, 1961.
14. L. D. Landau, *J. Phys.* (*USSR*), **10**, 25 (1946).
15. K. M. Case, Ann. Phys. (N.Y.), **7**, 349 (1959).
16. J. D. Jackson, *J. Nucl. Energy*, **C1**, 171 (1960).
17. O. Penrose, *Phys. Fluids*, **3**, 258 (1960).
18. H. E. Singhaus, *Phys. Fluids*, **7**, 1534–1540 (1964).
19. T. M. O'Neil and J. H. Malmberg, *Phys. Fluids*, **11**, 1754–1760 (1968).
20. V. K. Neil and A. M. Sessler, Rev. Sci. Instr., **36**, 429 (1965).
21. E. R. Harrison, Proc. Phys. Soc. (London), **79**, 317–325 (1962).
22. M. T. Vlaardingerbroeck, K. R. U. Weimer, and H. J. C. Nunnink, *Phillips Research Report*, **17**, 344–362 (1962).
23. E. G. Harris, *Phys. Fluids*, **7**, 1572–1577 (1964).
24. K. F. Lee, *Phys. Fluids*, **9**, 2435–2443 (1966).
25. E. A. Frieman, M. L. Goldberger, K. M. Watson, S. Weinberg, and M. N. Rosenbluth, *Phys. Fluids*, **5**, 196 (1962).
26. J. E. Simpson and D. A. Dunn, *J. Appl. Phys.*, **37**, 4201 (1966).
27. F. W. Crawford, *Int. J. Electronics*, **19**, 217 (1965).
28. D. L. Bobroff, H. A. Haus, and J. W. Kluver, *J. Appl. Phys.*, **33**, 2932 (1962).
29. H. M. Schneider, *Phys. Fluids*, **9**, 2299 (1966).
30. J. V. Parker, J. C. Nickel, and R. W. Gould, *Phys. Fluids*, **7**, 1489 (1964).
31. V. B. Gil'Denburg, *Sov. Phys. JETP*, **18**, 1359 (1964).
32. D. E. Baldwin and J. L. Hirshfield, *Appl. Phys. Letters*, **11**, 175–177 (1967).
33. D. E. Baldwin, D. M. Henderson, and J. L. Hirshfield, *Phys. Rev. Letters*, **20**, 314 (1968).
34. L. D. Pearlstein and D. Bhadra, *Phys. Fluids*, **12**, 213 (1969).
35. R. J. Briggs, and S. F. Paik, *Int. J. Electronics*, **23**, 163 (1968).
36. O. Buneman, *Phys. Rev.*, **115**, 503 (1959).
37. R. J. Briggs, MIT Quarterly Progress Report No. 85, 183 185 (1967).

Cusp Containment

IAN SPALDING

Department of Physics and Astronomy
University of Maryland
*College Park, Maryland**

* Permanent address: UKAEA Culham Laboratory, Abingdon, Berkshire, England.

Introduction

In this chapter we are primarily interested in the containment and thermonuclear possibilities exhibited by a class of open-ended magnetic bottles in which the confined plasma is virtually free of magnetic field. Ideally, the plasma is separated at a sharp interface from its confining magnetic field (B_0), so that at equilibrium the plasma pressure, p, is completely balanced by the external magnetic pressure, $B_0{}^2/8\pi$. The plasma parameter β, defined by the relation $\beta = 8\pi p/B_0{}^2$, therefore closely approaches its maximum value of unity throughout the bulk of the plasma.

The interest in these high β configurations arose from early Russian and American predictions (1, 2) that any field-free, infinite conductivity fluid bounded in equilibrium by a surface-current interface (the "free boundary") should have high gross stability, provided that the center of curvature of the interface is nowhere enclosed by the plasma. Since it is impossible to generate a smooth closed surface of this type, such configurations are characterized by the discontinuities, or "cusps," that occur where the surface curvature reverses (Figure 1*a*). Their high magnetohydrodynamic (MHD) stability is a consequence of the fact that the confining magnetic field increases in every direction away from the plasma. The cusp configuration is unique in being MHD stable at high β to arbitrary disturbances of finite amplitude (2).

The simplest three-dimensional cusp is generated by two axisymmetric coils connected in opposition (Figure 1*a*). However, as a consequence of the Bernoulli effect, high β plasma escaping from a linear theta-pinch coil (Figure 1*b*) will also flow down to a cusp at either end of the coil (3) and so many of the properties of high β cusps and linear theta pinches are closely related (4).

High β plasmas differ in several ways from their low β counterparts. First, the magnetic moment of the ions (and possibly of the electrons) is not a good adiabatic invariant, because the gyroradius in the interior of the plasma exceeds the scale length for variations in B. The containment (and scaling) in such bottles is therefore different from that exhibited by low β magnetic mirrors (5). In many respects our high β plasma behaves like a collision-free gas; the particles move linearly, but randomly, until reflected at the plasma interface. Most of the interesting physics therefore occurs at the sheath which separates the plasma from the vacuum field B_0. The plasma parameters in the sheath are normally such that $a_i > a_e > \lambda_D$, where a_e and a_i are the local electron and ion gyroradii, and λ_D is the electron Debye length. It will be shown that, in the absence of microinstabilities, the characteristic length of the sheath should therefore lie between a_i and a_e, depending upon the strength of the space-charge electric field in the boundary. The boundary

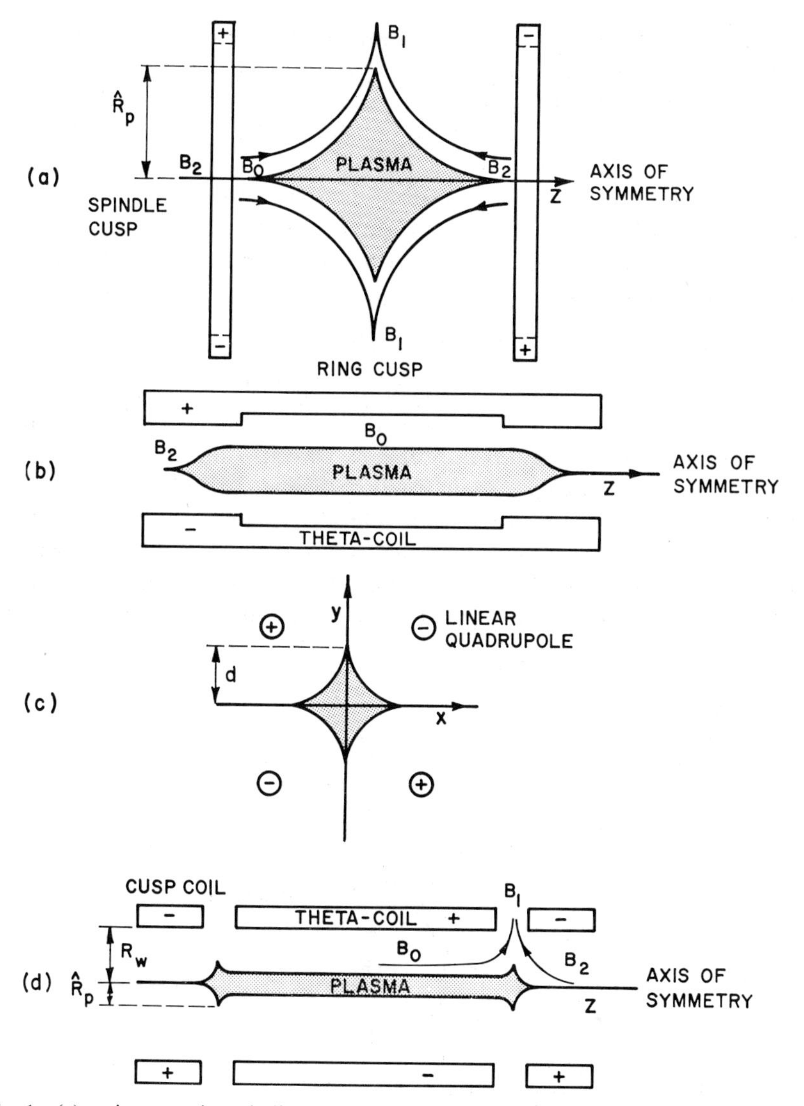

Fig. 1. (*a*) axisymmetric spindle-cusp; (*b*) axisymmetric θ pinch with mirrors; (*c*) two dimensional quadrupole; and (*d*) cusp-ended θ pinch ("long cusp").

layer between the earth's geomagnetic field and the (shock-heated) solar wind (Figure 2) also satisfies the conditions $a_i > a_e > \lambda_D$, so that the self-consistent structure and stability of the collision-free sheath is also a problem of some geophysical significance; indeed, this problem was first discussed by Chapman and Ferraro in 1932 (6, 7).

A second important difference from the low β problem is that it is not necessarily sufficient to make an electrostatic approximation when discussing the stability of the plasma boundary (8). Moreover, it becomes necessary to consider not only the Larmor drift V_D, due to pressure gradients, but also the magnetic drift V_M, due to field gradients (9). (Note that $V_M \sim \dfrac{\beta}{2(1-\beta)} \cdot V_D$.)

Finally, the density scale length in the sheath becomes as short as, or shorter than, the ion gyroradius, a situation that violates all MHD approximations. Similar boundary layer problems can also arise in collision-free shock theory (10), in high β toroidal devices (4), and in low β toroidal systems for which apertures or gross instabilities have removed the outermost layers of plasma. (Examples of this latter situation are provided by aperture-limited stellerators, and by the flute-unstable region bounding the $\int dl/B$ stability limit in multipoles (11).) To a large extent the instabilities (and,

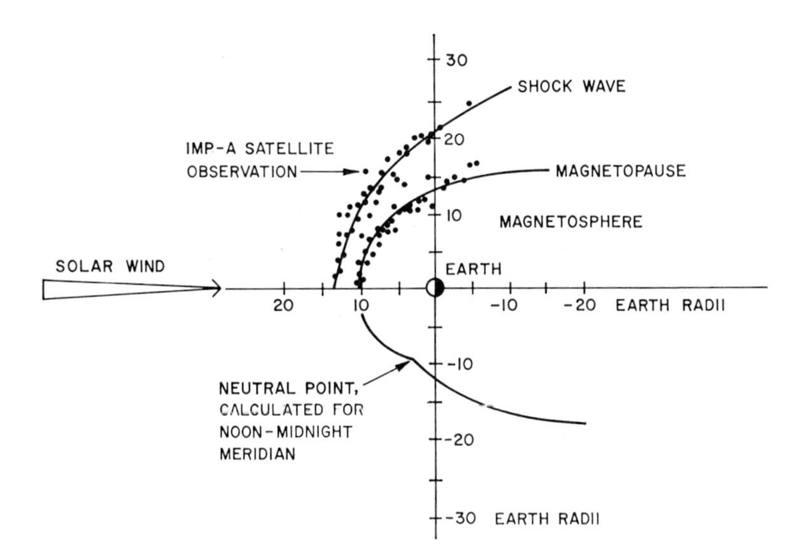

Fig. 2. The Bow shock, magnetopause, and neutral point. The full curves are calculated for the magnetosphere, the fitted curves for the shock boundary. (After Beard, Figs. 8 and 17 of Ref. 102.)

more particularly, the associated diffusion) to be expected at the collision-free sheath are still open questions.

The following approach has therefore been adopted in the following sections.

In Section I we summarize early theoretical work on equilibrium, gross stability, and quiescent confinement in cusp geometries. Much of this work has been discussed previously by Firsov (12) and Grad (13). Recent geophysical discussions of a possible small-scale lack of equilibrium that can develop in high β sheaths, and developments in models of collision-free confinement, are then outlined. Finally, current work on the microstability of the boundary is described briefly.

In Section II we discuss those experiments for which a meaningful comparison with the theory of Section I can be attempted. In general, some measurement of β and the time variation of plasma density (n) and temperature (T) will be regarded as the minimum diagnostics required for such a comparison. However, a representative survey of other attempts to form a cusp plasma are also listed. It is hoped that most experiments published between the earlier reviews of Artsimovitch (14) and Luk'yanov and Podgornyi (15) and the close of 1968 are included in this survey.

In Section III an assessment of the reactor feasibility of open-ended high$-\beta$ systems (16) is outlined. Combining the quiescent confinement models of Section I with some major practical and technological requirements of any fusion reactor we conclude that any open-ended, high$-\beta$ reactor having a mean (electrical) output less than 10 GW(E) must be pulsed. The only method currently envisaged for heating such a device is shock heating (collision-free); it has been argued that the capital cost of the reliable, fast, pulsed energy-storage units required to heat the plasma is likely to be so great that any pulsed reactor would be uneconomic if capacitor banks were used for this purpose (16, 17).

In Section IV we conclude by summarizing the main results of published cusp experiments and some points of contact with other active branches of plasma and geophysics.

I. Cusp Theory

A. *Gross Magnetohydrodynamic Equilibrium and Stability*

The prototype two-dimensional $\beta = 1$ cusp configuration is shown in Figure 1*c*. The magnetic field is generated by four line currents alternating in direction, coupled with plasma sheet currents that flow in alternate directions on adjacent segments. The shape of the equilibrium surface can be determined by solving Maxwell's equation $\nabla \times \mathbf{B} = 0$, $\nabla \cdot \mathbf{B} = 0$, using the following boundary conditions:

1. $\mathbf{B}$ reduces to appropriate values near all exterior conductors.
2. $\mathbf{n} \cdot \mathbf{B} = 0$ at a plasma surface having the unit normal vector $\mathbf{n}$.
3. $\mathbf{B}^2/8\pi = p =$ const at the plasma surface.

Using conformal mapping techniques, the limiting shape can be shown analytically (1, 11, 18) to be the hypercycloid

$$x^{2/3} + y^{2/3} = d^{2/3} \tag{1-1}$$

when the maximum plasma dimension d is much less than the coil spacing. More complicated three-dimensional equilibria, with coils of finite dimensions, are computed numerically (19).

It has been shown quite generally (20) that any equilibrium state is macroscopically stable, provided that the plasma pressure everywhere decreases in the direction of the principal curvature of the field lines. This theorem thus ensures gross stability to any cusp configuration in which a true equilibrium exists, i.e., one in which the plasma pressure is constant along the magnetic field, with uncontained plasma extending to infinity outside the cusps. Additionally, the free boundary model having $\beta = 1$ plasma has been shown by the energy principle (11) to be stable with respect to finite amplitude disturbances from equilibrium (2).

Rotation or any equivalent positive acceleration $\mathbf{g}$ associated with a body force that in the plasma reference frame is directed from the plasma toward the vacuum field tends to make the plasma MHD unstable to the Rayleigh-Taylor flute mode. However, the guiding center drift associated with a negative (cusp) radius of curvature $\mathbf{R}$ provides a stabilizing acceleration (11)

$$\mathbf{g_c} = \frac{\mathbf{R}}{R^2}\left(v_{\|}{}^2 + \frac{1}{2}v_{\perp}{}^2\right) \tag{1-2}$$

where $v_{\|}$ and $v_{\perp}$ are the particle velocities parallel and perpendicular to $\mathbf{B}$. It follows (21) that a $\beta \sim 1$ cusp plasma is MHD stable to any acceleration g, provided that

$$|B_0{}^2/8\pi R| > nm_i\, g \tag{1-3}$$

where m_i is the ion mass. In this stability condition is violated, instabilities having wave number k develop with a growth rate (11):

$$\gamma = \sqrt{(g - g_c)\,k} \tag{1-4}$$

Now, the mathematical idealization of an infinitely sharp plasma boundary is physically unrealistic, as discussed in Section I-B; it follows immediately that

1. There is an upper limit to the short wavelength growth rates in eq. 1-4. For example, Chandrasekhar (22) discusses an ideal fluid confined between plane boundaries that are separated by any arbitrary distance much greater than the density scale length a of the (exponentially varying) boundary; in a constant effective gravitational field G, the growth rate is given by

$$\gamma^2 = \left(\frac{G}{a}\right)\left(1 + \frac{1}{4a^2k^2}\right)^{-1}$$

and thus γ has a maximum growth rate $\gamma \max = \sqrt{G/a}$.

2. Plasma will be lost along the magnetic field lines. A self-consistent stability analysis should then consider a quasiequilibrium in which plasma escapes slowly along the field lines through the ring and point cusps (Figure 1*a*); such as analysis has not been attempted. However, Haas and Wesson note that the flow of plasma from a high β cusp is subsonic and that the flow energy density is therefore less than the energy density of the confining field; they conclude that in the fluid approximation the flow should not strongly effect the gross stability (23). Although large-scale motions of the plasma to the wall are therefore unlikely, small-scale turbulence, due to fine-scale effects not covered by MHD approximations, are not excluded at this stage of the theory (13).

B. *The Collision-free Sheath*

The current sheet of Section I-A implies an infinite current density that cannot occur in nature. An order of magnitude estimate of the real thickness of the layer can be made very readily for a plasma having equal ion and electron masses and mean particle velocities ($\bar{v}$), since space-charge electric fields can be ignored in this problem. The maximum current density j that can be sustained is approximately

$$\frac{(ne\bar{v})}{c}$$

so that the maximum field change (ΔB) in a layer of thickness δ is given by

$$\Delta B = B_0 = 4\pi \frac{(ne\bar{v})}{c} \delta \tag{1-5}$$

But for pressure balance

$$\frac{nm(\bar{v})^2}{3} \sim \frac{{B_0}^2}{8\pi} \tag{1-6}$$

Hence

$$\delta \sim \frac{3}{2}\left(\frac{m\bar{v}c}{eB_0}\right) \tag{1-7}$$

Thus the change in B from low values within the plasma to the vacuum value B_0 required for pressure balance cannot occur in a distance much less than a gyroradius. More rigorous treatments of the transition layer, in which the computed particle orbits give rise to precisely the current needed to produce the magnetic field gradient, are discussed below.

1. Plane Boundary, No Electric Field

Grad (24) has discussed the thinnest transition zone between a unidirectional magnetic field $B(x)\hat{\imath}_z$ at $x = +\infty$ and a collisionless plasma at $x = -\infty$ which is compatible with given plasma properties at $-\infty$. If all quantities are restricted to vary only in the x direction traversing the sheath, it is shown that a unique solution exists, corresponding to a given arbitrary velocity distribution at $-\infty$, provided that there are no "trapped" particles (i.e., particles that do not have access to minus infinity). Rather surprisingly, it is found that, for any nonsingular velocity distribution at minus infinity, B penetrates only a finite distance into the plasma, although the plasma extends to infinity, exponentially attenuated, in the magnetic field region (25).

For a Maxwellian distribution at minus infinity, B first becomes finite at a well-defined position X such that for $(x - X)$ small but positive,

$$\frac{B}{B_0} \sim 0.123\left(\frac{x - X}{\lambda}\right)^3 \tag{1-8}$$

where

$$\lambda^2 = \frac{m}{4\pi e^2 N_0} = \frac{c^2}{{\omega_{pe}}^2} = \frac{2mkT_0}{e^2 {B_0}^2}$$

B_0 is the vacuum B_z at $x = +\infty$, N_0, T_0 are the total density and temperature at $x = -\infty$, and ω_{pe} is the electron plasma frequency.

The collision-free scale length $\lambda = c/\omega_{pe}$ may thus be interpreted as the gyroradius of a particle evaluated using the field strength at $+\infty$ and a mean thermal speed at $-\infty$.

2. Rosenbluth Sheath

Let us now restrict the particle velocities at $-\infty$ to the single value v_0 in the x direction but let the ions and electrons have unequal masses ($m_i > m_e$). The electrons are bent more easily than the ions in the magnetic field, and so the ions try to get ahead of the electrons, producing a strong space-charge electric field $E(x)\hat{\imath}_x$ that ensures charge neutrality to the first order. Thus

$$n_i(x) = n_e(x) = n \tag{1-9}$$

where $n_i(x)$ is the net density of both in-going and reflected ions at x, and n_0 will signify the ion density inside the plasma. Particle conservation requires that

$$n_i(x)v_{xi}(x) = n_0 v_0 = n_e(x)v_{xe}(x)$$

or

$$v_{xi}(x) = v_{xe}(x) = v_x \tag{1-10}$$

Neglecting magnetic forces on the ions (for reasons discussed later), we find that the equation of motion for ions becomes

$$m_i(d/dt)v_x = m_i v_x(d/dx)v_x = eE(x) \tag{1-11}$$

and for electrons (having a charge $-e$),

$$m_e v_x(d/dx)v_x = -e\,[E(x) + (v_y/c)B] \tag{1-12}$$

$$m_e v_x(d/dx)v_y = +(e/c)v_x B \tag{1-13}$$

By neglecting terms in m_e/m_i, it follows from eqs. 1-11 and 1-12 that

$$m_i v_x(d/dx)v_x = -(e/c)v_y B \tag{1-14}$$

The presence of the ion mass in eq. 1-14 signifies that the ion momentum is transferred through the electric field $E(x)$ to the electron. Similarly, from eqs. 1-13 and 1-14 we have

$$(d/dx)(m_i v_x^2 + m_e v_y^2) = 0$$

i.e.,

$$\left(\frac{m_i}{m_e}\right)v_x^2 + v_y^2 = \left(\frac{m_i}{m_e}\right)v_0^2 \tag{1-15}$$

Thus eq. 1-15 shows that at the turning point, i.e., where $v_x = 0$, the electrons have acquired all the energy ($\frac{1}{2} m_i v_0^2$) initially possessed by the ions. Moreover, the electron orbits are elongated in the y direction, thus permitting the electrons to carry the diamagnetic current required for pressure balance.

Maxwell's equation can now be used to deduce the sheath thickness;

$$-\frac{\partial B}{\partial x} = -\frac{1}{2}\left(\frac{d}{d\eta}\right)B^2 = 4\pi j_y = 4\pi n\left(\frac{ev_y}{c}\right) \tag{1-16}$$

where $\eta(x)$ is the magnetic vector potential which we equate to zero inside the plasma. Equation 1-13 can be written in terms of η to give

$$m_e v_y = \left(\frac{e}{c}\right)\eta = p_y = 0 \tag{1-17}$$

where the constant of integration p_y (the y canonical momentum) is also zero, since both v_y and η are zero at $-\infty$.

From eq. 1-10, 1-15, 1-16, and 1-17 it follows that

$$\frac{d(B^2/8\pi)}{d\eta} = \frac{n_0 e^2}{m_e c^2} \cdot \frac{\eta}{[1 - (e^2\eta^2/m_i m_e v_0{}^2 c^2)]^{1/2}} \tag{1-18}$$

or

$$B^2/B_0{}^2 = 1 - [1 - (\eta/B_0 a)^2]^{1/2} \tag{1-19}$$

where

$$a = (m_i m_e)^{1/2} v_0 c/eB_0$$

The length parameter a of eq. 1-19 is the geometric mean of the ion and electron gyroradii, so that the boundary layer is thin compared to the ion gyroradius (see Figure 3) and our neglect of magnetic forces in eq. 1-11 is fully justified. The charge neutrality assumption of eq. 1-9 is also justified, provided that the Debye length is much smaller than a. This condition is satisfied when $(m_i v_0{}^2/m_e c^2)$ is small (26), i.e., provided that the ion kinetic energy is small compared to the electron rest energy of 510 keV.

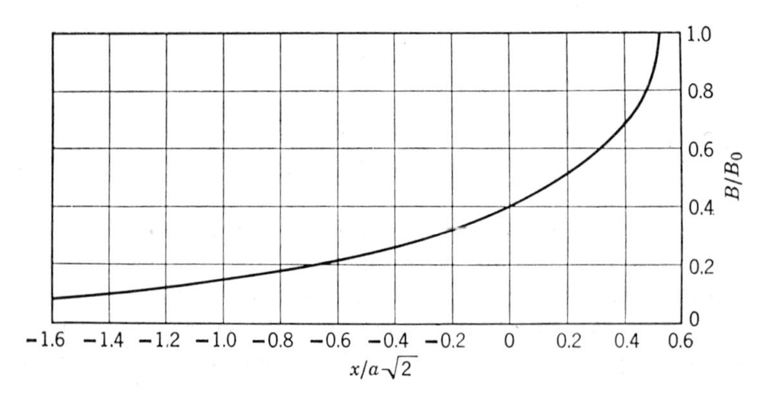

Fig. 3. The collision-free sheath, showing the normalized magnetic field B/B_0 as a function of (x/a), where a is the hybrid gyroradius of Eq. 1-19 (26).

This result, obtained by Rosenbluth (27), has been generalized by Bertotti (28) to include thermal motion in the direction of ∇B only. Bertotti also treats the charge neutrality assumption more rigorously.

3. Isotropic Plasma

Longmire (26) discusses the difficulty that arises when the relatively simple problem described above is extended to include a monoenergetic but isotropic velocity distribution deep within the plasma. It seems to be impossible to construct a self-consistent solution without requiring that some electrons be trapped in the boundary layer. Once trapped particles are permitted, the solutions are no longer unique, as the trapped electron distribution may be dictated either by physical intuition or by mathematical convenience. (Such solutions are discussed in refs. 29–32.)

In order to reduce the arbitrary nature of these solutions, Morse (33) investigated the time development of a sheath in which electrons are trapped in a predictable, although not unique, manner by a slowly rising magnetic field. Starting at time zero with a Maxwellian plasma having arbitrary ion and electron temperatures, he obtains numerical solutions that possess a scale length somewhat larger than the electron collision-free skin depth c/ω_{pe}. The width is almost independent of the ratio T_i/T_e over a wide range of values.

4. Two-dimensional Effects: Local Nonequilibrium in the Sheath

When the plasma parameters are permitted to vary in two dimensions, other complications arise. First, in the prototype two-dimensional cusp of Figure 1*c* we intuitively expect ions and electrons to escape along the magnetic field lines through a hole having a (minimum) half-width roughly equal to the hybrid gyroradius discussed in Section I-B-2. When $T_e \sim T_i$, electrons will attempt to escape more rapidly than ions, so that an ambipolar electric field must be developed parallel to B in order to maintain charge neutrality. An elementary computation suggests that the Maxwell-averaged ion loss rate is then multiplied by the factor $(1 + e\phi/kT)$, whereas the electrons are decelerated by the much greater factor $\exp(-e\phi/kT)$. Thus plasma will escape at a rate only slightly in excess of that expected for ions under conditions in which the plasma potential ϕ is zero.

Second, in a finite plasma, end-effects can become important. In particular, the transverse space-charge field E_x may be short-circuited by electrons flowing back and forth along B_z to any electron source that happens to exist at the ends of the plasma. The ions are then free to penetrate across the plasma sheath to a depth of up to an ion gyroradius, but provided there is no net motion of the plasma relative to B_z, a stationary equilibrium is still possible, as discussed in Section I-B-1. (It may be noted that, if $E_x = 0$, the ions carry their own diamagnetic current, in order to provide a pressure balance; there is then net rotation of the plasma. For this reason the short circuit is expected to propagate into plasma where E_x is finite from a region where $E_x = 0$ at a speed comparable to the local shear Alfvén velocity).

A third effect (34,35) arises when the collisionless plasma is moving with a velocity U parallel to B_z, and when there is simultaneous neutralization of the space charge in the sheath. Suppose that $(p = B_z^2/8\pi)$ across the plane interface, so that the MHD necessary and sufficient condition for a steady equilibrium is satisfied. The bulk motion of the ions parallel to B_z generates a local transverse magnetic field B_y as soon as the ions have penetrated more deeply than untrapped electrons into the neutralized boundary layer, as shown in Figure 4. This local magnetic field can be of order

$$B_y \sim B_z \cdot (U/v_i) \tag{1-20}$$

if the sheath is as thin as an ion gyroradius. If $U > v_i$, there appears to be no way of containing B_y, which tends to push the ions out into the plasma stream, so that there is a local nonequilibrium in the sheath that mixes field and plasma with a scale length of up to an ion gyroradius.

Lerche and Parker first discussed this "nonhydromagnetic" effect in relation to the confinement of the geomagnetic field by the solar wind.

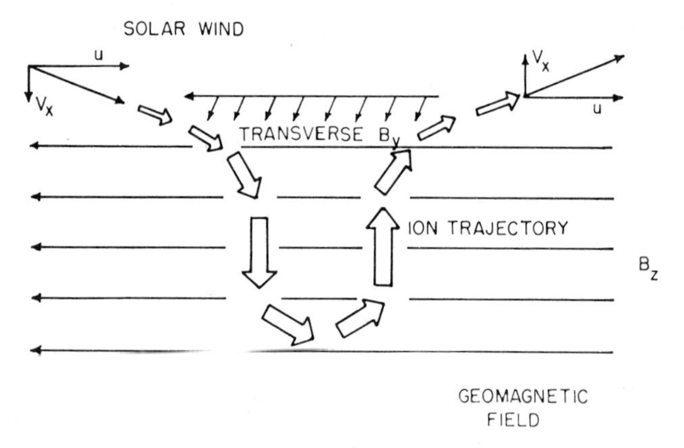

Fig. 4. A sketch of the incident ion trajectory for ions having a large velocity U parallel to B_z, a small velocity v_x normal to the field, and all space-charge electric fields neutralized. The velocity parallel to B_z is brought to zero by the Lorentz force in the transverse field B_y, so that beneath this transverse layer the structure of the boundary is essentially the same as for normal incidence of the ion beam. The sketch is based on Ref. 34.

Its importance depends on the exact degree of space-charge neutralization (36), but a similar effect is also possible in high β laboratory plasmas when the magnetic field lines terminate on the walls of the vacuum vessel.

C. *Microinstabilities*

Drift waves arising in a plane boundary layer between a plasma and its confining field have now been studied fairly extensively. When the waves (of frequency ω and wave number $\mathbf{k}$) show no variation along the magnetic field (i.e., $\mathbf{k} \cdot \mathbf{B} = 0$), any instabilities are driven entirely by the gradient drifts that propagate perpendicular to $\mathbf{B}$. Modes with $\mathbf{k} \cdot \mathbf{B} \neq 0$ can be related to the Alfvén and ion-acoustic waves of homogeneous plasma theory and can tap the thermal energy of the plasma by resonating with particles streaming along $\mathbf{B}$. (See, e.g, the review by Krall (8).) In most treatments it has been postulated that $\beta \ll \frac{1}{2}$, so that the magnetic and electrostatic drift modes decouple, and in many treatments it has also been assumed that the boundary is many ion gyroradii wide. Attention has been concentrated primarily on low-frequency, long wavelength modes that are expected to produce appreciable plasma transport across the field (i.e, $\omega < \Omega_i \ll \Omega_e$, where Ω_i and Ω_e are the ion and electron angular gyrofrequencies). However, we are also interested in instabilities producing only diffusion because an increased magnetic flux permeating the plasma volume would enhance the cusp end losses (to be discussed in Section I-D).

Since the stationary equilibria of the collision-free sheaths discussed in Section I-B do not appear to have been subjected to any systematic stability analysis, we shall merely select for discussion a few collision-free calculations that appear to be relevant at high β. It should be particularly noted that relatively infrequent particle collisions may introduce additional instabilities and strongly modify cyclotron or other resonances (8).

1. Low-Frequency ($\omega < \Omega_i$) Drift Instabilities

Mikhailovskii and Fridman (9) discuss the propagation of low-frequency waves in an inhomogeneous plasma of arbitrary β, in which the magnetic field $B(x)\hat{\imath}_z$, the density, and the temperature are assumed to vary only in the x direction. When $\beta > \frac{1}{2}$, many low-frequency, ion-acoustic modes become stable, but a drift (shear) Alfvén wave becomes unstable when the temperature gradient $\partial \ln T/\partial \ln n$ is sufficiently negative (37). It is assumed that $k^2 r_i^2 \ll 1$ and that the transverse wavelength $\lambda_\perp$ is less than the density scale length; the growth rate γ is then related to the magnetic and pressure-gradient drift frequencies, $\omega_M^* = k_\perp V_M$ and $\omega_p^* = k_\perp V_D$.

2. High-Frequency Drift Instabilities

At frequencies $\Omega_e \gg \omega > \Omega_i$, it is usually possible to ignore the effect of the magnetic field on the ions, which may be assumed to have straight-line orbits. Krall and Book (38) discuss the linear stability of ion-acoustic waves driven by (self-consistent) electron $\mathbf{E} \times \mathbf{B}$ and ∇B drifts across the magnetic field $B(x)\hat{\imath}_z$. Their discussion is concerned primarily with the structure of collision-free shocks in which the transit time of the shock front exceeds the electron gyro period but is less than the ion gyroperiod. An instability is predicted (for sufficiently large magnetic gradients) at wave numbers such that $\alpha_e{}^{-1} < k < \lambda_D{}^{-1}$. The number of resonant electrons is very sensitive to the electric field $\mathbf{E}$ (which must be self-consistent with the postulated equilibrium density and magnetic field distributions), but a similar treatment suggests that collision-free sheaths for which $(dn/dx \cdot dB/dx)$ is positive (as in some regions of the simple Rosenbluth sheath) and for which $T_e/T_i \gtrsim 1$ will also experience a rapidly growing instability having a growth rate $\sim \sqrt{\Omega_e \Omega_i}$. (See also references 108, 109.)

A more interesting situation is that in which $(dn/dx \cdot dB/dx)$ is everywhere *negative*, as in the equilibria discussed by Morse (33). At moderately low β (i.e., $\beta < \frac{1}{2}$) an electrostatic wave having $\Omega_i \ll \omega \ll \Omega_e$ and field components E_z, E_y has been shown by Krall (107) to be unstable when

$$\left(\frac{1}{n} \cdot \frac{dn}{dx}\right) a_i \sqrt{T_e/T_i} > 1 \tag{1-21}$$

with a maximum growth rate near the stability limit of

$$\gamma \sim \frac{\overline{kV_e{}^2}}{\Omega_e} \left(\frac{1}{n} \cdot \frac{dn}{dx}\right) \sim \omega_{pi} \tag{1-22}$$

Such an instability can produce field diffusion without appreciable particle transport across the field.

3. Bulk Velocity-Anisotropy

During the shock-heating and adiabatic compression of a high β plasma, the perpendicular pressure, $P_\perp$, may exceed the parallel pressure, $P_\parallel$. Although bulk velocity anisotropy is not an inevitable requirement of $\beta = 1$ containment (indeed, the opposite will be assumed in the discussion of Section I-D), the anisotropies that can arise during the formation of the plasma may well drive instabilities which lower the initial β of the plasma and so affect the subsequent confinement.

Thus Zayed and Kitsenko (39) discuss the excitation of electron cyclotron harmonics in a plasma having an anisotropic electron temperature. Longitudinal and transverse electromagnetic waves propagating at an angle to the magnetic field, and having typical wavelengths of the order of the electron gyroradius, are shown to be unstable. When $T_\perp \gg T_\parallel$, instability is predicted over the whole range of cyclotron harmonics to the electron plasma frequency.

At lower frequencies a fine-scale "mirror" instability with a destabilizing term proportional to $(P_\perp/P_\parallel - 1)$ can randomize the ion velocity distribution within a few ion gyroperiods, as demonstrated in a finite β digital computation by Morse (40). Finally, under conditions appropriate to an *escaping* plasma (i.e., $P_\parallel/P_\perp \gg 1$), a collision-free, fire-hose instability can develop (41).

D. *Cusp Confinement Models*

A rigorous theory of the cusp containment of quiescent plasma involves the computation of plasma currents and self-consistent, space-charge fields in three dimensions. We have noted that these electric fields may be controlled not only by the manner in which the plasma is created but also by its external environment during containment. It is therefore not entirely surprising that the most popular theoretical approach has been the construction of over-simplified but well-defined containment models. Recent work has largely obviated the need for detailed discussion of particle orbits and complex numerical solutions.

1. Collision-free Models

Taylor (42) used a statistical model to calculate the losses from a collision-free plasma under conditions in which all electric fields are neglected. It was postulated that a large volume of field-free Maxwellian plasma is separated from an axisymmetric cusp field by a thin boundary of arbitrary structure and that the equilibrium is a true steady state, in which all escaping particles are reintroduced into the containment zone, so that the steady-state Vlasov equation can be used. Accessible regions of phase space were calculated using two constants of the motion: the total particle energy and its canonical angular momentum p_θ. Taylor's analysis is only valid at very high values of β such that the local ion gyroradius (a_i) in the bulk of the plasma is larger than the maximum radius of $\beta = 1$ plasma ($\hat{R}_p$). Kaye (43) has extended this analysis to regimes in which the trapped (*parallel*) magnetic flux Ψ may be sufficiently large that $a_i < \hat{R}_p$. The flux of particles effusing through one-half of the ring cusp is

$$F_1 = \tfrac{1}{4}\, n\bar{v}\, 2\pi\hat{R}_p \cdot \delta_L \tag{1-23}$$

where $\bar{v}$ is the mean thermal speed $(8kT_i/\pi m_i)^{1/2}$ and the (half) hole size δ_L is shown to be

$$\delta_L = \left(\frac{3\pi}{8}\right)^{1/2} r_i \lambda_1^{-1}\left[1 + \sqrt{\frac{2}{3\pi}} \cdot \frac{\hat{R}_p}{r_i}(1-\beta)^{1/2}\right] \tag{1-24}$$

where r_i is the thermal gyroradius evaluated in the vacuum field B_0 close to the boundary (i.e., $r_i = m_i c \sqrt{3kT_i/m_i}/eB_0$), and λ_1 the mirror ratio (B_1/B_0) between B_0 and the maximum field B_1 on the median plane (Figure 1*a*).

Similarly, the particle flux through one point cusp is

$$F_2 = \tfrac{1}{4} n\bar{v} \cdot \pi \delta_p{}^2 \tag{1-25}$$

where

$$\delta_p{}^2 = \left(\frac{3\pi}{2}\right)^{1/2} \frac{r_i \hat{R}_p}{\lambda_2} \left[1 + \left(\frac{8}{27\pi}\right)^{1/2} \frac{r_i}{\hat{R}_p \lambda_2} + \left(\frac{2}{3\pi}\right)^{1/2} \frac{\hat{R}_p}{r_i} (1-\bar{\beta})^{1/2}\right] \tag{1-26}$$

Here $\lambda_2 = B_2/B_0$, and B_2 is the maximum field on the z axis. When $\bar{\beta} = 1$, Taylor's result is recovered and it is seen that, to order $(r_i/\hat{R}_p)$, $F_1 = F_2$ when $\lambda_1 = \lambda_2$. This result is not intuitively obvious and is a consequence of the conservation of canonical angular momentum; it implies that the end loss from a high$-\beta$ θ pinch should be comparable to that from a spindle cusp device; i.e., the diameter of the escape hole is roughly $3\sqrt{r_i \hat{R}_p/\lambda_2}$.

When $\hat{R}_p > a_i$, i.e., when $(1-\bar{\beta})^{1/2} > r_i/\hat{R}_p$, Kaye shows that the loss rate increases significantly; scaling as $(1-\bar{\beta})^{1/2}/\lambda$. Thus, at low $\bar{\beta}$, the collision-free model scales identically to the fluid model to be discussed below. It should be noted that, although $\bar{\beta}$ is well defined in a θ pinch in which the internal field can be uniform, in cusp geometries possessing a true field zero, eq. 1-24 and 1-26 are more precise when the term $(1-\bar{\beta})^{1/2}$ is replaced by $(\Psi/\pi\hat{R}_p{}^2 B_0)$.

The effect of a third invariant, the action integral $\oint p_r \cdot dr$ discussed previously by Grossman (44), has been reconsidered by Kaye (43). He shows that the loss rate at the line cusp is reduced to

$$F_1^* = \tfrac{1}{4} n\bar{v} \cdot 2\pi \hat{R}_p \cdot \delta_L^*$$

where

$$\delta_L^* = \left(\frac{3\pi}{8}\right)^{1/2} r_i \lambda_1{}^{-1} \left[\frac{2}{3\lambda_1^{1/2}} + \sqrt{\frac{2}{3\pi}} \cdot \frac{\hat{R}_p}{r_i} \cdot (1-\bar{\beta})^{1/2}\right] \tag{1-27}$$

and at the point cusp to

$$F_2^* = \tfrac{1}{4} n\bar{v} \cdot \pi \delta_p^{*2}$$

where

$$\delta_p^{*2} = \left(\frac{3\pi}{2}\right)^{1/2} \frac{r_i \hat{R}_p}{\lambda_2} \left[\frac{2}{3\lambda_2^{1/2}} + \left(\frac{8}{27\pi}\right)^{1/2} \frac{r_i}{\hat{R}_p \lambda_2} + \left(\frac{2}{3\pi}\right)^{1/2} \frac{\hat{R}_p}{r_i} (1-\bar{\beta})^{1/2}\right] \tag{1-28}$$

Thus when $\lambda_1 = \lambda_2$, $F_1^* = F_2^*$ (to order $r_i/\hat{R}_p$). When $(1-\bar{\beta})^{1/2} < (r_i/\hat{R}_p) \ll 1$, the additional adiabatic invariant, which is related to the generalized magnetic moment of Berkowitz (45), reduces the loss rate by a factor $\frac{2}{3}\lambda^{-1/2}$ over the nonadiabatic results of eqs. 1-24 and 1-26. This enhanced mirror effect is due to the exclusion of a certain class of particles having high longitudinal velocities from the (adiabatic) sheath, and so is sensitive to weak transverse electric fields. By assuming that magnetic field lines are equipotentials, and that the electric field is linear at the point cusp, Kaye estimates that the effect will vanish when $E > kT_i/er_i$, a result that should not be sensitive to the exact shape of the electric field profile. (The effect also vanishes if an ion makes a collision during the time it spends in the sheath, i.e., if $1_{ii} \ngtr \lambda^{3/2}L$, where 1_{ii} is the ion–ion mean free path and L is the length of the plasma.) In the much stronger electric fields associated with a $\beta = 1$ plasma having a sheath width Δ of less than an ion gyroradius, i.e., $r_i \gg \Delta > (r_i r_e)^{1/2}$, Kaye computes a flux through the line ($q = 1$) or point ($q = 2$) cusp of

$$F_q = \tfrac{1}{4} n\bar{v} \cdot 2\pi\hat{R}_p \cdot \Delta\lambda_q^{-1} \tag{1-29}$$

The mean-free path condition is then relaxed to $1_{ii} > L_N$, where L_N is the effective length of the magnetic nozzle.

A fundamental postulate of these models is the assumption that there exists a high β (nonadiabatic) region remote from the cusps which supplies isotropic plasma to the nozzles; this assumption may be valid in cusp geometries that possess true field zeros but is not necessarily true in a simple θ pinch unless $L > 1_{ii} > L_N$. It is therefore convenient to summarize those regimes in which eq. 1-23–1-29 are valid (43):

1. In a strong transverse sheath field $E > kT_i/er_i$, eq. 1-29 can be applied to to cusps when

$$(1-\bar{\beta})^{-1/2} > \frac{\hat{R}_p}{r_i}\left(\frac{\hat{R}_p}{\Delta}\right)^{1/2} > \frac{1_{ii}}{L_N} > 1$$

   and to θ pinches when, additionally, $1 \gg 1_{ii}/L$.
2. In weak electric fields $E < kT_i/er_i$,
   a. Equations 1-27 and 1-28 can be applied to cusps only, provided that

$$(1-\bar{\beta})^{-1/2} \gtrsim \frac{R_p}{R_i} \gg \frac{1_{ii}}{\lambda^{3/2}L} > 1$$

   (Kaye has also shown that for $\bar{\beta} \to 1$, eq. 1-27 gives the effective hole size δ_L^* at a two-dimensional linear cusp.)

b. Equations 1-23 to 1-26 can be applied to cusps only, provided that

$$(1 - \bar{\beta})^{1/2} \gtrsim \frac{R_p}{R_i} \gg \frac{1_{ii}}{\lambda L} > 1$$

(For cusps or θ pinches having $L > 1_{ii} > L_N$, eq. 1-23 to 1-26 give a lower limit to the losses which are enhanced by collisions in the sheath.)

c. Equations 1-23 to 1-26 can be applied to cusps when

$$\frac{\hat{R}_p}{r_i} > (1 - \bar{\beta})^{1/2} \gg \frac{1_{ii}}{L_N} > 1$$

and to θ pinches when, additionally, $1 \gg 1_{ii}/L$.

Kilb (46) discusses the difficulty of formulating self-consistent, collision-free equilibria for the θ pinch when $E = 0$. For $\beta > \frac{1}{3}$, equilibrium can only be found if the distribution is almost isotropic; in the absence of randomization by nonadiabatic (cusp) effects, velocity space instabilities are expected to provide the required isotropy. (See also refs. 47 and 48.) Morse (49) has discussed a particular ion distribution that is in equilibrium and is stable to ion–ion collisions (i.e., the model is valid when $L > 1_{ii} > L_N$). In this model the electrons are cold and completely neutralize the ion space charge (i.e., $E = 0$); the plasma is assumed to rotate as a rigid body with the ions carrying the diamagnetic current. For this rigid rotor distribution function Kaye (43) derives a loss rate given by

$$F_2^+ = \frac{n\bar{v}}{4} \cdot \pi \hat{R}_p{}^2 \left(\frac{1 + \sqrt{1 - \beta_A}}{2\lambda_2} \right) \tag{1-30}$$

where β_A is the local β on the axis of symmetry. Unless radial temperature gradients are introduced, β_A completely specifies the sharpness of the sheath in this model, the maximum $\bar{\beta}$ being 0.80 when $\beta_A = 1.00$.

Finally, Kaye notes that the $\beta \sim 1$ collision-free cusp losses discussed above can imply a strong asymmetry of the loss cone for p_θ positive or negative. When $E = 0$ there is a preferential loss at the point cusp of ions rotating in the countergyro direction that leads to an angular momentum loss of order $\hat{R}_p\sqrt{m_i kT}$ per escaping particle. He estimates that an observable rotational instability will therefore develop within a few sound transit times along any open-ended, high β device lacking a magnetic plane of symmetry. This estimate of the onset of the instability includes an approximate treatment (50) of finite gyroradius stabilization but not the additional stabilization afforded by eddy currents in the coils surrounding the plasma (51). The net loss of angular momentum is reduced by transverse electric fields (which tend to equalize the ion and electron angular momentum loss rates), by lowering β (which reduces the importance of cusp losses), or by diffusion.

2. One-Fluid Plasma

Wesson (52) has studied the transient one-dimensional (axial) flow of a one-fluid plasma from a long straight θ pinch of uniform β and length L. High β plasma is lost rapidly behind the wavefront of an area wave that propagates inward from the ends of the pinch; however, the wave can propagate quite slowly, and the loss rate is then quite small. As $\beta \to 1$ the particle flux through one end of the pinch approximates to the steady-state loss rate derived earlier by Taylor and Wesson (53)

$$F_2' = nc_s \cdot \pi \hat{R}_p{}^2 \cdot G(\lambda) \cdot (1 - \beta)^{1/2} \tag{1-31}$$

where c_s is the sound speed $(\gamma P/\rho)^{1/2}$. It can be shown that, for $\gamma = 2$,

$$G(\lambda) = \{6\lambda \sin[\tfrac{1}{3} \sin^{-1}(1/\lambda)] - 2\}^{1/2} \tag{1-32}$$

where $G(\lambda)$ is a monotonically decreasing function of λ that tends to $(\lambda_2 \cdot \sqrt{3})^{-1}$ for $\lambda_2 > 2$. (Equation 1-32 is also a good approximation for $G(\lambda)$ when $\gamma = \frac{5}{3}$.) Defining a particle containment time τ_N by

$$2F_2' = \frac{n \cdot \pi \hat{R}_p{}^2 L}{\tau_N} \tag{1-33}$$

it follows that

$$\tau_N = \left(\frac{L}{2C_s}\right) \cdot \frac{1}{G(\lambda) \cdot (1 - \beta)^{1/2}} \tag{1-34}$$

When $\beta \sim 1$, τ_N becomes unrealistically large because sheath losses are not included in the model. At lower values of β, τ_N is roughly $(2\pi)^{-1/2}$ of that predicted by the collision-free model (eqs. 1-25 to 1-26) but scales with β, L, T in the same manner (see Figure 5).

3. Diffusion Model

A slightly more realistic model can be constructed for a $\bar{\beta} \sim 1$ plasma when diffusion transports plasma across a sheath of width Δ more rapidly than it is lost along the magnetic field by the cusp losses discussed above. The sheath will broaden until the cross-field diffusion is balanced by sonic flow out of the sheath (54). For example, in a cylindrical plasma, where $\hat{R}_p \gg \Delta > r_i$,

$$2nc_s \cdot 2\hat{R}_p \Delta \cdot G(\lambda) \sim 2\pi\, R_p L D \frac{n}{\Delta} \tag{1-35}$$

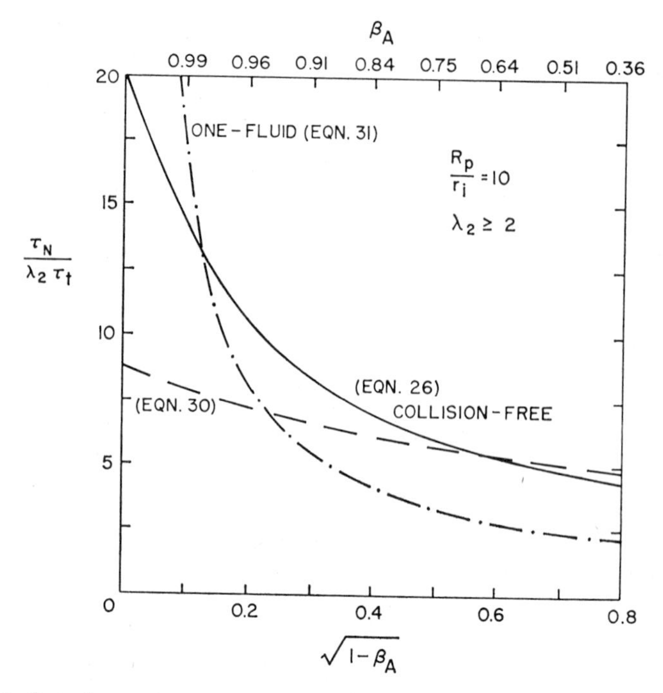

Fig. 5. θ pinch containment as a function of $\beta_A \cdot \lambda_2$ is the mirror ratio; the particle containment time τ_N is expressed in units of the ion transit time $\tau_t = L/2\sqrt{m_i/3kT_i}$. The various models are identified by the numbers of the relevant equations. (After Ref. 43.)

where D is the effective diffusion coefficient. Hence

$$\Delta^2 = \left(\frac{L}{2c_s}\right)\frac{D}{G(\lambda)} \tag{1-36}$$

and

$$\tau_N = (\pi\hat{R}_p{}^2 Ln)\left(2\pi\hat{R}_p LD\frac{n}{\Delta}\right)^{-1} \sim \hat{R}_p \cdot \left(\frac{L\lambda}{5DC_s}\right)^{1/2} \tag{1-37}$$

i.e., strong diffusion will cause a different scaling of τ_N with L, λ, and T. It may be noted that, when $\Delta \sim r_i$, particle transport by binary ion–ion collisions should be considered; Simon (55) shows that, if the transverse ambipolar electric fields are short-circuited (i.e., $E = 0$), the binary diffusion coefficient $(\beta c^2/8\pi\sigma)$ is enhanced by a factor of order $(m_i/m_e)^{1/2}\,(r_i/\Delta)^2$. In the absence of such a short-circuit, like-particle collisions can be neglected (56).

4. Energy Loss

The mean energy carried out per ion in eqs. 1-23 to 1-28 is $5kT_i/2$, to order (r_i/R_p), a result expected if each ion effuses through a hole characteristic of its own gyroradius (43). The adiabatic fluid model, with $\gamma = \frac{5}{3}$, also happens to yield the same result. The ion energy containment time τ_{iE} is thus related to τ_N by

$$\tau_{iE} = 0.6\,\tau_N \tag{1-38}$$

The electron thermal flux along the magnetic field can exceed that due to the ions when $T_e \sim T_i$, the actual value being determined by the magnitude of the positive potential established between the plasma and the wall of the vacuum vessel (cf. Section I-B-4). Using a one-dimensional model in which no secondary ions are produced in the sheath region, we can write the total (ion plus electron) energy flux per unit area flowing into the wall (Q) as

$$Q = \tfrac{1}{4}\, n\bar{v}_e \cdot 2kT_e \cdot E \tag{1-39}$$

where E is a thermal transmission factor through the sheath. The term E ranges from a minimum value of $(\pi m_e/8m_i)^{1/2}\,[5 - \ln(2\pi m_e/m_i)]$ for the classical Langmuir sheath in which the wall is perfectly absorbing and nonemitting (57) to a maximum of $[0.33 + 2.2\,(m_e/m_i)^{1/2}]$ for an electron-emitting wall (58). It follows that, in the absence of a fully developed sheath, the total thermal flux may exceed the ion thermal flux by a factor of order $0.33\sqrt{m_i/m_e}$. Electron collisions will tend to reduce this energy loss until, in the collisional limit, the thermal conduction coefficient (59) can be used to calculate the energy loss.

II. Cusp Experiments

A. *The Creation of a Cusped Plasma*

The establishment of a plasma with properties approaching those of the free boundary equilibria discussed in the early part of Section I proved to be a major experimental problem. Two principal experimental approaches may be distinguished:

1. Plasma is injected from an external field-free region into a quasi-static, cusped confining field. (The creation of energetic plasma by a suitable plasma gun is then relatively easy, but the injection process and capture is difficult to describe theoretically and is complicated to analyze experimentally.)

2. Plasma is produced in situ and is then heated further by the rapid application of a confining cusp field. (Heating by adiabatic compression in

geometries possessing a vacuum field zero is relatively inefficient, so that high voltage pulsed techniques are required to catch and shock-heat plasmas of reasonable volume. At the expense of some experimental complexity the final state of the plasma is then more amenable to ab init. calculations in which the initial heating is computed using one or two-dimensional computer codes and the subsequent containment phase is described using suitable confinement models.)

A third possible approach, laser heating of lithium hydride and solid or liquid hydrogen targets suspended, in vacuo, at the center of a steady cusp field is still in its infancy. The technique is attractive because highly ionized plasma can be created within the confining field and well away from the walls of the vacuum (60); however, the dimensions of such experiments will tend to be small because the energy of present-day Q-switched laser pulses is much less than the energies currently produced in electrodynamically heated plasmas.

The injection approach was very popular in the period between 1958 to 1961, and these early experiments have been adequately reviewed by Artsimovitch (14) and Yu Luk'yanov and Podgornyi (15). The diagnostic techniques used were relatively rudimentary; laser techniques, in particular, had not been sufficiently developed to permit the spatially resolved measurements that are now both possible and desirable. Some more recent injection experiments in cusped geometry are listed in Table I, with a brief summary of the main experimental parameters. Many of these plasmas are either highly collisional or of low average β. Detailed results from the experiments will be discussed in Sections II-B and C.

Two other groups of injection experiments are indirectly related to the containment problem. In the first of these, high β plasma is injected into a converging (dipole) magnetic field. The plasma is analyzed downstream, where it is not necessarily high β. Thus Jensen (61) discusses the injection of a helium plasma having $n_e \sim 3 \times 10^{14} \mathrm{cm}^{-3}$ and $T_e \sim 3$ to 10 eV, from a conical pinch gun into a field of 0 to 2.5 kG. He shows that, during injection, a high β plasma of relatively low conductivity can trap magnetic flux; this subsequently distorts the confining cusp field in the manner discussed by Artsimovitch (14). Similarly, Ashby and Avis (62) have demonstrated the experimental importance of wall short-circuiting effects in a low $-\beta$ plasma column, and also the existence of an electrostatic streaming instability having a frequency greater than ω_{ci} in the column (63). Whether such a short circuit can propagate under arbitrary vacuum and wall conditions into a high β plasma that is escaping from the confinement region at Alfvénic velocities, and whether the observed instability is significantly modified by higher values of β and by electric fields, is not answered by these experiments.

In the other group of experiments a cusped field is used either to guide a plasma along a magnetic duct or to accelerate it by traveling wave techniques. Thus Tuckfield and Scott (64) used a quasistatic linear hexapole duct to reduce the contamination of helium plasma injected from an inductive gun; Oleson et al. purified a hydrogen plasma by radial injection into a spindle cusp (65); and Schindler (66) accelerated a hydrogenic plasma to a final energy of 21 keV in an axisymmetric, three-staged, pulsed cusp.

In Table II we list an alternative set of experiments in which the plasma is heated and subsequently contained by a rising cusp field. Watteau (75) described qualitatively the adiabatic compression of a plasma preheated by a Z-pinch, while Allen and co-workers described quantitatively the adiabatic compression of a 4-eV plasma formed by colliding shock waves (76–78), and of a plasma formed by gun plasmas colliding in vacuo with a directed energy of 200 eV (71,79). A shock-heated spindle cusp experiment has been described by Belitz and Kugler (80) and Chalice, the pulsed cusp filled by two θ-pinches by Karol, Carr, Harvey, and Yevick (81). A recent Culham experiment, "Centaur," uses two pulsed cusps to terminate the ends of a very high β shock-heated θ-pinch (82–84). Finally, we list some parameters of the Garching experiments, Lupus and Spinne. In these experiments (85,86) a toroidal θ-pinch is used to form a high β plasma and a toroidal equilibrium is achieved by the superposition of a pulsed hexapole field. The hexapole introduces cusp losses through six parallel slits running along the major circumference of the torus. (This circumference is tabulated as L in Table II.)

One toroidal experiment that does not conveniently fit into either of the categories listed in Tables I and II is the Polytron. In this experiment multiple cusp coils are aligned with their axes forming the major circumference of a torus. An axial current I_z is induced in the plasma, producing an azimuthal Hall current by interaction with the radial cusp field B_r, and ions are thus preferentially accelerated along the axis of the torus (87). If the directed ion velocity exceeds the ion thermal speed, it has been predicted that the ion loss rate through the ring cusps will be reduced; direct measurements of the particle and energy containment times are not yet available, however.

B. *Experiments on Gross Equilibrium and Stability*

Framing photography of the first injection (15) and spindle-cusp compression (75–77) experiments provided striking qualitative evidence for the gross stability of cusped equilibria; this evidence seemed particularly encouraging in view of the violent instabilities often observed in simple Z-pinches. Although Rayleigh-Taylor instabilities were sometimes observed during the establishment of the equilibrium (e.g., when eq. 1-3 was violated (77), no large-scale breakup of the plasma during a containment time of up

TABLE I

Typical Injection—Cusp Experiments[a]

References	Plasma source; confinement geometry	Diameter D (cm)	Length L(cm)	B (max) (kG)	n_e cm^{-3}	W'' keV	T_e eV	Quoted β near axis
67	Single-pulse coaxial gun; axisymmetric quadrupole and octupole	90	120	4.5	10^{12}–10^{14}	5×10^{-2}	15	?
68	12 conical Z-pinch guns; axisymmetric triple cusp (radial injection)	20	45	1.9	7.5×10^{14}	$>5 \times 10^{-3}$	4.5	<1
69	Coaxial gun; spindle-cusp	53	53	12	$\sim 3 \times 10^{13}$	13	Nonthermal	$<1/2$
70, 71	2 θ-Pinch (single pulse) guns; spindle cusp	25	230	3.2	$\sim 10^{15}$	2.4×10^{-1}	20	$\gg 0.90$ in core
72	Conical Z-pinch; spindle cusp	40	40	4	$(3–10) \times 10^{15}$	~ 1	?	~ 1
73	Titanium guns; spindle cusp (radial injection)	12	12	3.9	$\sim 8 \times 10^{15}$	5×10^{-2}	>5	~ 1
74	2 multiple-pulse coaxial guns; spindle cusp	17	15	3.9	10^{13}	2×10^{-2}	6	~ 1

[a] Axial injection unless radial injection at ring cusps is specifically noted. W'' is the injected energy in keV.

TABLE II
Recent Compression—Cusp Experiments

Refs.	Description	D (cm)	L (cm)	B kg	Rise time (μsec)	$\hat{n}$ (cm^{-3})	$\hat{Te}$ eV	β_A
75	Adiabatic spindle cusp	11	8	25	4.5	?	?	?
76–78	Ditto (shock preheat)	20	20	24	15	$2{\cdot}5 \times 10^{16}$	15	0.98
79	Ditto (gun preheat)	20	20	34	15	10^{16}	70	~1·0
80	Shock-heated spindle cusp	10·5	13	70	1·1	10^{17}	120	?
81	Linear θ-cusp-θ pinch	5	2·5	27	1	$\sim 3 \times 10^{16}$	100–180	?
82–84	Shock-heated linear cusp-θ-cusp pinch	19	50	60	2·1	$1{\cdot}5 \times 10^{16}$	150	0.99 ± 0.01
85	Shock-heated toroidal hexapole	6	163	10	3·0	3×10^{16}	50	0·8
86	Shock-heated toroidal hexapole	6	163	21	3·0	$3{\cdot}5 \times 10^{16}$	93	0·4
		6	163	10·5	3·0	$1{\cdot}4 \times 10^{16}$	62	1·0

to 10 sound transit-times (78) was observed. Confinement in simple θ-pinches (i.e., those without significant reversed trapped field) has tended to be rather shorter, when measured on a sound-transit time scale (τ_s), and so a true equilibrium has not always been established; nevertheless, simple θ pinches are often observed to be remarkably stable. This is somewhat surprising, since ideal MHD theory applied to a sharp boundary of the form $e^{i(m\theta + kz + \omega t)}$ would predict an interchange instability having a growth rate of order $(\sqrt{2m'\tau_s})$.

Several stabilizing effects have been discussed:

1. Eddy current stabilization by a cylindrical conducting wall of radius R_w is predicted by Haas and Wesson (88) to stabilize the gross ($m = 1$) mode in a long θ-pinch plasma having $\beta > [1 + (R_p/R_w)^2]^{-1}$ everywhere along the pinch.
2. Finite density gradients (22,88) and finite gyroradius or viscous effects (89) are expected to stabilize, or inhibit growth rates, of the finer scale (high m number) perturbations in either cusp or θ-pinch geometries.
3. "Line-tying" of magnetic field lines by a conducting wall or cold plasma is known to stabilize low β mirror plasmas. However, some fairly direct experiments (90) suggest that this effect should not be dominant in high β laboratory plasmas.

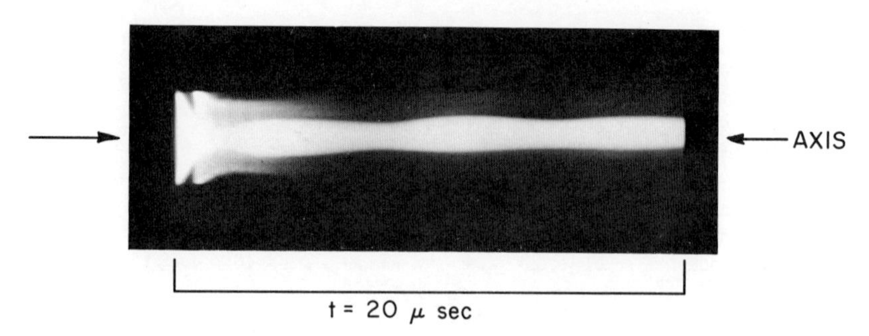

Fig. 6. Streak photograph illustrating the $m = 1$ stability of a high $\bar{\beta}$ mirrored θ pinch (84). $B_0 = 16.5$ kG, 0.02 torr D_2, $Z = 6.0$ cm, mirror ratio $\lambda_2 = 2.25$. The periodic modulation of the plasma radius correlates with a low-amplitude ripple in the confining field B_0.

Perhaps the most quantitative test of MHD stability theory is therefore provided by the Centaur experiment (Figure 1*d*), in which the plasma is known to be confined for times much greater than $(L/2c_s)(2m)^{-1/2}$ and which has a density profile that is known to be relatively "sharp." In both a simple θ-pinch (82,83) and in a θ-pinch having a mirror ratio variable from 1 to 4 (84), the plasma was shown to be $m = 1$ stable for at least 6 sound-transit times (Figure 6). In both cases the measured density profile and radially averaged $\bar{\beta}$ satisfy the Haas-Wesson criterion for stability at the mid-plane of the experiment for over 6 μsec. The axial β of the unmirrored θ-pinch was determined by Faraday rotation of an argon-ion laser beam to be $\beta_A = 99 \pm 1\%$ from $t = 2$ to 6 μsec, a value that also satisfied the Haas-Wesson stability criterion (Figure 7). At lower initial values of β, an $m = 1$ instability was observed (82), with a growth rate that agreed with theory within the experimental error of a factor of 2. Line-tying therefore appears to be an unlikely explanation for the observed stability of high β cusps or θ-pinches. Finally, the application of cusp curvature at either end of the device was shown to stabilize $m = 1$ drifts occurring later in the low β phase of the discharge, but not low amplitude $m = 2$ modes which were always observed, with rotation, during the final decay of the plasma. (Stereoscopic streak photography identified the rotation as having the same sense as the ion gyro direction in the θ-pinch field of the θ-cusp experiment).

Most of the experiments in Table 2 and Refs. 68 and 70 of Table 1 provide additional qualitative evidence for the stabilizing effects of cusp curvature. It is interesting to note that no quantitative comparison between the theoretical shape and any actual high β plasma equilibrium has been published.

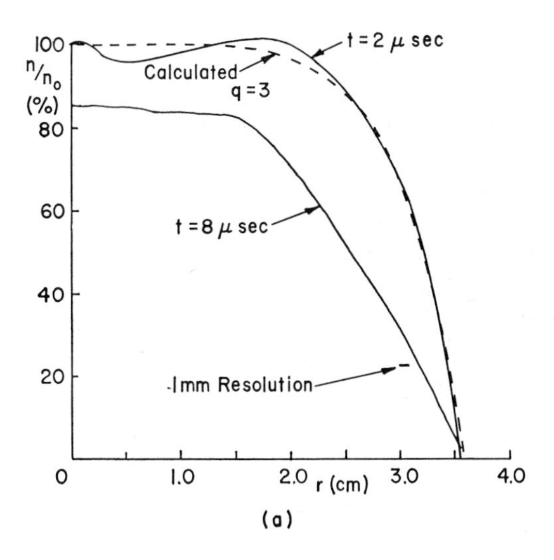

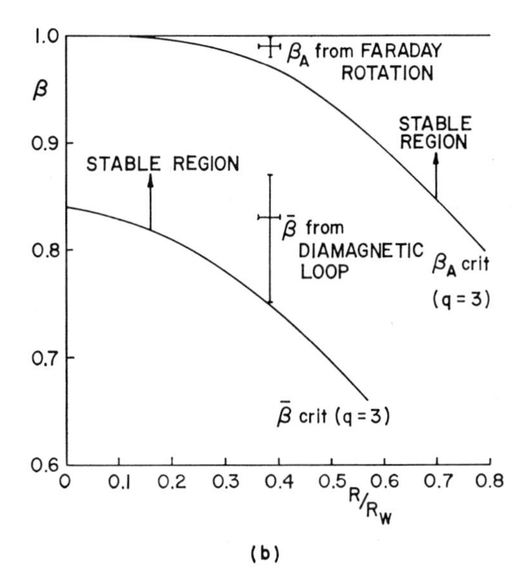

Fig. 7. (*a*) Radial density profiles in a stable high $\bar{\beta}$ θ pinch. (*b*) Haas-Wesson MHD stability criterion fitted to the experimental data of Fig. 7*a*. The measured value of axial β_A lies above the $m = 1$ stability boundary, calculated for the density profile measured at $t = 2$ μsec (83).

C. *Cusp Containment Experiments*

In many early experiments the plasma conductivity was so low that binary collisions (76–78) or low β effects determined the loss rate. However, in the Spice II experiment (79), multichannel, time-resolved measurements of the (full) width of the stagnation pressure profile (Figure 8) at the ring cusp revealed a hole only one ion gyrodiameter (i.e., $\sim 2r_i$) wide, with approximately equal particle losses through the point cusp. These measurements implied that $\bar{\beta}$ was high (a fact directly verified during the early phase of the compression by internal magnetic probe measurements); spectroscopic measurements of n_e and T_e suggested that the hole size could hardly be explained by binary collisions alone. It was concluded that either wall

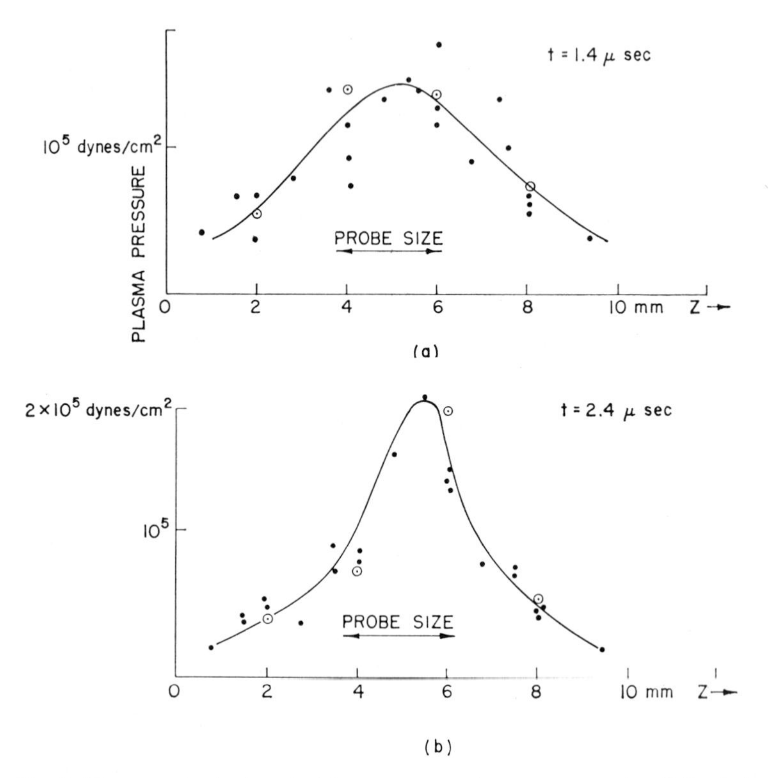

Fig. 8. Time resolved piezoelectric pressure probe measurements of the hole size at the ring cusp of a spindle cusp compression experiment. The ringed points show the stagnation pressure measured in a single discharge at times $t = 1.4$ and 2.4 μsec after initiating the compression. Other points were measured on successive discharges at various axial distances Z and at a radius of 20 cm near the central plane (79).

short-circuiting of the space charge electric field in the collision-free sheath or anomalous diffusion were responsible for the measured containment time of 5 to 10 ion transit times. These hole-size measurements were tested for self-consistency by following numerically the adiabatic compression of a plasma contained in a magnetic bottle that had an escape hole $2r_i$ wide at the ring cusp, with comparable losses through the point cusps. The loss rate thus obeys eqs. 1-23 and 1-25 for a $\beta \sim 1$ plasma. The measured density and electron temperature agreed semiquantitatively (i.e., within a factor of 2) with the time variation predicted by the model (79). Rather better experimental agreement has recently been found by Belitz and Kugler (80), using an identical model to predict the variation of electron density, ion and electron temperature, and hydrogen continuum emission at 4976 Å during the containment phase of a shock-heated cusp experiment in which the plasma is a factor 10 denser and factor 2 hotter than in the earlier experiment (Figure 9).

Attempts have been made to distinguish between anomalous diffusion and wall short-circuiting effects by measuring the mean electric field in the sheath of a high β injection cusp experiment (70). In this experiment an electric field somewhat greater than (kT_i/er_i) was measured in a sheath roughly an ion gyroradius wide (Figure 10), but diagnostic difficulties prevented a quantitative comparison of its magnitude with the transverse ion pressure gradient (91). There is little, if any, direct experimental evidence for the importance of wall short-circuiting effects in high β experiments.

It has been noted that sharp sheaths are not necessarily indicative of low diffusion rates; e.g., diffusion at a rate comparable to that given by the Bohm formula could account for the maintenance of an ion gyrodiameter hole in the Spice II experiment (57). To reduce this ambiguity, an experiment has been designed in which $\delta_B > r_i > \delta_c$, where δ_B and δ_c are the skin depths of Section I-D-3, evaluated using the Bohm and classical diffusion coefficients, respectively (82,83). Before this experiment is discussed further, we compare in Table III the sheath thickness Δ measured at the plasma boundary in five, recent, high β compression experiments. It will be noted in Table III that $r_i \sim \delta_B$ for both the Spice II and Julich experiments; it is therefore possible that the good agreement between these experiments and the predictions of a model in which plasma effuses through an ion gyrodiameter hole may be fortuitous.

The extremely small escape hole deduced indirectly from spectroscopic measurements of the time variation of number density in early Spinne experiments was tentatively attributed to experimental errors produced by an influx of neutrals from the wall (85). In later experiments on the same device (86), the line density of a somewhat hotter plasma was observed to decay

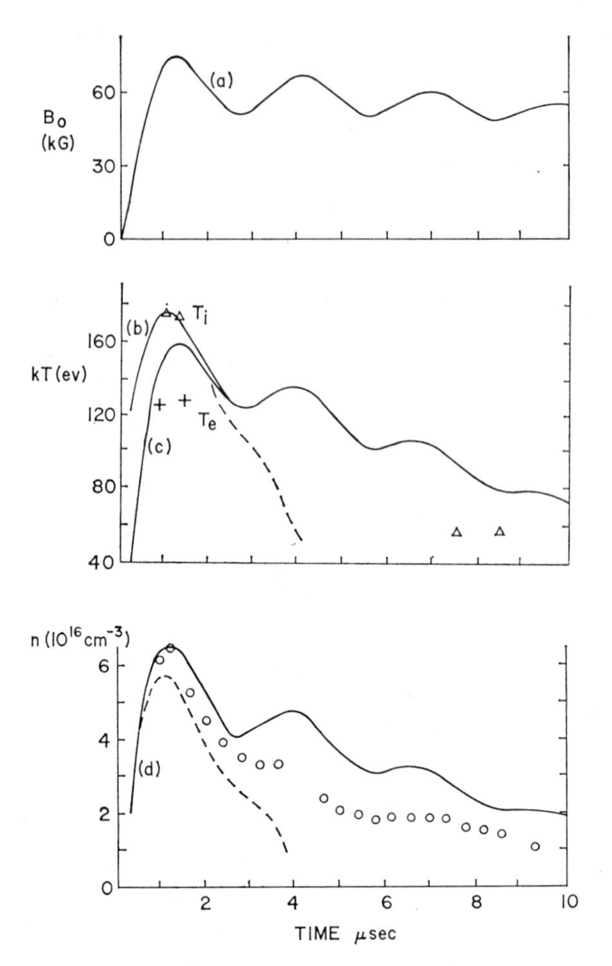

Fig. 9. A shock-heated, spindle cusp experiment (80). The main compression field B_0, and the time variation of T_i, T_e, and n predicted by Eqs. 1-23 to 1-26 are plotted (for $\bar{\beta} \to 1$) as the continuous curves in Figs. 9*a*, *b*, *c*, and *d*, respectively. (The dashed curve neglects the effective mirror ratio λ, and so gives a poor fit to the experimental points.)

in about 4 μsec, i.e., in roughly the sound transit time along field lines connecting plasma to the wall, so that any distinction between "cusp" and "diffusion" losses on the basis of the line density measurements would appear to be difficult. In the Centaur experiment particles are contained either in a simple θ-pinch ($\lambda_2 = 1$) or a cusp-ended θ-pinch ($\lambda_1 = 1$, $\lambda_2 = 3.5$) for roughly six sound-transit times along the (half) length of the device, a

TABLE III

References	79	80	85	83	83
Experiment	Spice II	Julich	Spinne	Centaur (θ)	Centaur (Cusp)
Gas	H_2	H_2	H_2	D_2	D_2
Time of measurement (μsec)	4	1	4	3	3
B_0 (kg)	~ 7	~ 35	10	14	14
Mirror ratio	$\lambda_1 = \lambda_2 \sim 2{\cdot}5$	$\lambda_1 = \lambda_2 \sim 2$	$\lambda_1 \sim 1{\cdot}2$	$\lambda_2 = 1$	$\lambda_1 = 1$ $\lambda_2 = 3{\cdot}5$
n_e (cm^{-3})	10^{16}	10^{17}	3×10^{16}	$1{\cdot}2 \times 10^{16}$	$1{\cdot}2 \times 10^{16}$
T_i (eV)	>65	180	>40	~ 280	~ 280
T_e (eV)	65	120	40	140($\pm$20)	140
R_p(cm)	~ 4	~ 2	~ 0.8	3·4	3·4
r_i(eq. 1-24) mm	2	0·7	1	3	3
δ_B(eq. 1-35) mm	4	0·7	0·7	5	5
Δ (measured) mm	2[a, b]	0·7[b]	0·02[b]	8[a, b]	8[a, b]

[a] Δ measured spectroscopically or by pressure probes.
[b] Δ estimated from particle containment time.

well-defined equilibrium is established, and some test of the alternative loss mechanisms appears possible. Also n_e and T_e were measured on the axis by Thomson scattering of ruby light with an accuracy of $\pm 15\%$ and a resolution of ~ 1 cm; the radial density profile at four different axial positions was deduced from calibrated streak photographs and photoelectric measurements of the optical continuum, while $\int n_e \, dz$ was determined interferometrically to provide an independent check on the density measurements. The radially averaged $\bar{\beta}$ was deduced from diamagnetic measurements; under carefully controlled preheat conditions (i.e., critically damped z-pinch with -120 G Bz bias) $\bar{\beta}$ was $83^{+5}_{-8}\%$ and $\beta_A = 99 \pm 1\%$.

The sheath width Δ determined directly from measurements of the e-folding length of the (flat-topped) density profile was consistent with a value deduced less directly from measurements of the particle containment time, the plasma dimensions, and the sonic velocity; it was also significantly greater than either the ion-gyroradius or the binary resistive skin depth (computed using the MHD code). The authors concluded that the initial thickness and, more particularly, the maintenance of such a wide sheath around the high β plasma core were suggestive of anomalous diffusion; fluctuations in light emission at frequencies near the ion gyrofrequency (and

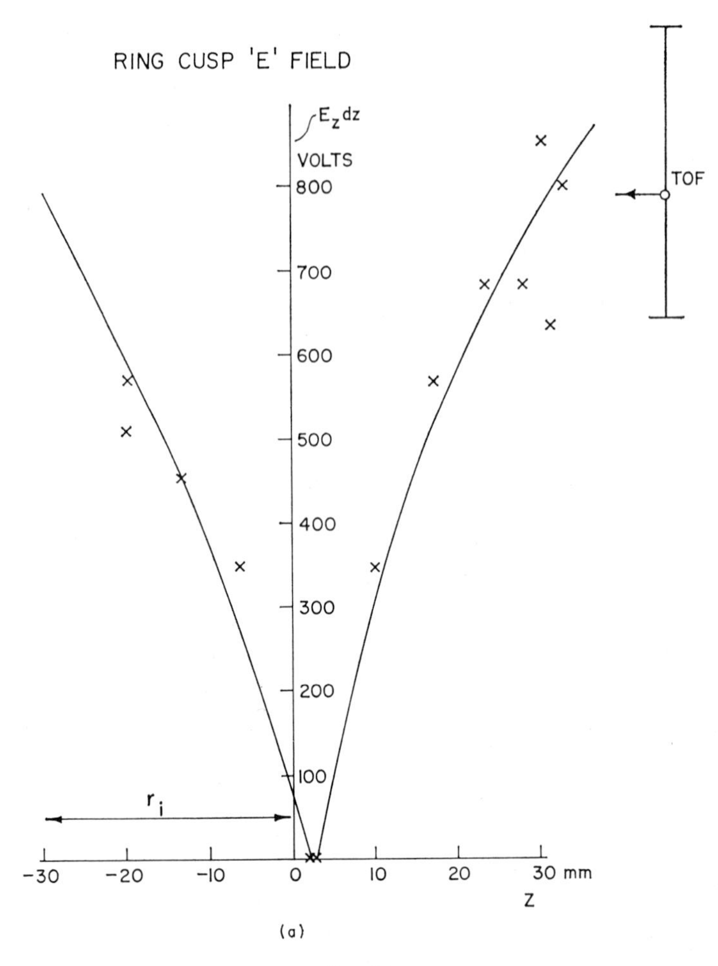

Fig. 10*a*

of electiic fields in an associated injection experiment, see Figure 10*b*) were considered significant (83). In order to test the diffusion model of eq 1-37, they then measured the particle containment time in a simple θ-pinch as the mirror ratio λ_2 was varied from one to four. (In all cases the column was stable for at least 10 μsec due to the high initial β of the plasma). Within the estimated error of $\pm 20\%$, τ_N scaled as $(\lambda_2)^{1/2}$ (Figure 11), which is in agreement with the diffusion model, provided that neither R_p, L, nor T a restrong functions of the mirror ratio. (The radial temperature variation has not yet been checked experimentally.)

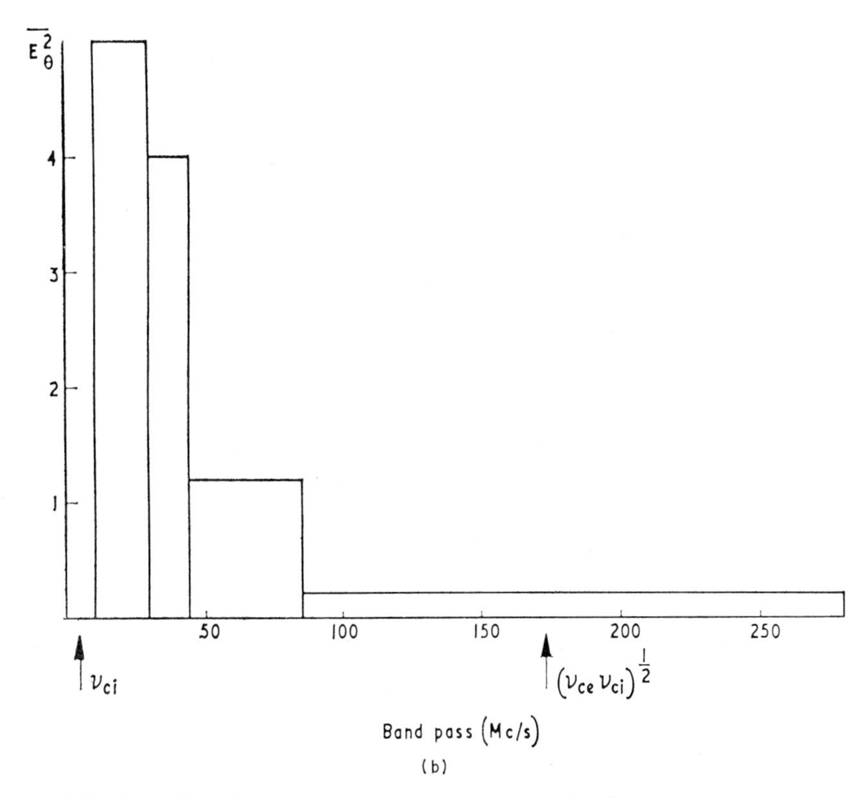

Fig. 10. (*a*) $\int E_z \cdot dz$ measured from the midplane of a high β injection cusp experiment. The directed energy of the ions was determined by the time flight (TOF). From Ref. (91.) (*b*) Histogram of the electric field fluctuation E_θ measured as a function of frequency ν in the sheath of high β plasma spear (70) flowing in a uniform guide field B_z. (Spalding, unpublished.)

D. *Related Experiments*

Particle losses in θ pinches were reviewed in 1967 (89). Mach-Zender interferograms taken on the first half-cycle of the Scylla IV discharge over a range of filling pressures permitted the effective radius δ_p of the point cusp to be estimated from the particle containment time, the plasma volume, and the ion velocity. At 50 m torr of D_2, for example, it was found that $\hat{R}_p \sim 10$ mm, $\bar{R}_p \sim 6$ mm, $\delta_p \sim 4.3$ mm, and $r_i \sim 3$ mm; i.e., the hole size was somewhat less than the plasma radius and greater than the gyroradius. Although $\bar{\beta}$ was not measured in this experiment, the density profiles suggest

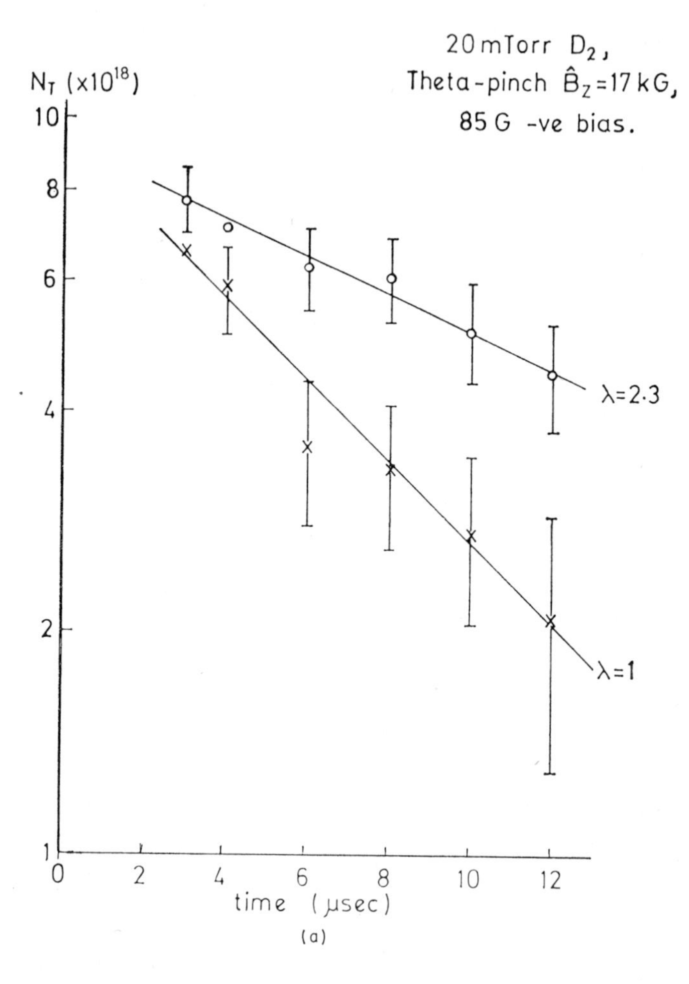

Fig. 11*a*

that $\bar{\beta} < \frac{1}{2}$. (Thus it is probable that in neither the Scylla 4 nor the Centaur experiment is $\bar{\beta}$ sufficiently high that the first cusp term in eq. 1-26 completely dominates the losses.)

Anomalously rapid diffusion of magnetic field into high β plasma has also been clearly demonstrated by measurements of the time variation of the radial density profile at the center of an 8-meter long θ-pinch (92). During the implosion phase, i.e., long before any disturbance can conceivably propagate from the ends of the device, the number density on the axis varies in a manner that gives good agreement with computer calculations employing

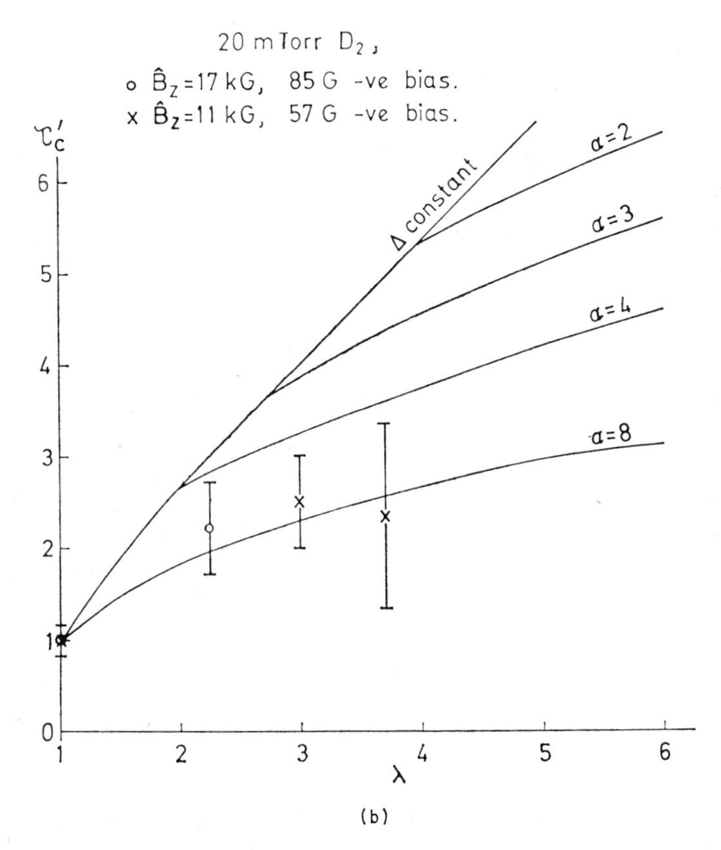

Fig. 11. (*a*) Total number of electrons N_T as a function of time t for two high $\bar{\beta}$ θ pinches having mirror ratios $\lambda = 1$ and 2.3, respectively. (*b*) Particle containment time τ_c' in a high $\bar{\beta}$ θ pinch as a function of the mirror ratio λ. (τ_c' is normalized to the containment time when $\lambda = 1$.) The curves are derived from Eq. 37, with increasing α representing increasing anomalous diffusion (84).

an effective electron-ion collision frequency due to Sagdeev (93). (This calculation does *not*, however, include the effect of magnetic fields on the ion sound instability.) During the quiescent phase of the discharge the diffusion appears to be classical; however, at this time $\beta_A \sim \frac{1}{2}$, the plasma density gradient is low, and so direct extrapolation of these results to a $\beta = 1$ collision-free sheath is not possible.

El-Khalafawy et al. (94) have examined the microwave radiation emitted during the first stages of a θ pinch implosion. Strong emission was observed at all electron cyclotron harmonics up to the electron plasma frequency; the

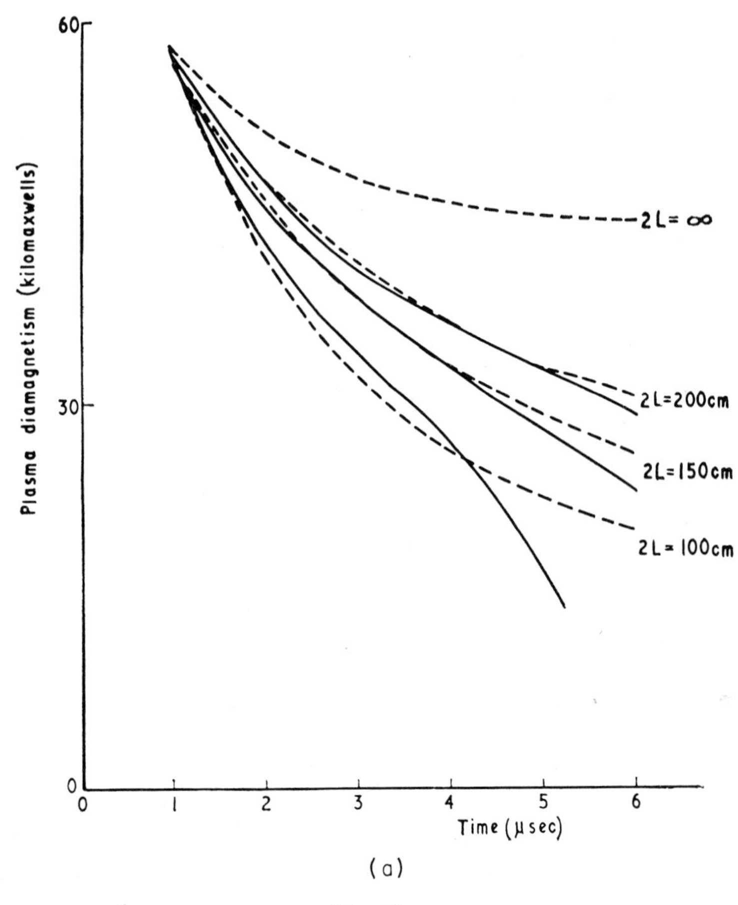

Fig. 12*a*

emission was strongest at angles between 35° and 55° to the magnetic field direction, and was well correlated with the signal produced by an X-ray probe measuring the perpendicular electron energy. Thus, in this experiment, the electron cyclotron instability discussed by Zayed and Kitsenko (39) has been excited by a relatively weak beam of high-energy rotating electrons.

Energy losses attributed to ion and electron thermal conduction have been discussed briefly in the collisional mirror experiments of refs. 95 and 71.

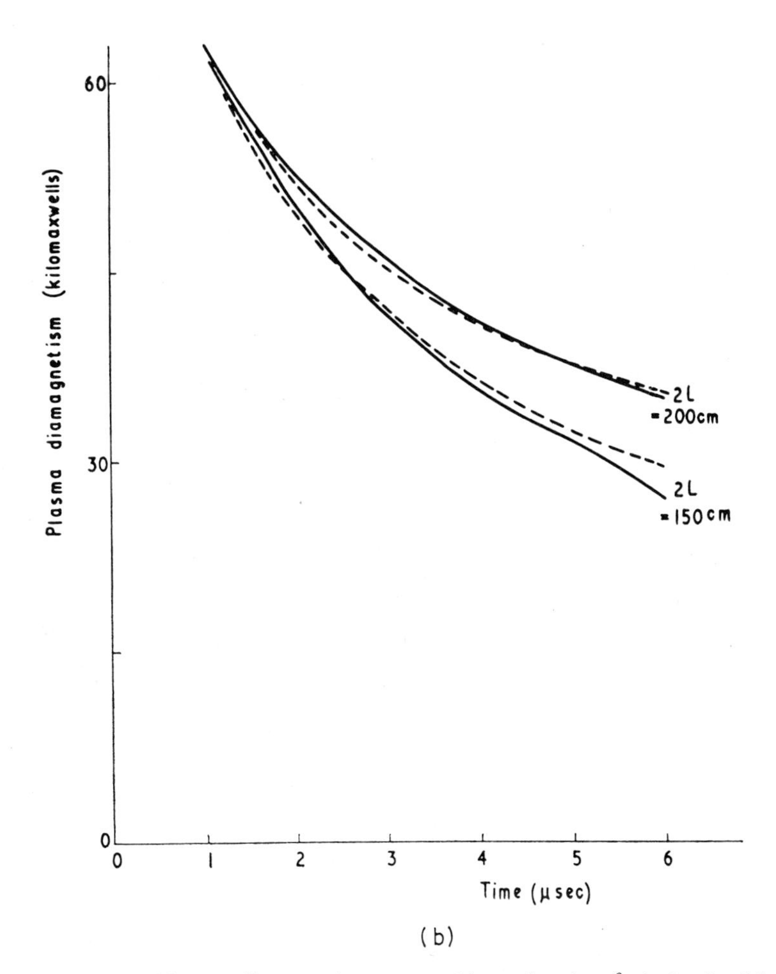

Fig. 12. The mid-plane diamagnetism measured in a mirrorless θ pinch of coil length ($2L$) is shown as a continuous line at various times after initiating the discharge. The filling pressures are (*a*) 43 m torrs and (*b*) 61 m torrs. The plasma has a low $\bar{\beta}$. The broken lines are computed using an electron thermal conduction model (96).

In the later experiment the time variation of the local electron temperature in a $\beta \sim \frac{1}{2}$ mirror plasma formed by colliding two gun plasmas was measured to an accuracy of $\pm 15\%$ by Thomson scattering and was adequately explained by a thermal conduction model (71). Very extensive θ-pinch scaling experiments at comparable β have been described by Green et al. (96) (See Figure 12.) More recently, electron conduction losses have been investigated in low β mirror experiments (97).

III. Cusp Reactors

It is useful to make some formal assessment, however naïve, of cusp reactor possibilities for the following reasons: first, certain cusp geometries create technological problems that appear to be insurmountable, however optimistically one extrapolates current engineering practice. Second, similar assessments for toroidal, inertial, and low β open-ended systems then permit a semiquantitative comparison of the advantages and disadvantages of the competing systems. In particular, the relative importance of various physical problems becomes clearer.

In this spirit the effusion model with zero transverse electric field (eqs. 1-23, 1-25, and 1-38), which should represent an overestimate of cusp losses from a strictly collision-free quiescent plasma, has been used to calculate the main parameters of a DT reactor satisfying the following practical or technological constraints (16):

1. The mean electrical output should preferably not exceed 10 GW (E).
2. The Lawson product $n\tau \sim 10^{15}\text{cm}^{-3}$ (at $kTi = 12$ keV). (The power circulating internally can then be less than 17% of the electrical output and the tritium scavenging efficiency does not need to be excessive, 17.)
3. The mean energy flux of 14 MeV fusion neutrons incident on the vacuum wall should not exceed 550 W/cm^{-2} (98). (In a (DT) reactor this factor is not likely to vary by more than a factor of 10.)
4. To avoid excessive thermal stresses due to free–free X radiation absorbed on the inner skin of a pulsed (cylindrical) DT reactor (16), $n_e B_0(R_p{}^2/R_w) < 8 \times 10^{21}$ G/cm^{-2}.
   (This criterion avoids one criticism often made of high-density pulsed systems.)

Any quasispherical spindle cusp geometry proves to be impracticable because of the intense radiation flux absorbed on the inner wall of the vacuum vessel; this flux is intense because of the large plasma volume required for adequate confinement. Even if the sheath could be reduced by space-charge electric fields to its limiting value of a few electron gyroradii (current theoretical and experimental work suggests that this is unlikely), and the wall radius were simultaneously increased to reduce its thermal loading, the device would still be unattractive because of its enormous output and the magnetic energy required.

To reduce the wall loading problem, a cusp-ended cylindrical θ pinch was also considered. It was shown that with $\bar{\beta} \sim 1$, $R_p = 10$ cm, $R_w = 60$ cm $B_0 = 100$ kg, $B_1 = B_2 = 600$ kg, $kT_i = kT_e = 12$ keV and $n \sim 10^{16}$ cm^{-3}, the thermal loading may be acceptable during a pulse of duration of 0.1 sec.

The cylindrical geometry has the additional advantage that in situ shock heating of the plasma can be envisaged; the confining field B_0 was provided by a super conducting coil, and the local cusp field B_1, B_2 by pulsed force-free coils. A typical length and output of 1 km and 10 GW (E) (Figure 13) were deduced for long cusp geometry, but the fusion products are not contained sufficiently well to permit a sustained (i.e., self-heated) reaction, unless the output power and length of the device are increased by more than an order of magnitude. It was therefore concluded that such reactors must be pulsed.

In this geometry the crucial physical problem is shown to be the microstability of the high β plasma boundary (even with an ion gyroradius sheath),

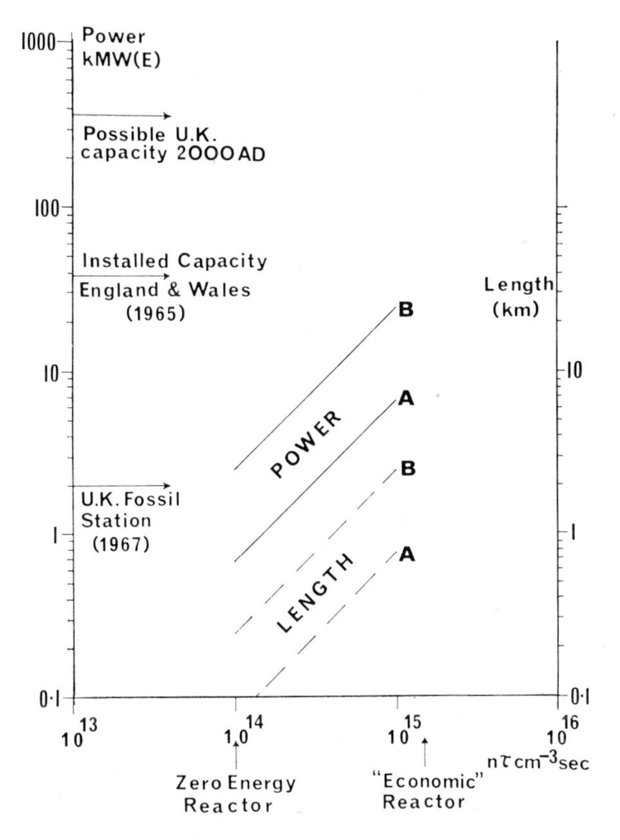

Fig. 13. Mean output power and length of a hypothetical long-cusp reactor. Curve A assumes collision-free high β containment given by Eqs. 1-27 and 1-28. Curve B assumes collision-free high β containment given by Eqs. 1-23 to 1-26.

since the containment time decreases if diffusion is significantly faster than the classical value, i.e., if it exceeds 10^{-5} of the Bohm value; the high β approximation also fails if the internal field exceeds about 3% of the confining field. Technical problems arise concerning the electical insulation and thermal conductivity of the vacuum wall, and the capital cost (per kW) of the reliable, low-inductance, pulsed-energy source needed for heating the plasma. It should be particularly noted that minor changes in the assumed loss rate (e.g., if a mirrored θ pinch proved stable enough to permit a slightly shorter reactor) do not strongly modify the economics nor the microstability requirements of the sheath. However, if RF stoppering (99) or magnetically rough walls (100) were to reduce the cusp losses sufficiently for a sustained reactor to be practicable, the entire system is likely to be very different from that discussed in this chapter. Finally, any increase in the thermal loading allowed at the vacuum wall permits an increase in the plasma radius and so eases the restriction on the permissible rate or diffusion in the high β sheath. However, the reactor output is then increased.

IV. Summary and Conclusions

Both the one-fluid (53) and the collision-free model (43) of cusp containment predict a very strong dependence of the end losses upon the average β of the plasma; unless an appreciable volume of the plasma is virtually field-free, the particle containment in relatively dense plasmas cannot exceed a few ion transit times for any geometry where $\lambda_1 = \lambda_2 \sim 1$. At low densities the experimental situation may be complicated by conventional mirror confinement in the outer, adiabatic, regions of the device. These outer regions can become flute-unstable if plasma is lost more rapidly in the non-adiabatic region near the field zero; theoretical estimates of the containment time in this low $\bar{\beta}$ situation would then be extremely difficult to make (13).

In a few of the cusp injection experiments described in this article, the injection time has been comparable to, or greater than, the estimated confinement time, so that direct measurement of the latter quantity is difficult. In more experiments the electron temperature of the injected plasma has been lowered by expansion or impurity radiation, to such an extent that binary collisions are sufficient to account for the low $\bar{\beta}$ and rapid losses. In one experiment the energy of the injected ions was so high that the gyroradius was comparable to the equilibrium radius of the plasma and $\bar{\beta}$ was again low (69). Finally, in one experiment plasma "spears" having a well-defined, high β core with $\beta_A \gg 90\%$ (70) and an electron temperature of 20 eV were collided in a spindle cusp, and electric fields of order kT_i/er_i in a sheath $\sim 1 r_i$ wide were measured (91). In all injection experiments known to the author the relatively poor confinement can be directly attributed to the low $\bar{\beta}$ achieved. In a few of these experiments the low $\bar{\beta}$ itself is less easily explained; it is significant that strong fluctuations were observed by Karr and Osher (69) during the injection phase of their experiment.

Compression experiments have produced electron temperatures sufficiently high that classical diffusion should permit very high values of $\bar{\beta}$ and contribute negligibly to the end loss (80–82). Moreover, the higher densities achieved in compression experiments have been well suited to spectroscopic, interferometric, laser scattering, and Faraday rotation diagnostic techniques. The gross ($m = 1$) stability of high β cusps and θ-pinches appears to be in quantitative agreement with the predictions of ideal MHD theory (88); see also ref. 4. However, the origin of the rotation observed (after most of the plasma has already escaped) in θ-cusp experiments (83) and in simple θ-pinches (4) has not yet been explained quantitatively; it may be significant that rotation has not been reported in spindle-cusp geometries possessing a magnetic plane of symmetry (43). Haines (101) has reviewed several rotation mechanisms involving residual asymmetries in the confining magnetic field that could be important.

In many θ pinch experiments $r_i \sim \hat{R}_p$, $\bar{\beta} < \frac{1}{2}$, the plasma decays in roughly the sound transit time, and so tests of cusp containment have not been possible. In several cusp experiments $\hat{R}_p \gg r_i$, (79,80,83), $\bar{\beta}$ appears (79,80) or is measured (83) to be high, and a containment time of several sound transits is observed. However, in all cusp or θ pinch experiments mentioned in this chapter in which the heating has been predicted from the Hain-Roberts computer code, or in which T_e has been measured directly, the average β is lower than that computed using Spitzer resistivity. The containment (and sheath width) observed in the Centaur θ-cusp and θ-pinch experiments almost certainly excludes wall short-circuiting or nonequilibrium effects as the sole cause of plasma loss: either β is lowered anomalously by a few percent throughout the plasma volume during the initial formation of the plasma, or anomalous diffusion occurs at the collision-free sheath, which is a few ion gyroradii wide, throughout the equilibrium phase of the discharge (82,83). The density profile measured in this sheath roughly satisfies the Krall instability criterion of eq. 1-21, and so should be marginally unstable. Circumstantial evidence for continual diffusion is also provided by the $\lambda^{1/2}$ mirror scaling observed in this experiment (Figure 11). Bodin et al (92) detected anomalously rapid diffusion in an 8-meter θ-pinch at times long before end effects could affect their measurements. (In their experiment the electron conduction time and the averaged Alfvén speed in the sheath are of the order of 500 μsec and 20 μsec, respectively.) They reported numerical agreement with Sagdeev's ion-sound effective collision frequency (92), but found almost classical diffusion later in the discharge (when $\beta_A < 1$, and the density gradients are less steep).

It appears probable, therefore that, collision-free sheaths narrower than a few ion gyroradii (i.e., the sheaths discussed in Section I) are unstable, although much more experimental work is required to identify positively the microinstability and to measure the resulting diffusion coefficient over a wider range of plasma parameters. As discussed in Section III, such a conclusion would invalidate the reactor scaling in Figure 13. In the design of steady-state, collision-free shock experiments, and in discussions of the magneto sheath (see, e.g., Refs. 34–35; 102–104) it is often postulated that the current sheath is collision-free and behaves like a perfect (non leaky) piston. It is apparent that this assumption may be invalid. This assumption is also implicit in a paper by Spreiter and Summers (105), who calculate the loss rate of protons from the solar wind at the neutral points into the earth's magnetosphere. The (theoretical) cusp hole size used in their computation is considerably smaller than that actually measured in the laboratory experiments discussed in Section I.

Finally, it should be remarked that the microstability of the collision-free sheath is a problem of consequence to some recent high β toroidal proposals, especially if a high average β is required for equilibrium or gross stability of the configuration (106).

Acknowledgments

The author's interest in this subject has been enlivened and stimulated by his close association with Dr. T. K. Allen. He is particularly indebted to Dr. A. S. Kaye and Professor N. A. Krall for the communication of results in advance of publication, and wishes to thank Drs. R. J. Bickerton and R. S. Pease for their active interest in this work.

References

1. S. I. Braginskii and B. B. Kadomtsev, *Plasma Physics and the Problems of Controlled Thermonuclear Reactions*, Vol. 3, Pergamon Press, 1959, p. 356.
2. J. Berkowitz, H. Grad, and H. Rubin, *Proc. Second U.N. Conference on the Peaceful Uses of Atomic Energy, Geneva*, **31**, 177 (1958).
3. J. Wesson, *Plasma Physics, and Controlled Nuclear Fusion Research*, Vol. 1 IAEA, Vienna, 1966, p. 223.
4. H. A. B. Bodin, *Methods of Experimental Physics*, Vol. 9A, Academic Press, New York, 1970.
5. V. D. Sivukhin, *Reviews of Plasma Physics*, Vol. 5, Consultants Bureau, New York, 1970, p. 497.

6. S. Chapman and U.C.A. Ferraro, *J. Terr. Magnetism Atmos. Elec.*, **37**, 147 (1932).
7. F. L. Scarf, *Advances in Plasma Physics*, Vol. 1, Interscience, New York, 1968.
8. N. A. Krall, *Advances in Plasma Physics*, Vol. 1, Interscience, New York, 1968.
9. A. B. Mikhailovskii and A. M. Fridman, *Zh. Eksp. Teor. Fiz.* **51**, 1430 (1966).
10. C. K. Chu and R. A. Gross, *Advances in Plasma Physics*, Vol. 1, Interscience, New York, 1968.
11. D. J. Rose and M. Clark, *Plasmas and Controlled Fusion*, M.I.T Press and Wiley, New York, 1961.
12. O. B. Firsov, *Plasma Physics and the Problem of Controlled Thermonuclear Reactions*, Vol. 3, Pergamon Press, 1959, p. 386.
13. H. Grad, *Prog. Nucl. Energy*, Series 11, **2**, 189 (1963).
14. L. A. Artsimovitch, *Controlled Thermonuclear Reactions*, Gordon and Breach, New York, 1964, p. 368.
15. S. Yu Luk'yanov and I. M. Podgornyi, *J. Nucl. Energy*, Part C, **4**, 57 (1962).
16. I. J. Spalding, *Nuclear Fusion*, **8**, 161, (1968).
17. R. Carruthers, P. A. Davenport, and J. T. D. Mitchell, Culham Laboratory Report CLM-R85, 1967.
18. J. Berkowitz, K. O. Friedrichs, H. Goertzel, H. Grad, J. Killeen, and E. Rubin *Proc. Second U.N. Conference on the Peaceful Uses of Atomic Energy, Geneva*, **31**, 171 (1958).
19. E. Dowd, New York University Report NYO-10435, 1964.
20. K. Hain, R. Lüst, and A. Schlüter, *Z. Naturforsch.* **12a**, 833 (1957).
21. H. Grad, New York University Report NYO-7969, 1957.
22. S. Chandrasekhar, *Hydrodynamic and Hydromagnetic Stability*, Oxford University Press, London, 1961, p. 435.
23. F. Haas and J. Wesson, *Phys. Fluids*, **9**, 2472 (1966).
24. H. Grad, *Phys. Fluids*, **4**, 1366 (1961).
25. G. Schmidt and D. Finkelstein, *Phys. Rev.* **126**, 1611 (1962).
26. C. L. Longmire, *Elementary Plasma Physics*, Interscience, 1963, p. 100.
27. M. Rosenbluth, in R. K. M. Landshoff, ed., *Magnetohydrodynamics: A Symposium*, Stanford University Press, Stanford, California, 1957.
28. B. Bertotti, *Ann. Phys. (N. Y.)*, **25**, 271 (1963).
29. J. Hurley, *Phys. Fluids*, **6**, 83 (1963).
30. R. B. Nicholson, *Phys. Fluids*, **6**, 1581 (1963).
31. A. Sestero, *Phys. Fluids*, **7**, 44 (1964).
32. S. H. Lamm, *Phys. Fluids*, **10**, 2454 (1967).
33. R. L. Morse, *Phys. Fluids*, **8**, 308 (1965).
34. E. N. Parker, *J. Geophys. Res.*, **72**, 4365 (1967).
35. I. Lerche, *J. Geophys Res.*, **72**, 5295 (1967).
36. V. C. A. Ferraro and C. M. Davies, *J. Geophys. Res.*, **73**, 3605 (1968).
37. A. B. Mikhailovskii and A. M. Fridman, *Zh. Tekh. Fiz.*, **37**, 1782 (1967).
38. N. Krall and D. Book, *Phys. Rev. Letters*, **23**, 574 (1969).
39. K. E. Zayed and A. B. Kitsenko, *Plasma Physics*, **10**, 673 (1968).
40. R. L. Morse, Proc. APS Topical Conf. on Pulsed High Density Plasmas LA-3770, Paper F3, 1967.
41. R. L. Morse, *Phys. Fluids*, **10**, 1017 (1967).
42. J. B. Taylor, Culham Laboratory Report CLM-R58, 1966.
43. A. S. Kaye, Thesis, Oxford University, Oxford, England, 1968, to be published.
44. W. B. Grossman, *Phys. Fluids*, **9**, 2478 (1966).

45. J. Berkowitz, New York University Report NYO-2536, 1959.
46. R. W. Kilb, Proc. APS Topical Conf. on Pulsed High Density Plasmas LA-3770, Paper B5, 1967.
47. J. P. Friedberg and W. Grossman, *Phys. Fluids*, **11**, 2476 (1968).
48. R. L. Morse, *Phys. Fluids*, **11**, 1558, (1968).
49. R. L. Morse, *Phys. Fluids*, **9**, 2536 (1966).
50. J. B. Taylor, *J. Nucl. Energy*, Part C, **4**, 401 (1962).
51. F. A. Haas and J. A. Wesson, *Phys. Fluids*, **10**, 2245 (1967).
52. J. A. Wesson, *Plasma Physics and Controlled Nuclear Fusion Research*, Vol. 1, IAEA, Vienna, 1966, p. 223.
53. J. B. Taylor and J. A. Wesson, *Nuclear Fusion*, **5**, 159 (1965).
54. N. J. Phillips, *Proc. Phys. Soc.* (*London*), **77**, 965 (1961).
55. A. Simon, *Phys. Rev.*, **100**, 1557 (1965).
56. A. N. Kaufman, *Phys. Fluids*, **1**, 252 (1958).
57. G. D. Hobbs and I. J. Spalding, Culham Laboratory Report CLM-R57, 1966.
58. G. D. Hobbs and J. A. Wesson, *Plasma Physics*, **9**, 85 (1967).
59. L. Spitzer, *Physics of Fully Ionized Gases*, 2nd ed., Interscience, New York, 1962, p. 144.
60. A. F. Haught and D. H. Polk, *Phys. Fluids*, **9**, 2047 (1966).
61. V. O. Jensen, *Phys. Fluids*, **11**, 240 (1968).
62. D. E. T. Ashby and B. E. Avis, *J. Nucl. Energy*, Part C, **8**, 1 (1966).
63. D. E. T. Ashby and A. Paton, *J. Nucl. Energy*, Part C, **9**, 359 (1967).
64. R. G. Tuckfield and F. R. Scott, *Phys. Fluids*, **8**, 1197 (1965).
65. N. L. Oleson, J. F. Steinhaus, and W. L. Barr, *Phys. Fluids*, **9**, 2056 (1966).
66. H. Schindler, *Z. Naturforsch.*, **21a**, 351 (1966).
67. L. L. Gorelik and N. G. Koval'skii, *Dokl. Akad. Nauk SSSR*, **147**, 576 (1962).
68. V. G. Zykov, I. A. Smelanenko, V. T. Tolok, and K. D. Sinel'nikov, *Zn. Tekhn. Fiz.*, **35**, 62 (1965).
69. H. J. Karr and J. E. Osher, *Phys. Fluids*, **9**, 750 (1966).
70. T. K. Allen, A. J. Cox, and I. J. Spalding, *Proc VIIth Int. Conf. Phenomena in Ionized Gases*, Vol. 2, Belgrade, 1966, p. 171.
71. T. K. Allen, A. J. Cox, and I. J. Spalding, *Plasma Physics and Controlled Nuclear Fusion Research*, Vol. 2, IAEA, Vienna, 1966, p. 427.
72. H. Ito, T. Isshimura, K. Hirano, A. Ozaki, K. Shinano, S. Goto, *Plasma Physics, and Controlled Nuclear Fusion Research*, Vol. 2, IAEA, Vienna, 1966, p. 419.
73. L. T. M. Ornstein, C. A. J. Hugenholtz, and H. Avander Laan, Rijnhuizen Report 66-34, 1966.
74. Y. Nakano, K. Hirano, J. Irisawa, and S. Mikoshiba, *J. Inst. Elec. Engrs.* (*Japan*), **87**, 118 (1968).
75. J. P. Watteau, *Phys. Fluids*, **4**, 607 (1961).
76. T. K. Allen and R. J. Bickerton, *Nature*, **191**, 794 (1961).
77. T. K. Allen, R. W. P. McWhirter, and I. J. Spalding, *Nuclear Fusion*, **1**, 67 (1962 Suppl.).
78. T. K. Allen and I. J. Spalding, *Phys. Fluids*, **8**, 2032 (1965).
79. T. K. Allen, K. Doble, T. J. L. Jones, R. M. Payne, and I. J. Spalding, *Phys. Fluids*, **9**, 1394 (1966).
80. H. J. Belitz, E. Kugler, *Plasma Physics and Controlled Nuclear Fusion, Research*, Vol. 1, IAEA, Vienna, 1966, p. 287 (and private communication).
81. J. Karol, W. Carr, R. Harvey, G. Yevick, *Bull. Am. Phys. Soc.*, Series 2, **13**, 1490 (1968).

82. T. K. Allen, H. Coxell, M. Hill, and I. J. Spalding, Proc. APS Topical Conf. on Pulsed High Density Plasmas, LA-3770, Paper G4, 1967.
83. I. J. Spalding, M. J. Eden, A. D. R. Phelps, and T. K. Allen, *Plasma Physics and Controlled Nuclear Fusion Research*, Vol. 2, IAEA, Vienna, 1969, p. 639.
84. I. J. Spalding, M. J. Eden, M. Hill, A. S. Kaye, A. D. R. Phelps, and T. K. Allen, *Bull. Am. Phys. Soc.*, Series 2, **13**, 1550 (1968) (to be published).
85. G. V. Gierke et al., *Plasma Physics and Controlled Nuclear Fusion Research*, Vol. 1, IAEA, Vienna, 1966, p. 331.
86. J. Junker and W. Lotz, Garching Report IPP 1/82, 1968.
87. A. E. Dangor, G. J. Parkinson, R. M. Dunnett, M. G. Haines, and R. Latham, *Plasma Physics and Controlled Nuclear Fusion Research*, Vol. 1, IAEA, Vienna, 1969, p. 255.
88. F. A. Haas and J. A. Wesson, *Phys. Fluids*, **10**, 2245 (1967).
89. F. L. Ribe, *Nuclear Fusion*, **7**, 81 (1967).
90. R. L. Bingham, L. M. Goldman, and R. W. Kilb, *Plasma Physics and Controlled Nuclear Fusion Research*, Vol. 2, IAEA, Vienna, 1969, p. 667.
91. F. Insinger and I. J. Spalding, First European Meeting on Fusion, Munich, 1966 (unpublished).
92. H. A. B. Bodin, J. McCartan, A. A. Newton, and G. H. Wolf, *Plasma Physics and Controlled Nuclear Fusion Research*, Vol. 2, IAEA, Vienna, 1969, p. 533.
93. R. Z. Sagdeev, *Proc. Symp. on Appl. Math.*, April 1965, American Math. Soc., 18, 281 (1967).
94. T. A. El-Khalafawy, V. A. Souprunenko, A. M. Ternopol, M. A. Bourham, A. M. Yousef, and L. G. El-Hak, *Plasma Physics*, **10**, 117 (1968).
95. P. Evrard, J. Jacquinot, C. Leloup, J. P. Poffe, and P. Waelbrook, Euratom Report EUR-CEA-FC 274, 1964.
96. T. S. Green, D. L. Fisher, A. H. Gabriel, F. J. Morgan, and A. A. Newton, *Phys. Fluids*, **10**, 1663 (1967).
97. D. W. Mason, J. W. Hill, and G. Francis, *Plasma Physics and Controlled Nuclear Fusion Research*, Vol. 2, IAEA, Vienna, 1969, p. 239.
98. W. G. Homeyer, MIT Research Report TR-325, 1965. (Also available as AD-619669.)
99. J. Teichmann and R. Klima, *Phys. Letters*, **11**, 231 (1964).
100. J. L. Tuck, *Plasma Physics and Controlled Nuclear Fusion Research*, Vol. 2, IAEA, Vienna, 1969, p. 595.
101. M. G. Haines, *Advances in Physics*, **14**, 167 (1965).
102. D. B. Beard, *Reports on Progress in Physics*, **30**, 409 (1967).
103. L. Lees, *AIAA J.*, **2**, 1576 (1964).
104. I. Lerche and E. N. Parker, *Astrophys. J.*, **140**, 731 (1967).
105. J. R. Spreiter and A. L. Summers, *Planet. Space Science*, **15**, 787 (1967).
106. A. A. Blank, H. Grad, and H. Weitzner, *Plasma Physics and Controlled Nuclear Fusion Research*, Vol. 2, IAEA, Vienna, 1969, p. 607.
107. N. A. Krall, private communication.
108. C. N. Lashmore-Davies, *J. Phys. A: Gen. Phys.*, **3**, L40 (1970).
109. J. P. Boris, J. M. Dawson, J. H. Orens, and K. V. Roberts, *Phys. Rev. Letters*, **25**, 706 (1970).

Relativistic Beam Equilibria*

GREGORY BENFORD AND DAVID L. BOOK

Lawrence Radiation Laboratory, University of California
Livermore, California

Introduction

Beams of high-velocity electrons have long been of interest to plasma physicists. The first theoretical studies of them were made by Bennett (1) and Alfvén (2). In recent years experimental advances have made possible for the first time the laboratory study (3–6) of the current limitations long assumed to

* This work was performed under auspices of the U.S. Atomic Energy Commission.

hold for electron beams; this has in turn stimulated the theoretical search for new forms of equilibria.

Relativistic electron beams have been developed primarily because they afford a mechanism for transport of high energies into a small area. This makes them useful tools for the nondestructive study of shock waves in solids, as well as X-ray emission. Electron beams are much more useful in this way than lasers, since they can carry higher energy densities and they penetrate materials to much greater depths. It also seems that ions may be accelerated inside such beams; total proton fluxes of the order of 10^{14} particle at energies of 1 or 2 MeV have been noted (7).

The higher the beam current, the more useful it is as a tool. For this reason currents exceeding 10^4 A are desirable, and this is the regime in which theoretical estimates have placed bounds on the maximum beam current. In this article we shall consider whether such current limitations may be exceeded. We shall also present a catalog of the known analytically tractable equilibria for beams that are infinite along their z-axis and have cylindrical symmetry.

The range of possible equilibria for a beam varies as the medium around the beam changes. It is possible to inject a beam into a totally evacuated chamber, in which case the coulomb repulsion of the beam quickly blows it apart in the transverse direction. If the beam is injected into an ionized gas, however, a back current may be induced in the plasma by the changing magnetic field at the head of the beam. It is important to recognize that this back current will not persist forever, but instead diffuses away as ohmic heating in the plasma absorbs the energy of the induced magnetic field. This is illustrated in Figure 1. As the beam enters the ionized medium, there exists a region in which charge neutralization occurs, called the head of the beam. In this space the rising self-magnetic-field of the beam induces a changing axial electric field in the plasma, causing a back current to flow, which in turn decays due to collisions within the plasma. The back current dies away inversely as the distance from the head of the beam with a characteristic distance

$$D = \frac{V_z a^2 \sigma}{c^2} = V_z \tau$$

where V_z is the beam velocity, a the radius of a uniform beam, σ the plasma conductivity, and c is the velocity of light; τ is the resistive diffusion time. This distance D, in which the back current is important and charge neutralization has already occurred, may be quite long (several meters) in present experiments. The equilibrium calculations that follow assume (for mathematical convenience) that the beam is infinite along z. This means that the beam radius a is small compared to the beam length L, so that $a/L \ll 1$. If we

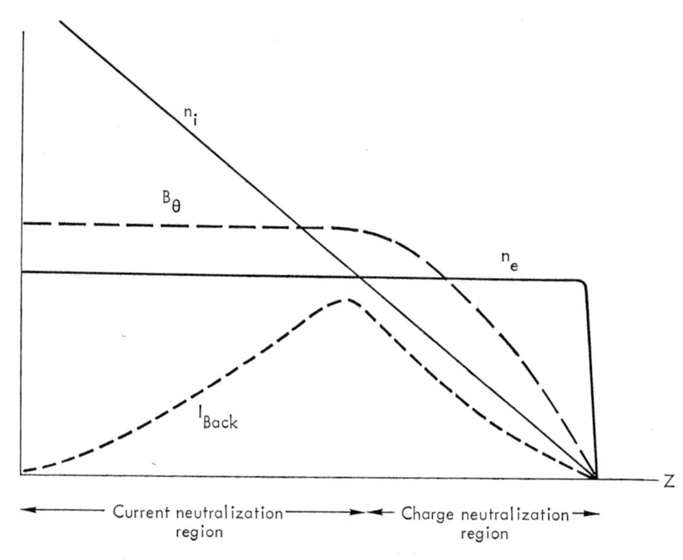

Fig. 1.

wish to include back currents in the background plasma, we must assume as well that $a/D \ll 1$.

The neglect of z-dependence means that the time scale of the transient effects associated with charge and current neutralization is either very short, so that the effects are nil, or very long, so that transient fields and currents are constant. Thus, depending on the model being studied, we take the time for formation of the equilibrium to be much longer or much shorter than the transient times of the problem. Theoretical descriptions of the transient behavior in plasma are for the most part complicated and invoke many approximations. We do not review such treatments (3,8,9) in the present paper.

Streaming electrons in the presence of ions will necessarily undergo collision with the ions. Momentum transfer will slow the beam and change the beam equilibrium. We shall neglect this time-dependent effect. This approximation is valid in two situations: (1) in the early stages of the beam lifetime, the beam has come to equilibrium but few electron-ion collisions have occurred; and (2) if the beam is kept from decelerating by a small applied electric field E_z, a steady-state balance can be struck in which energy gained from E_z is lost through electromagnetic radiation (due to transverse electron oscillations) and the momentum loss through collisions is compensated by E_z(10). Thus the basic beam parameters do not change. If E_z is much smaller than the equilibrium fields, it may be neglected as a first approximation.

Actual beam experiments confine the beam to the interior of a conducting cylinder, the drift tube. For a perfectly circular beam cross section these walls have no effect. Only when a circular equilibrium is perturbed, as in stability theory, will the fields associated with the conducting walls become important. Even then, such effects may be neglected if the drift tube radius R is very much larger than the beam radius, $a/R \ll 1$.

A. *Current Limitations*

A good deal of recent beam research is devoted to the production of high currents. The first derivation of an upper limit to beam currents was that of Alfvén (2), who treated a cylindrically symmetric model of uniform current density, monoenergetic particles that were charge-neutralized by a cold background plasma. He integrated the equations of motion for particle trajectories are found a family of curves like those shown in Figure 2. For low currents, particle motion is approximately sinusoidal. The type of particle trajectory is determined by how much current is included within the maximum radius reached by a particle. Alfvén found that pinch forces limited the current to

$$I_A = 17{,}000\ \beta\gamma\ \mathrm{A}.$$

Here $\beta = (V_z)/c$ and $\gamma = \sqrt{1 - (v/c)^2}$. If the included current is much less than I_A, a particle executes nearly sinusoidal motion (curve 1). As the current increases, though, the particle gains in transverse velocity (curve 2)

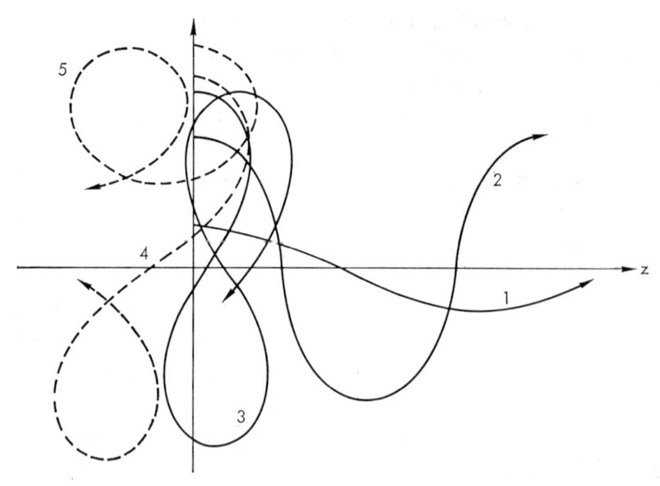

Fig. 2.

until the included current reaches I_A, when the particle crosses the axis normal to it. If the current is increased still more, the average particle drift velocity begins to decrease (curves 3, 4, and 5). This current limitation is not dependent on the dimensions of the beam.

Lawson (11) later considered this same model without assuming charge neutralization and found the current to be limited to

$$I_L = 17{,}000\ \beta^3\gamma/(\beta^2 + f - 1)$$

where f is the fraction of neutralization. If f and β^2 could be maintained constant along the beam, very large currents could be produced. But since ions are produced by collisions between the beam and background plasma, f rises as the beam passes through any one point. Lawson's limit may be obtained by requiring that a particle's Larmor orbit in the maximum self-field of the beam equal the beam radius. This means that a particle cannot reverse its direction along z by orbiting in the self-field. This immediately suggests a method for increasing beam currents if the beam can be made to depart from a constant density. Assuming that the beam is hollowed out, we find that a current limitation will arise only when particles can execute a Larmor orbit within the thickness of the cylindrical shell that the beam forms. We shall see later that such hollowing of a beam can be described by several equilibria and does lead to higher limiting currents.

Another way to propagate arbitrarily high currents is by the production of a reverse current in the plasma medium through which the beam passes. This "magnetic neutralization" arises from large numbers of electrons from a background plasma that counterstream slowly within the beam, so that the net current density is much reduced. Then the self-field of the beam is small and transverse motion proportionally reduced.

A final method for increasing beam currents is by application of an external field B_z much stronger than the self-field B_θ. Electrons tend to spiral about magnetic field lines. The addition of B_z to B_θ changes these lines from circles to helices. When $B_z \gg B_\theta$, the electrons drift predominantly in the z-direction and the beam is "forced" to propagate, even though $I > I_A$. All of these methods of increasing beam current will be examined in the equilibria that follow.

All of the beam models in this paper (excluding two Astron cases) are infinite along z ($a/L \ll 1$). This means, among other things, that the fractional neutralization f is considered constant. Experimentally, however, this is not always true. Collisions between the beam and the background gas produce a background plasma density n_i which is continually building up during an experiment. For gas pressures around 0.5 torr, for example, n_i grows approximately according to

$$n_i = n n_0 \sigma_I L$$

where n_0 is the gas density, n the beam density, σ_I the appropriate effective ionization cross section (a function of the beam energy), and L the length of beam that has passed. In air at 0.5 torr, a 1 MeV beam (6) of density 2×10^{11} electrons/cm^3, with $\sigma_I = 2 \times 10^{-18}$ cm^2, gives

$$f = n_i/n = 4L$$

where L is in meters. After a quarter of a meter has passed, $f = 1$. If the beam diameter is more than a few centimeters, the models we treat obviously do not apply. At lower gas pressures, however, the region over which f is approximately constant may be many meters.

Beams may be conveniently classified by the magnitude of ν/γ, where

$$\nu = \frac{Ne^2}{4\pi mc^2} \equiv Nr_0$$

N is the number of positive and negative charges per centimeter of the beam length, and r_0 the classical electron radius. For $\nu/\gamma \gg 1$ we must have many electrons inside every r_0 of beam length; this represents a very high current, indeed. Generally, the larger ν/γ, the more unstable a beam becomes. (The condition that magnetohydrodynamic instability theory apply to beams is that $\nu/\gamma \gg 1$ (12).)

Using a uniform beam model that is not self-consistent but does represent most of the features of a beam contained in a pipe, Spence, Ecker, and Yonas (13) have estimated that perpendicular motion is given by

$$\frac{\langle V_\perp{}^2 \rangle}{\langle V_z{}^2 \rangle} = \left(\frac{\nu}{\gamma}\right)_{\text{net}} (1 - f_m)$$

where

$$f_m = \frac{I_0}{I}$$

and I_0 is the back current. Intuition tells us that a current limitation occurs when transverse motion becomes dominant. In the absence of back currents, this means $\nu/\gamma \geq 1$. It is important to note that no current limitation exists for nonrelativistic beams, e.g.; consider a uniform neutralized beam. An equilibrium will exist if, at the surface ($r = a$)

$$\frac{1}{c}\mathbf{j} \times \mathbf{B} = \nabla P_\perp$$

where $P_\perp$ is the transverse beam pressure, approximately given by

$$P_\perp \simeq \frac{mV_\perp{}^2}{2} n\gamma$$

Then $j \simeq P_\perp c/aB_\theta$, where $B_\theta(r = a)$ has the form

$$B_\theta \simeq \frac{2\pi V_z ne}{c} a$$

The total current $I = \pi a^2$ will be

$$I \simeq \frac{m\gamma V_\perp{}^2 c^2}{4eV_z} = I_A \cdot \left(\frac{V_\perp{}^2}{2V_z}\right)$$

Lowering $2V_z$ below $V_\perp$ permits currents arbitrarily larger than I_A. But high-current beams are useful because of their high energy, and beams with small V_z/c must necessarily have low-energy content. Thus they are not now of great interest.

B. *Basic Equations*

In constructing a model of a relativistic beam, we proceed as follows:

1. Specify geometry; the direction of current flow implies a degree of symmetry.
2. Write an equilibrium distribution function f_0 in terms of the constants of motion appropriate to this geometry.
3. Calculate current density j and charge density $-nq$ from f_0, as functionals of the potentials.
4. Solve for the vector potential $\mathbf{A}$ and scalar potential ϕ by means of

$$\nabla \times \nabla \times \mathbf{A} = \frac{4\pi j}{c} \tag{I-1}$$

and

$$-\nabla^2 \Phi = 4\pi nq \tag{I-2}$$

The mathematical techniques employed make use of two properties of these equations. First, even if these equations are coupled, it is often possible to find a linear combination of Φ and the component of $\mathbf{A}$ in the $\mathbf{j}$ direction (the only one that is important) which decouples the eqūations. Second, the equation

$$\frac{d^2\Phi}{dx^2} = f(\Phi)$$

can be integrated after multiplication by $d\Phi/dx$:

$$\frac{1}{2}\left(\frac{d\Phi}{dx}\right)^2 = \int f(\Phi)\, d\Phi$$

In particular, the solution of the equation

$$\frac{d^2\Phi}{dx^2} = e^{-\Phi} \tag{I-3}$$

is

$$\Phi = \ln \frac{\cosh^2[\eta(x - x_0)]}{2\eta^2} \tag{I-4}$$

where η and x_0 are arbitrary.

The general solution of

$$\frac{\partial^2\Phi}{\partial x^2} + \frac{\partial^2\Phi}{\partial y^2} = e^{-\Phi}$$

the two-dimensional form of eq. (1-3), is also known (14–16). It is

$$e^{-\Psi} = 8\,\frac{(\partial g^2/\partial x) + (\partial g^2/\partial y)^2}{(1 + g^2 + h^2)^2}$$

where $f(z) = f(x + iy) = g(x, y) + ih(x, y)$ is an arbitrary, single-valued analytic function, and g and h are its real and imaginary parts. This formula can be useful for describing beams with noncircular cross sections. There is, unfortunately, no known general solution for the three-dimensional analog of eq. (I-3).

$$\frac{\partial^2\Phi}{\partial x^2} + \frac{\partial^2\Phi}{\partial y^2} + \frac{\partial^2\Phi}{\partial z^2} = e^{-\Phi}$$

The degree of neutralization and the parameters in f_0 are at our disposal in constructing the model. In addition the general solution (eq. I-4 or its equivalent) can be chosen so as to shape the profile. We can sometimes specify a solution with a homogeneous externally imposed, magnetic field in the direction of current flow. Finally, the classical ($mc \to \infty$) or ultra-relativistic ($mc \to 0$) limit may be taken to simplify the results. We consider systems with axial and azimuthal symmetry (except for two models in Section II). In terms of cylindrical coordinates r, θ, z, all quantities are independent of θ and z. Thus

$$H = c(p^2 + m^2c^2)^{1/2} + q\phi \tag{I-5}$$

$$P_z = p_z + \frac{q}{c} A_z \tag{I-6}$$

and

$$P_\theta = r(p_\theta + \frac{q}{c} A_\theta) \tag{I-7}$$

are constants of motion.

C. *Boundary Conditions*

No equilibrium is specified by f_0 alone; the boundary conditions on ϕ and $\mathbf{A}$ must be given. The solutions usually satisfy automatically the requirement that the field go to zero as r approaches infinity. The relevant potential (ϕ or $\mathbf{A}_z$ or a linear combination of the two) will satisfy a second-order differential equation, usually in terms of r only. The general solution contains two constants of integration. It will sometimes turn out that one field (e.g., B_θ) diverges $\sim 1/r$ as $r \to 0$, unless one of the parameters has a specific value. If the parameter takes on any other value, a line charge or current at $r = 0$ has to be present to provide a source for the field. This is essentially a way of adjusting boundary conditions to fit the solution. It is consistent with the field equations, which in cylindrical coordinates are undefined at $r = 0$.

If a model describing a beam contained in a conducting tube is desired, E_z must vanish at the wall. (Throughout most of this article E_z will vanish, anyway, because the beam is uniform in the z-direction.) An infinite cylindrically symmetric equilibrium with $B_r \neq 0$ may be cut off at any arbitrary radius and a conducting wall placed there. If particles undergo specular reflection at the boundaries, the equilibrium solution is good inside the walls. This property is not particularly useful in practical application, however, because specular reflection does not occur in plasmas in most collisions with walls.

I. Axial Current Models

A. *Diffuse Beams*

1. Generalized Bennett Distribution

Consider the distribution function

$$f_0 = \bar{K} \exp\left[-\frac{a}{c} H + \alpha P_z\right] \tag{1-1}$$

As Buneman (7) has noted, the argument of the exponential can be written (using the summation convention) in the covariant form $a_\mu P_\mu$, where P_μ is the canonical 4-momentum and a_μ is a constant 4-vector with nonvanishing components in the z- and time-like directions. By going to a new inertial frame, appropriately chosen, we can transform this expression to $a'P'_0$, where P'_0 is the total energy in the new coordinate system. In this frame, the distribution is Maxwellian.

In any frame of reference, there will be particles moving at all velocities between $-c$ and $+c$. Thus eq. 1-1 defines a model that differs significantly from most experiments, in which all electrons move in the positive z-direction. A true Maxwellian can only arise experimentally after the passage of enough time to permit randomization through collisions. Nevertheless, many attractive simplifications result from assuming the simple form of eq. 1-1.

If two or more species described by equilibrium distributions of the form eq. 1-1 are present, it is in general impossible to transform them all to rest simultaneously. We are interested, in fact, in situations in which electrons drift with relativistic speeds through a stationary background of ions in the laboratory frame. We shall assume that this drift is in the positive sense ($\alpha > 0$).

The distribution function f_0 is invariant under Lorentz transformation (7). (Since d^3r and d^3p, the differential volumes in coordinate and momentum space, are *not* invariant, this implies that the number of particles in a differential volume of phase or coordinate space is not an invariant either.) Because f_0 transforms as a scalar, we require the invariance of both $\bar{K}$ and of $a_\mu P_\mu$ separately. It is this which shows that a_μ does indeed transform as a well-defined 4-vector. The condition that eq. 1-1 give rise to finite particle densities on integration over momentum is $a > |\alpha|$. This inequality is preserved under Lorentz transformation; hence, in all frames, $a_\mu a_\mu = a^2 - \alpha^2 > 0$ (a_μ is a time-like vector) and $a > 0$.

Following the outline given earlier, we next calculate the particle density and current associated with eq. 1-1. Defining the current 4-vector $j_\mu = (nqc, j_x, j_y, j_z)$, where $n(r) = \int d^3p f_0$ and $\mathbf{j}(r) = q \int d^3p\ \mathbf{v}\ f_0$, with $\mathbf{v} = (p^2 + m^2c^2)^{-1/2}\mathbf{p}c$, we find

$$j_\mu = qc\bar{K}e^{-\psi}Qa_\mu \tag{1-2}$$

where the quantities

$$\psi = \frac{q}{e}(a\phi - \alpha A_z) = \frac{q}{c}a_\mu A_\mu \tag{1-3}$$

and

$$Q\frac{4\pi(mc)^2}{a^2 - \alpha^2}K_2[mc(a^2 - \alpha^2)^{1/2}] \tag{1-4}$$

are Lorentz invariants. In the ultrarelativistic limit ($mc \to 0$), the small-argument expansion for the modified Bessel function K_2 allows us to write instead of eq. 1-4

$$Q \to \frac{8\pi}{(a^2 - \alpha^2)^2}$$

Now, if we were to omit the contribution of the ions entirely, Ampere's law and Poisson's equation would then take the form ($q = -e$)

$$\frac{1}{r}\frac{d}{dr}\left(r\frac{dA_\mu}{dr}\right) = -\frac{4\pi}{c}j_\mu = 4\pi e\bar{K}Qa_\mu e^{-\psi} \tag{1-5}$$

Multiplying by e/ca_μ, we obtain an equation for ψ:

$$\frac{1}{r}\frac{d}{dr}\left(r\frac{d\psi}{dr}\right) = -\frac{4\pi e^2}{c}\bar{K}Q(a^2 - \alpha^2)e^{-\psi} = Ge^{-\psi} \tag{1-6}$$

Writing $x = \log r$ and $\Phi = \psi - 2\ln r - \ln G$, we transform eq. 1-6 into eq. 1-3. The general solution of eq. 1-6 is thus

$$\psi = \ln\left\{\frac{Gr^2}{8\eta^2}\left[\left(\frac{r}{r_0}\right)^\eta + \left(\frac{r_0}{r}\right)^\eta\right]^2\right\} \tag{1-7}$$

But as a result of neglecting the ions, G is *negative*. Hence the general solution ψ breaks down, and we must reconsider the steps leading to eq. 1-5. Physically, this reflects the fact that two charges with a relative velocity v have an electrostatic repulsion larger than the magnetic attraction by a factor of $(v/c)^{-2}$. (Equation 1-2 says that $\langle v_z/c\rangle = \alpha/a$.) If a beam is to exist in equilibrium, at least partial charge neutralization is necessary.

The simplest way to handle the ions is to assume that they form a cold immobile background with a density everywhere proportional to $n_e(r)$. In general, this is what we shall do. However, it is more rigorous to start with the ion equilibrium distribution function f_{0i}. Integration over momentum yields the ion density $n_i(r)$. Unless n_i has the same dependence on r as does n_e, the solution of Poisson's equation would be very difficult. This condition can always be met, however.

We illustrate these remarks by treating the ions in detail for the model under discussion. Specifically, we assume that *both* species satisfy an equation of the form 1-1:

$$f_{0j} = \bar{K}_j \exp[-a_\mu^{(j)}p_\mu^{(j)}] \qquad j = i, e \tag{1-8}$$

Hence eq. 1-5 is replaced by

$$\frac{1}{r}\frac{d}{dr}\left(r\frac{dA_\mu}{dr}\right) = 4\pi\sum_{j=i,e} q_j\bar{K}_jQ_ja_\mu^{(j)}e^{-\psi_j} \tag{1-9}$$

where

$$\psi_j = \frac{q_j}{c}a_\mu^{(j)}A_\mu \tag{1-10}$$

and

$$Q_j = \frac{4\pi(m_jc)^2}{a_j^2 - \alpha_j^2}K_2[m_jc(a_j^2 - a_j^2)^{1/2}]$$

By adding $(q/c)a^{(l)}$, to eq. 1-9 we get

$$\frac{1}{r}\frac{d}{dr}\left(r\frac{d\psi_l}{dr}\right) = -\frac{4\pi}{c}\sum_j q_j q_l \overline{K}_j Q_j a^{(j)} a^{(l)} e^{-\psi j} \qquad l = i, e \qquad (1\text{-}11)$$

When can we solve the system defined by eq. 1-11? (We shall defer discussion of boundary conditions for the time being.)

The condition that the linear transformation 1-10 be invertible is

$$\begin{bmatrix} a^{(i)} & \alpha^{(i)} \\ a^{(e)} & \alpha^{(e)} \end{bmatrix} = a^{(i)} a^{(e)} \left(\frac{\alpha^{(e)}}{a^{(e)}} - \frac{\alpha^{(i)}}{a^{(i)}}\right) \neq 0 \qquad (1\text{-}12)$$

If this does not hold, the vectors $a_\mu^{(i)}$ and $a_\mu^{(e)}$ are parallel,

$$a_\mu^{(i)} = \kappa a_\mu^{(e)} \qquad (1\text{-}13)$$

for some constant κ which must be positive. The system 1-10 reduces to a single equation,

$$\frac{1}{r}\frac{d}{dr}\left(r\frac{d\psi_e}{dr}\right) = -\frac{4\pi e^2}{c} a_\mu^{(e)} a_\mu^{(e)} [\overline{K}_e Q_e \exp(-\psi_e) - \kappa \overline{K}_i Q_i \exp(\kappa \psi_e)]$$

which, however, appears intractable by analytic methods. We note that when eq. 1-13 holds, both species are at rest in the same frame, although they may have different temperatures there. Thus this is a case of no interest in a discussion of beams.

If eq. 1-12 holds, though, we can find solutions to eq. 1-11 by assuming that electrons and ions have the same radial profiles:

$$\exp(-\psi_j) = \sigma_j \frac{8\eta^2}{r^2}\left[\left(\frac{r}{r_0}\right)^\eta + \left(\frac{r_0}{r}\right)^\eta\right]^{-2} \equiv \sigma_j \exp(-\psi_0)$$

Making use of the fact that, when $G = 1$, ψ of eq. 1-7 equals ψ_0, we find on substituting in eq. 1-11 that

$$1 = \sigma_e M_{ee} + \sigma_i M_{ei}, \qquad 1 = \sigma_e M_{ie} + \sigma_i M_{ii} \qquad (1\text{-}14)$$

where

$$M_{lj} = -\frac{4\pi}{c} q_l q_j \overline{K}_j Q_j a^{(j)} \cdot a^{(l)}$$

The solution of eq. 1-14 is

$$\sigma_e = \frac{1}{\Delta}\begin{bmatrix} 1 & M_{ei} \\ 1 & M_{ii} \end{bmatrix} = \frac{M_{ii} - M_{ei}}{\Delta}$$

$$\sigma_i = \frac{1}{\Delta}\begin{bmatrix} M_{ee} & 1 \\ M_{ie} & 1 \end{bmatrix} = \frac{M_{ee} - M_{ie}}{\Delta}$$

where the determinant is

$$\Delta = \begin{bmatrix} M_{ee} & M_{ei} \\ M_{ie} & M_{ii} \end{bmatrix} = \left(\frac{4\pi e^2}{c}\right)^2 \overline{K}_e \overline{K}_i Q_e Q_i \begin{bmatrix} a^{(e)} \cdot a^{(e)} & -a^{(e)} \cdot a^{(i)} \\ -a^{(i)} \cdot a^{(e)} & a^{(i)} \cdot a^{(i)} \end{bmatrix}$$

This determinant vanishes if and only if

$$[a^{(e)} \cdot a^{(e)}][a^{(i)} \cdot a^{(i)}] - [a^{(e)} \cdot a^{(i)}]^2 = -(\alpha_e a_i - \alpha_i a_e)^2 = 0$$

which we have assumed by eq. 1-12 not to be the case. So there results

$$\sigma_e = \frac{c}{4\pi e^2} \frac{a_i^2 - \alpha_i^2 + a_e a_i - \alpha_e \alpha_i}{\overline{K}_e Q_e(\alpha_e a_i - \alpha_i a_e)^2}$$

$$\sigma_i = \frac{c}{4\pi e^2} \frac{a_e^2 - \alpha_e^2 + a_e a_j - \alpha_e \alpha_i}{\overline{K}_i Q_i(\alpha_e a_i - \alpha_i a_e)^2}$$

The particle densities are

$$n_e(r) = \overline{K}_e Q_e a_e e^{-\psi_e} = \frac{c a_e}{4\pi e^2} \frac{a_i^2 - \alpha_i^2 + a_e a_i - \alpha_e \alpha_i}{(\alpha_e a_i - \alpha_i a_e)^2} \cdot e^{-\psi_0} \quad (1\text{-}15)$$

$$n_i(r) = K_i Q_i a_i e^{-\psi_i} = f n_e(r) \quad (1\text{-}16)$$

where f, the degree of neutralization, is defined to be

$$f = \frac{a_i a_e^2 - \alpha_e^2 + a_e a_i - \alpha_e \alpha_i}{a_e a_i^2 - \alpha_i^2 + a_e a_i - \alpha_e \alpha_i} \quad (1\text{-}17)$$

Inverting eq. 1-10, we find that

$$A_z = +\frac{c}{e} \frac{a_i + a_e}{\alpha_e a_i - \alpha_i a_e} \ln\left\{\frac{r^2}{8\eta^2}\left[\left(\frac{r}{r_0}\right)^\eta + \left(\frac{r_0}{r}\right)^\eta\right]^2\right\} + \text{const} \quad (1\text{-}18)$$

$$\phi = +\frac{c}{e} \frac{\alpha_i + \alpha_e}{\alpha_e a_i - \alpha_i a_e} \ln\left\{\frac{r^2}{8\eta^2}\left[\left(\frac{r}{r_0}\right)^\eta + \left(\frac{r_0}{r}\right)^\eta\right]^2\right\} + \text{const} \quad (1\text{-}19)$$

Hence

$$B_\theta = -\frac{c}{e} \frac{a_i + a_e}{\alpha_e a_i - \alpha_i a_e} \frac{2}{r}\left\{1 + \eta \frac{(r/r_0)^\eta - (r_0/r)^\eta}{(r/r_0)^\eta + (r_0/r)^\eta}\right\} \quad (1\text{-}20)$$

$$E_r = -\frac{c}{e} \frac{\alpha_i + \alpha_e}{\alpha_e a_i - \alpha_i a_e} \frac{2}{r}\left\{1 + \eta \frac{(r/r_0)^\eta - (r_0/r)^\eta}{(r/r_0)^\eta + (r_0/r)^\eta}\right\} \quad (1\text{-}21)$$

These results become clearer if we work in the rest-frame of the ions ($\alpha_i = 0$). Then eq. 1-11 becomes for $l = e$

$$\frac{1}{r}\frac{d}{dr} r \frac{d\psi}{dr} = \frac{4\pi e^2}{c} \overline{K} Q a^2 \left(\frac{\alpha^2}{a^2} + f - 1\right) e^{-\psi} \equiv G e^{-\psi} \quad (1\text{-}22)$$

(Species labels have been dropped and from now on all unlabeled quantities that would be species-dependent refer to electrons.) From eq. 1-17, as a_i varies from 0 to ∞, f goes from $1 - \alpha^2/a^2$ to 1. So in contrast to eq. 1-6, G in eq. 1-22 is always positive. In terms of f, eqs. 1-20 and 1-21 become

$$B_\theta = -\frac{c}{e}\frac{2\alpha}{[\alpha^2 + (f-1)a^2]r}\left\{1 + \eta\frac{(r/r_0)^\eta - (r_0/r)^\eta}{(r/r_0)^\eta + (r_0/r)^\eta}\right\} \tag{1-23}$$

$$E_r = -\frac{c}{e}\frac{2a(1-f)}{[\alpha^2 + (f-1)a^2]r}\left\{1 + \eta\frac{(r/r_0)^\eta - (r_0/r)^\eta}{(r/r_0)^\eta + (r_0/r)^\eta}\right\} \tag{1-24}$$

Equation 1-15 becomes

$$n(r) = \frac{c}{4\pi e^2}\frac{a}{\alpha^2 + (f-1)a^2}\frac{8\eta^2}{r^2}\left[\left(\frac{r}{r_0}\right)^\eta + \left(\frac{r_0}{r}\right)^\eta\right]^{-2} \tag{1-25}$$

while

$$j(r) = \frac{c}{4\pi e}\frac{\alpha}{\alpha^2 + (f-1)a^2}\frac{8\eta^2}{r^2}\left[\left(\frac{r}{r_0}\right)^\eta + \left(\frac{r_0}{r}\right)^\eta\right]^{-2} \tag{1-26}$$

The character of the solutions is determined by competition between self-pinching (the α^2/a^2 term) and electrostatic repulsion (the $1-f$ term). We see that, if $a_i \to \infty$ or $a_e \to \infty$, $f \to 1$ and $\phi \to$ const. This is easily understood when we recall our initial assumptions. From eq. 1-8, $a_i \to \infty$ means zero ion temperature. But if the ions are cold and stationary, and if the system is in equilibrium, the total force (electrostatic plus magnetic) on an individual ion has to vanish. The magnetic force vanishes because $v = 0$; therefore, the electrostatic force also vanishes, and ϕ must be constant. For this to happen, charge neutralization must be complete, so $f = 1$. A similar argument applies in the case of cold electrons.

The solutions 1-23 to 1-26 have a number of interesting properties. First, it is clear that r_0 is simply a scale length associated with radial variation in the beam profile. The second constant η is a shaping parameter (Figure 3). For $\eta < 1$, $n(r)$ and $j(r)$ diverge at $r = 0$; for $n > 1$, n and j vanish at $r = 0$, and the profile is hollowed out near the axis. For $n = 1$, n and j are finite on the axis. This is the solution first found by Bennett (1).

The form of B_θ and E_r also reflects different choices of η. For $\eta > 1$ ($\eta < 1$), they diverge $\simeq 1/r$ to $+\infty$ $(-\infty)$ as $r \to 0$. If $\eta = 1$ (the Bennett distribution), both vanish at the axis. Temporarily, let us suppose $f = 1$, so that E_r vanishes, and examine the behavior of B_θ.

As $r \to \infty$, $B_\theta \to -[2(1+\eta)c/re\alpha]$. This is just the field produced by a current $I_{\text{total}} = [(1+\eta)c^2/e\alpha]$. As $r \to 0$, $B_\theta \to -[2(1-\eta)c/re\alpha]$. This is the field that would be produced by a current $I_0 = [(1-\eta)c^2/e\alpha]$ on the axis. But from the form of $j(r)$, our solution clearly lacks such a current. The

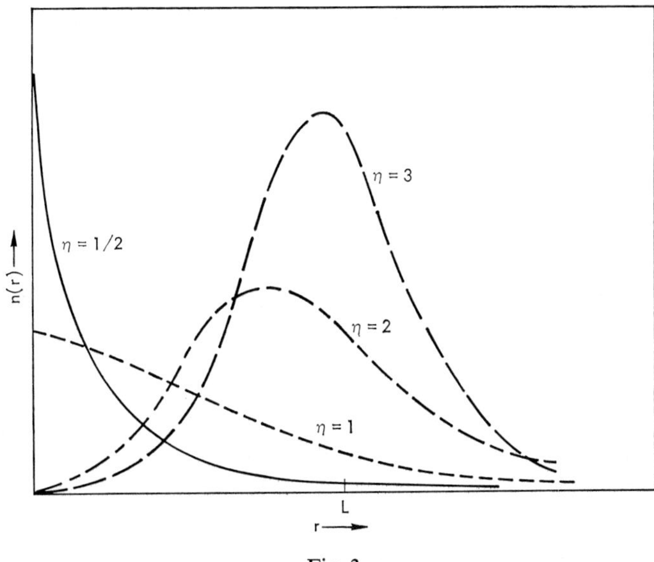

Fig. 3.

explanation is that the axial current must be added as a boundary condition. Without it, the solution would not be self-consistent. Only for the original Bennett model ($\eta = 1$) does this axial current vanish.

Physically, the additional current is required to balance internal forces in the beam if $\eta \neq 1$. When $\eta = 1$, self-pinching just compensates for the tendency of the thermal motion to make the electrons fly apart. If the beam were compressed toward the axis, it would tend to push apart again. By putting an additional current on the axis parallel to that of the streaming electrons, an attractive force that can hold the beam together is provided. If the beam is hollowed out, an oppositely directed axial current is needed to keep it from pinching back toward the Bennett configuration. Thus I_0 and I_{total} are parallel for $\eta < 1$ and antiparallel for $n > 1$.

If $f \neq 1$, similar considerations relate E_r and an axial line charge λ. For general η the charge per unit length is

$$\lambda = \frac{-(1-f)}{\alpha^2/a^2 + f - 1} \frac{(1-\eta)c}{ea} \tag{1-27}$$

while the axial current becomes

$$I_0 = \frac{-\alpha}{\alpha^2 + (f-1)a^2} \frac{(1-\eta)c^2}{e} \tag{1-28}$$

In the Introduction we noted that back currents persist for a length $D = v_z a^2 \sigma / c^2$ behind the head of a uniform beam. If the beam is hollowed out, however, the resistive diffusion time τ of the induced B_θ is changed. For a uniform beam, τ is proportional to a^2, the beam radius. But, for a hollow beam, τ is proportional to $a\Delta$, where Δ is the thickness of the cylinder of electrons; the appropriate diffusion length is

$$D' = \frac{v_z a \Delta \sigma}{c^2}$$

The condition for the applicability of a model with no z-dependence is then $D' \gg a$ which is somewhat more stringent.

It is somewhat remarkable that $\overline{K}_j$ appears in none of the macroscopic quantities calculated from f_{0j}. We could just as well set $\overline{K}_j = 1$. This is because $\eta(r)$ and $j(r)$ depend exponentially on ϕ and A_z. The arbitrary additive constants in these potentials give rise to an indeterminate constant, multiplying $\overline{K}_j$ in j_μ; hence the magnitude of $\overline{K}_j$ is indeterminate and physically meaningless.

By the integration of $n(r)$, we find the number of electrons per unit length:

$$N = 2\pi \int_0^\infty n(r) r \, dr = 2\eta \frac{c}{e^2 a} \left[\left(\frac{\alpha}{a} \right)^2 + f - 1 \right]^{-1} \tag{1-29}$$

This quantity is independent of r_0 but scales with η. From eq. 1-29 or from the difference between I_{total} and I_0 above, we find the current carried by the electron beam:

$$I_B = \frac{2\eta c^2 \alpha}{e a^2} \left[\left(\frac{\alpha}{a} \right)^2 + f - 1 \right]^{-1} \tag{1-30}$$

Provided that $f \neq 1$, $I_B \to 0$ as $\alpha \to 0$. If $f = 1$, $I_B \to \infty$ as $\alpha \to 0$. This may appear odd until we note that, in the same limit, $N \to \infty$ as α^{-2}, so that $I_B / N \to 0$.

As $f \to 1 - \alpha^2/a^2$, I_B diverges. This follows from the presence of the factor $[\alpha^2 + (f - 1)a^2]^{-1}$ in eq. 1-26. In this limit, the neutralizing ions barely suffice to reduce electrostatic repulsive forces to where they can be balanced by the tendency of currents to pinch. For the pinch effect to succeed in holding the beam together under these conditions, very large currents are needed.

How can such boundary conditions as eqs 1-27 and 1-28 be met? In the simple case $f = 1$, a current at $r = 0$ is required—in other words, a current-carrying wire at the center of the beam. Such a wire must have a radius small compared with r_0, since our solution breaks down inside the wire. There are no experiments with this design using relativistic beams, though some plasma experiments (19) have hard core conductors.

Another interpretation (20) is possible, however, if $\eta > 1$. If the beam is strongly hollowed out, and if the background through which it passes is ionized (and therefore conducting) only near the axis, a return current can flow there. There is no reason to expect a return current to flow in an equilibrium model, since experimentally it arises from time-varying effects that induce an E_z at the head of the beam (the approximations necessary for back currents to be present in a thin beam equilibrium were discussed in the Introduction), nor is there yet experimental evidence for return currents localized near the axis. Still, it is attractive to interpret I_0 in this fashion, since partial or total current neutralization is an important feature of many beam experiments (21,22).

As discussed earlier, the Lawson-Alfvén limit predicts a maximum beam current of the order

$$I_A = \frac{mc^3}{e} \beta_z \gamma$$

We have already found $\beta_z = 1/[n(r)c] \int d^3p v_z f_0 = \alpha/a$; γ can be found in the same fashion:

$$\gamma = \frac{1}{n(r)mc} \int d^3p(p^2 + m^2c^2)^{1/2} f_0 = 4\pi(mc)^2 \left\{ \frac{a^2 + \alpha^2}{(a^2 - \alpha^2)^2} \right.$$

$$\left. \times K_2[mc(a^2 - \alpha^2)^{1/2}] - \frac{mca^2}{(a^2 - \alpha^2)^{3/2}} K_2'[mc(a^2 - \alpha^2)^{1/2}] \right\} \quad (1\text{-}31)$$

In the ultrarelativistic limit, γ assumes the simple form

$$\gamma \simeq (mca)^{-1} \frac{3a^2 + \alpha^2}{a^2 - \alpha^2} \gg 1 \quad (1\text{-}32)$$

Hence

$$I_A \cong \frac{c^2}{e} \frac{\alpha}{a^2} \frac{3a^2 + \alpha^2}{a^2 - \alpha^2}$$

Comparison with eq. 1-30 shows that

$$\frac{I_B}{I_A} = 2\eta \left[\left(\frac{\alpha}{a} \right)^2 + f - 1 \right]^{-1} \frac{a^2 - \alpha^2}{3a^2 + \alpha^2} \quad (1\text{-}33)$$

We see that eq. 1-33 can exceed one by an arbitrary amount, merely by making η large enough. The second factor can also be large if $f \simeq 1 - \alpha^2/a^2$. But as noted in the Introduction, f increases along the course of the beam from front to back in actual experiments. We cannot expect a model that depends on marginal neutralization for current enhancement to give a good fit to such experiments.

It is of interest to calculate the energy associated with particle motion, magnetic fields, and the electrical fields in this model. The kinetic energy per unit length of beam is

$$W_K = N(\gamma - 1)mc^2 \simeq \frac{4\eta c^2}{e^2} \frac{1}{\alpha^2 + a^2(f-1)} \frac{a^2 + \alpha^2}{a^2 - \alpha^2}$$

in the ultrarelativistic limit, by eq. 1-32.

The magnetic energy density is $B^2/8\pi$. It varies as r^{-2} for large r and, except when $\eta = 1$, for small r as well. Consequently, the integral

$$2\pi \int r\, dr\, B^2/8\pi$$

is divergent at both limits. One should expect this, since B looks like the field of a straight wire carrying current I_{total} as $r \to \infty$, and the boundary current I_0 is all-important for $r \to 0$. To make the magnetic field energy per unit length W_B finite, we introduce cutoffs at $r = r_1$ and $r = r_2 > r_1$. Then

$$W_B = \frac{1}{4}\int_{r_1}^{r_2} r\, dr B^2 = \frac{c}{e^2} \frac{\alpha_2}{[\alpha^2 + (f-1)a^2]^2}$$

$$\times \left\{(1+\eta^2)\ln(\rho_2/\rho_1) + 2\ln\frac{\rho_2{}^\eta + \rho_2{}^{-\eta}}{\rho_1{}^\eta + \rho_1{}^{-\eta}} - \eta\left[\frac{\rho_2{}^\eta - \rho_2{}^{-\eta}}{\rho_2{}^\eta + \rho_2{}^{-\eta}} - \frac{\rho_1{}^\eta - \rho_1{}^{-\eta}}{\rho_1{}^\eta + \rho_1{}^{-\eta}}\right]\right\}$$

where $\rho_2 = r_2/r_0$, $\rho_1 = r_1/r_0$. If $\rho_2 \gg 1 \gg \rho_1$,

$$W_B \simeq \frac{c^2}{e^2} \frac{\alpha^2}{[\alpha^2 + (f-1)a^2]^2} (1+\eta^2)\ln(\rho_2/\rho_1) + 2\eta^2 \ln(\rho_2 \rho_1)$$

Likewise, the electric field energy per unit length is

$$W_E = \frac{1}{4}\int_{r_1}^{r_2} E^2 r\, dr = (1-f)^2\, a^2/\alpha^2\, W_B$$

For a neutralized $\eta > 1$ beam with I_0 carried in a small conducting channel or wire ($\rho_1 \simeq \frac{1}{10}$) and boundary by a cylinder ($\rho_2 \simeq 5$),

$$\frac{W_B}{W_K} \sim \eta \frac{a^2 - \alpha^2}{a^2 + \alpha^2}$$

so for very hollow beams ($n \gg 1$), most of the energy resides in the magnetic fields. However, for $\eta > 1$, W_B/W_K is not sensitive to the inner and outer boundaries of the beam. This contrasts with the double delta-function model of Section I-B where a 10% change in the beam radius doubles the field energy.

It can easily be verified that the beam electrons traverse orbits in which all three degrees of freedom are coupled, even when $f = 1$. Consequently, to do a stability analysis of the linearized Vlasov-Maxwell equations by the method of integration along unperturbed orbits is hopelessly difficult. However, since f_0 is independent of P_θ, there is no complication associated with postulating the existence of a uniform external field B_{z0}, derived from

$$A_\theta = \frac{B_{z0}\, r}{2} \tag{1-34}$$

If $B_{z0} \gg B_\theta$, integration over unperturbed orbits becomes practical.

2. Related Models

If we assume charge neutrality from the start, then $\phi = 0$. Now only eq. I-1 need be solved. The energy no longer depends on r through ϕ:

$$H = (p^2 + m^2c^2)^{1/2}c$$

We can accordingly permit f_0 to have a more general dependence on H,

$$f_0 = K(H) \exp(\alpha P_z) \tag{1-35}$$

The only restrictions on $K(H)$ are $K(H) > 0$ and convergence of the integral defining the current:

$$\begin{aligned} j(r) &= -ec \int d^3p v_z\, K(H) \exp(\alpha P_z) \\ &= -\frac{4\pi ec}{\alpha^2} \exp\left(\frac{-\alpha e}{c} A_z\right) \\ &\quad \times \int_0^\infty \frac{p\, dp}{(p^2 + m^2c^2)^{1/2}} [\alpha p \cosh(\alpha p) - \sinh(\alpha p)] K(H) < \infty \end{aligned}$$

We now consider some possible models of this type.

a. *Weighted Maxwellian*

Let

$$K(H) = \overline{K}(H/E_0)^\mu \exp(-aH/c)$$

For $\mu = 0$, eq. 1-35 reduces to eq. 1-1. For μ a positive integer, we have

$$j(r) = -\overline{K} ec\alpha \exp\left(-\frac{e}{c}\alpha A_z\right)(-c/E_0)^\mu\, \partial^\mu Q/\partial a^\mu$$

where Q is given by eq. 1-4; $n(r)$ is found analogously.

The current, density, and magnetic field profiles have exactly the same radial dependence as in the previous section; only the normalization of the density changes. Thus

$$n(r) = \frac{c}{4\pi e^2\alpha^2} \frac{(\partial/\partial a)^\mu (aQ)}{(\partial/\partial a)^\mu Q} \frac{8\eta^2}{r^2} \left[\left(\frac{r}{r_0}\right)^\eta + \left(\frac{r_0}{r}\right)^\eta\right]^{-2}$$

$$j(r) = \frac{-c^2}{4\pi e\alpha} \frac{8\eta^2}{r^2} \left[\left(\frac{r}{r_0}\right)^\eta + \left(\frac{r_0}{r}\right)^\eta\right]^{-2}$$

$$B_\theta(r) = \frac{-2c}{e\alpha r}\left[1 + \eta \frac{(r/r_0)^\eta - (r_0/r)^\eta}{(r/r_0)^\eta + (r_0/r)^\eta}\right]$$

Again, a current I_0 on the axis is required if $\eta \neq 1$.

$$I_0 = \frac{(\eta - 1)c^2}{\alpha e}$$

b. *Weighted Energy Shell*

We take $K(H) = \overline{K}\delta(H - E_0)$ in (42), where $E_0 > mc^2$. Now all electrons in the beam have the same energy. Then, writing $p_0 = (E_0{}^2 - m^2c^4)^{1/2}/c$, we find

$$j(r) = \frac{-4\pi e\overline{K} \exp\left[-(e/c)\alpha A_z\right]}{\alpha^2} \left[\alpha p_0 \cosh(\alpha p_0) - \sinh(\alpha p_0)\right]$$

When $p_0 \gg mc$, this becomes

$$j(r) \simeq -\frac{2\pi e\overline{K}p_0}{\alpha} \exp\left(\alpha p_0 - \frac{e}{c}\alpha A_z\right)$$

When $p_0 \ll mc$, it becomes

$$j(r) \simeq -\frac{4\pi}{3}\overline{K}e\alpha p_0{}^3 \exp\left(-\frac{e}{c}\alpha A_z\right)$$

In either case, the methods of Section I-A-1 can be applied and the resulting behavior of $n(r)$, $j(r)$, and $B(r)$ is identical with that already studied.

c. *Semirelativistic Maxwellian*

The chief difficulty in carrying out the integrations leading to $j(r)$ and $n(r)$ is that $j(r)$ contains an awkward factor $(p^2 + m^2c^2)^{-1/2}$. If we introduce this factor in f_0 to cancel the denominator of v_z, $j(r)$ is easier to calculate but $n(r)$ may not be expressible in closed form. This device was used by Marx (23) in his "relativistic rigid rotor" (see Section II-C). It works for axial currents as well.

We take for f_0 a nonrelativistic displaced Maxwellian, multiplied by $(p^2 + m^2c^2)^{1/2}$

$$f_0 = \overline{K}(p^2 + m^2c^2)^{1/2} \exp\left(-\frac{p^2}{2\chi^2} + \alpha P_z\right)$$

Then

$$j(r) = -\overline{K}ec(2\pi)^{3/2}\, a\chi^5 \exp\left(\frac{1}{2}\alpha^2\chi^2 - \frac{e}{c}\alpha A_z\right)$$

This leads again to an equation for A_z of the type already seen. The solutions for $B_\theta(r)$ and $j(r)$ are the same as eqs. 1-23 and 1-26. As anticipated, $n(r)$ cannot be obtained in closed form except in the ultrarelativistic limit.

3. Plane Orbit Case

Since P_θ is a constant of motion, it is possible to build into f_0 an arbitrary dependence on this variable. In general, this will give rise to azimuthal currents j_θ, as well as j_z. Consequently, by Eq. I-7 P_θ will have an electromagnetic term, and will have to solve a system of equations coupling A_θ, A_z, and ϕ. But if f_0 is an even function of P_θ and we assume no external A_θ, there will be solutions with j_θ (and therefore A_θ) vanishing. Here we discuss some simple examples that are related to the Bennett distribution.

Suppose that all the beam electrons have the same canonical angular momentum $\overline{P}_\theta$, but otherwise are Maxwellian, i.e.,

$$f_0 = \overline{K} \exp\left[-\frac{a}{c}H + \alpha P_z\right] \delta(P_\theta - \overline{P}_\theta) \qquad (1\text{-}36)$$

Individual electrons in such a beam follow helical paths. If $\alpha \to 0$, the helices collapse into circles about the axis. For a discussion of this case, see Section II.

If $\overline{P}_\theta = 0$, the azimuthal current j_θ vanishes. Further, each electron orbit lies in a plane through the beam axis. The orbits are therefore considerably simpler than in the preceding cases. Since eq. 1-36 contains an exponential dependence on A_z and ϕ, we expect Maxwell's equations to have a general solution resembling 1-7. But the electrons feel no centrifugal force, so that a weaker boundary current I_0 at $r = 0$ should suffice to hold them together.

To see that this is so, we integrate eq. 1-36 to find current and number densities,

$$n(r) = \frac{2\pi\overline{K}e^{-\psi}}{r}\frac{a}{(a^2-\alpha^2)^{1/2}}\left[\frac{mc}{(a^2-\alpha^2)^{1/2}} + \frac{1}{a^2-\alpha^2}\right]\exp\left[-mc(a^2-\alpha^2)^{1/2}\right] \qquad (1\text{-}37)$$

$$j(r) = -(\alpha/a)\, ecn(r) \qquad (1\text{-}38)$$

where we have again written $\psi = e/c(\alpha A_z - a\phi)$. Also, γ can be found analogously to eq. 1-31; in the ultrarelativistic limit,

$$\gamma \doteqdot (mca)^{-1} \frac{2a^2 + \alpha^2}{a^2 - \alpha^2}$$

Substituting eqs. 1-37 and 1-38 in Maxwell's equations, and assuming a partial neutralization f produces by an ion background,

$$n_i(r) = f n_e(r)$$

we get

$$\frac{1}{r}\frac{d}{dr}\left(r\frac{d\psi}{dr}\right) = \frac{1}{rL} e^{-\psi} \tag{1-39}$$

where

$$\frac{1}{L} = \bar{K} \frac{8\pi^2 e^2}{c} \frac{\alpha^2 + (f-1)a^2}{(a^2 - \alpha^2)^{1/2}} \left[\frac{mc}{(a^2 - \alpha^2)^{1/2}} + \frac{1}{a^2 - \alpha^2}\right] \cdot \exp\left[-mc(a^2 - \alpha^2)^{1/2}\right] \tag{1-40}$$

Setting $\Phi = \psi - \ln(r/L)$ and $x = \ln r$, we again obtain eq. I-3. Thus the general solution of eq. 1-39 is

$$\psi = \ln\left\{\frac{r}{8\eta^2 L}\left[\left(\frac{r}{r_0}\right)^\eta + \left(\frac{r_0}{r}\right)^\eta\right]\right\}$$

with η and r_0 arbitrary.

From the general solution we see that

$$n(r) = \frac{e}{4\pi e^2} \frac{a}{\alpha^2 + (f-1)a^2} \frac{8\eta^2}{r^2} \left[\left(\frac{r}{r_0}\right)^\eta + \left(\frac{r_0}{r}\right)^\eta\right]^{-2} \tag{1-41}$$

and

$$j(r) = -ec(\alpha/a)n(r), \tag{1-42}$$

identical with eqs. 1-25 and 1-26. It follows that eqs 1-29 and 1-30 also hold. However,

$$B_\theta = -\frac{c}{e} \frac{2\alpha}{[\alpha^2 + (f-1)a^2]r} \left[\frac{1}{2} + \eta \frac{(r/r_0)^\eta - (r_0/r)^\eta}{(r/r_0)^\eta + (r_0/r)^\eta}\right] \tag{1-43}$$

$$E_r = -\frac{c}{e} \frac{2a(1-f)}{[\alpha^2 + (f-1)a^2]r} \left[\frac{1}{2} + \eta \frac{(r/r_0)^\eta - (r/r_0)^\eta}{(r/r_0)^\eta + (r_0/r)^\eta}\right] \tag{1-44}$$

which differ from eqs. 1-23 and 1-24 in having one-half instead of one inside the brackets.

The explanation for this is related to the feature of zero electron angular momentum. Equations 1-43 and 1-44 show that for $\eta \leq \frac{1}{2}$, the radial force on an electron is always directed toward $r = 0$. Thus every electron passes through the axis. For $\eta > \frac{1}{2}$, there are two classes of electrons: those passing through the axis those localized between a minimum and a maximum radius. The existence of this second class is easily shown. Let $f = 1$, so that $E_r = 0$. From eq. 1-43, $B_\theta = 0$ ar $r = r^*$, where

$$r^* = r_0 \left[\frac{2\eta - 1}{2\eta + 1}\right]^{1/2\eta}$$

Thus an electron with $r = 0$ placed at $r = r^*$, feeling no forces, will stream with uniform velocity along z; if slightly displaced from $r = r^*$, it will oscillate radially about this point between some r_1 and r_2.

As we have seen, the beam current is given by eq. 1-26. Equation 1-43 however, shows that the total current is

$$I_{\text{total}} = \frac{-\alpha}{\alpha^2 + (f-1)a^2} \frac{(\frac{1}{2} + \eta)c^2}{e} \tag{1-45}$$

while at $r = 0$ there is a current needed as a boundary condition:

$$I_0 = \frac{-\alpha}{\alpha^2 + (f-1)a^2} \frac{(\frac{1}{2} - \eta)c^2}{e} \tag{1-46}$$

This is to be compared with eq. 1-28. Equation 1-46 vanishes for $\eta = \frac{1}{2}$, which corresponds to divergent particle densities at $r = 0$. This is the "natural" equilibrium configuration, analogous to Bennett's solution $\eta = 1$ in Section I-A-1. It is the equilibrium that a beam having f_0 given by eq. 1-36 with $\bar{P}_\theta = 0$ would seek if no axial boundary current were provided. This natural equilibrium is cusped at $r = 0$ by comparison with that of Section I-A-1 as a result of the fact that all electrons pass through the axis. If they had an ordinary thermal spread in angular momenta, centrifugal forces would pull them into the profile of the Bennett distribution.

With this exception, the model behaves much like those already examined. In particular, we can exceed the Lawson-Alfvén limit by choosing η large enough. The ratio of I_B to I_A, analogous to eq. 1-33, is

$$I_B/I_A = 2\eta[(\alpha/a^2)^2 + f - 1]^{-1} (a^2 - \alpha^2)(2a^2 + \alpha^2)^{-1}$$

Just as in the preceding section, if $f = 1$, we can take f_0 to have an arbitrary dependence on energy. The only result of doing this is to change the formula 1-40 for the characteristic length L. Except for normalization factors, eqs. 1-41 to 1-44 retain their form, as does the boundary condition 1-46.

Although it would be of interest to obtain solutions for equilibria with general dependence on P_θ, e.g.,

$$f_0 = \overline{K} \exp\left(-\frac{a}{c} H + \alpha P_z - \Lambda^{-2} P_\theta^{\,2}\right)$$

or

$$f_0 = \overline{K}\delta(H - E_0)\delta(P_\theta^{\,2} - \overline{P}_\theta^{\,2}) \exp(\alpha P_z)$$

the field equations cannot be solved in closed form. It is possible, though, to show that, in these cases as well, hollowing out of the beam leads to currents exceeding I_A (24).

B. *Beams with Sharp Boundaries*

1. Double Delta-Function Model

This model, introduced by Mjolsness (5) and investigated at length by Rostoker and Hammer (3), assigns to all beam electrons the same energy E_0 and the same canonical z-momentum $\overline{P}_z$. The distribution function,

$$f_0 = \overline{K}\delta(H - E_0)\delta(P_z - \overline{P}_z) \tag{1-47}$$

is not separable into position- and momentum-dependent factors, so that the streaming velocity $\langle v_z \rangle$ is a function of position. This distribution can give rise to a sharply bounded cylinder of electrons. If the beam propagates in a uniform longitudinal external field (eq. 1-34), the cylinder rotates almost rigidly about its axis.

We assume that the arguments of the delta functions have one zero each, and

$$\left[\left(\overline{P}_z + \frac{e}{c} A_z\right)^2 + m^2c^2\right]^{1/2} c - e\phi < E_0 \tag{1-48}$$

an assumption to which we shall return after finding explicit solutions for ϕ and A_z. Then

$$n(r) = \frac{2\pi\overline{K}}{c^2} (E_0 + e\phi) \tag{1-49}$$

$$j(r) = -2\pi\overline{K}e\left(\overline{P}_z + \frac{e}{c} A_z\right) \tag{1-50}$$

Thus without any restriction on the degree of neutralization f, eqs. I-1 and I-2 decouple automatically. They take the form

$$\frac{1}{r}\frac{d}{dr}\left(r \frac{dA_z}{dr}\right) = \frac{8\pi^2 e\overline{K}}{c}\left(\overline{P}_z + \frac{e}{c} A_z\right) \tag{1-51}$$

$$\frac{1}{r}\frac{d}{dr}\left(r \frac{d\phi}{dr}\right) = \frac{8\pi^2 e\overline{K}(1 - f)}{c} (E_0 + e\phi) \tag{1-52}$$

where we have made the usual assumption about the ion background, $n_i(r) = fn_e(r)$. We anticipate that, just as in Section I-A, f must exceed some minimum value or electrostatic repulsion will dominate self-pinching. However, we shall see shortly that this imposes only a trivial limitation on the model.

To solve eq. 1-51, we write $\chi = (eA_z/c\bar{P}_z) + 1$ and $\rho = r/L$, where $L^{-2} = 8\pi^2 e^2 \bar{K}/c^2$. We can set $\bar{K} = n_0 c^2/2\pi E_0$, where n_0 is a density scale factor; so L, the scale length on which A_z varies, is the usual collisionless skin depth c/ω_p. Then χ satisfies

$$\frac{d^2\chi}{d\rho^2} + \frac{1}{\rho}\frac{d\chi}{d\rho} - \chi = 0 \tag{1-53}$$

the equation for modified Bessel functions of order zero. Similarly, to solve eq. 1-52 we write $\chi' = e\phi/E_0 + 1$ and $\rho' = r\sqrt{1-f}/L$. Then $\chi'(\rho')$ satisfies eq. 1-53 also, and we have, in general,

$$\frac{eA_z}{c\bar{P}_z} + 1 = C_1 I_0\left(\frac{r}{L}\right) + D_1 K_0\left(\frac{r}{L}\right) \tag{1-54}$$

$$\frac{e\phi}{E_0} + 1 = C_2 I_0\left(\frac{r\sqrt{1-f}}{L}\right) + D_2 K_0\left(\frac{r\sqrt{1-f}}{L}\right) \tag{1-55}$$

C_1, C_2, D_1, D_2 arbitrary.

Since the functions $I_0(\rho)$ and $K_0(\rho)$ diverge at $\rho = \infty$ and $\rho = 0$, respectively, eqs. 1-54 and 1-55 will not satisfy eq. 1-48 everywhere. When the inequality fails, $n(r) = 0 = j(r)$ and eq. 1-53 is replaced by

$$\frac{1}{\rho}\frac{d}{d\rho}\left(\rho\frac{d\chi}{d\rho}\right) = 0 \tag{1-56}$$

i.e., $\chi = C' + D' \ln \rho$.

A number of possibilities are open. If $f = 1$, then $\phi = \text{const}$ and the density profile is flat; further, $\gamma = (1 + p^2/m^2c^2)^{1/2}$ becomes the same for all particles. Clearly, $n(r)$ and $j(r)$ are, in general, not proportional. Indeed, they could be proportional only if $f = 0$ (ions wholly absent).

Let us suppose $D_1 = 0 = D_2$. Then, following Hammer and Rostoker (3), we write $C_1 = 1 = C_2$. We can assume this even in the general case, for the C's and D's appear in f_0 only multiplying $\bar{P}_z$ and E_0:

$$f_0 = \bar{K}\delta\{(p^2 + m^2c^2)^{1/2}c - E_0[C_2 I_0(r\sqrt{1-f}/L) + D_2 K_0(r\sqrt{1-f}/L)]\}$$
$$\cdot\,\delta\{p_z - \bar{P}_z[C_1 I_0(r/L) + D_1 K_0(r/L)]\}$$

and consequently appear the same way in quantities expressed as moments of f_0 ($n(r)$, $j(r)$, $\gamma(r)$), or in terms of these. Thus there is no loss of generality incurred in absorbing C_1 into $\bar{P}_z$ and C_2 into E_0 and redefining D_1 and D_2 accordingly.

With the above choice of coefficients, we have $\phi(r = 0) = 0 = A_z(r = 0)$. Then if $\gamma(r = 0)$ is denoted by γ_0, we find

$$mc^2\gamma_0 = E_0 \qquad \text{and} \qquad \bar{P}_z = m\gamma_0 V_z$$

where $V_z = v_z(r = 0)$. Now,

$$\phi = \frac{mc^2\gamma_0}{e}\left[I_0(r\sqrt{1-f}/L) - 1\right] \tag{1-57}$$

$$A_z = \frac{mcV_z\gamma_0}{e}\left[I_0(r/L) - 1\right] \tag{1-58}$$

These equations hold inside a region $r \leq a$, with a determined by substituting eqs. 1-57 and 1-58 in eq. 1-48:

$$\beta_z^{\,2}\gamma_0^{\,2}I_0^{\,2}(a/L) + 1 = \gamma_0^{\,2}I_0^{\,2}(a\sqrt{1-f}/L) \tag{1-59}$$

where $\beta_z = V_z/c$. The solutions for $r > a$ are found by smoothly connecting the solutions of eq. 1-56 to eqs 1-57 and 1-58:

$$\phi = \frac{mc^2\gamma_0}{e}\left[I_0\left(\frac{a\sqrt{1-f}}{L}\right) - 1 + \frac{a\sqrt{1-f}}{L}\,I_1\left(\frac{a\sqrt{1-f}}{L}\right)\ln r/a\right] \tag{1-60}$$

and

$$A_z = \frac{mcV_z\gamma_0}{e}\left[I_0\left(\frac{a}{L}\right) - 1 + \frac{a}{L}\,I_1\left(\frac{a}{L}\right)\ln\frac{r}{a}\right] \tag{1-61}$$

It follows that

$$\begin{aligned} E_r = -\frac{\partial\phi}{\partial r} &= -\frac{mc^2\gamma_0\sqrt{1-f}}{eL}\,I_1\left(\frac{r\sqrt{1-f}}{L}\right), \qquad r \leq a \\ &= -\frac{mc^2\gamma_0\sqrt{1-f}}{eL}\,I_1\left(\frac{a\sqrt{1-f}}{L}\right)\frac{a}{r}, \qquad r > a \end{aligned} \tag{1-62}$$

and

$$\begin{aligned} B_\theta = -\frac{\partial A_z}{\partial r} &= -\frac{mcV_z\gamma_0}{eL}\,I_1\left(\frac{r}{L}\right), \qquad r \leq a \\ &= -\frac{mcV_z\gamma_0}{eL}\,I_1\left(\frac{a}{L}\right)\frac{a}{r}, \qquad r > a \end{aligned} \tag{1-63}$$

From eqs. 1-57, 1-58, and the definition of γ, we have for $r \leq a$

$$\frac{n(r)}{n_0} = \frac{\gamma(r)}{\gamma_0} = I_0\left(\frac{r\sqrt{1-f}}{L}\right) \tag{1-64}$$

and

$$\langle v_z(r) \rangle = V_z \frac{I_0(r/L)}{I_0(r\sqrt{1-f}/L)} \tag{1-65}$$

(see Figure 4).

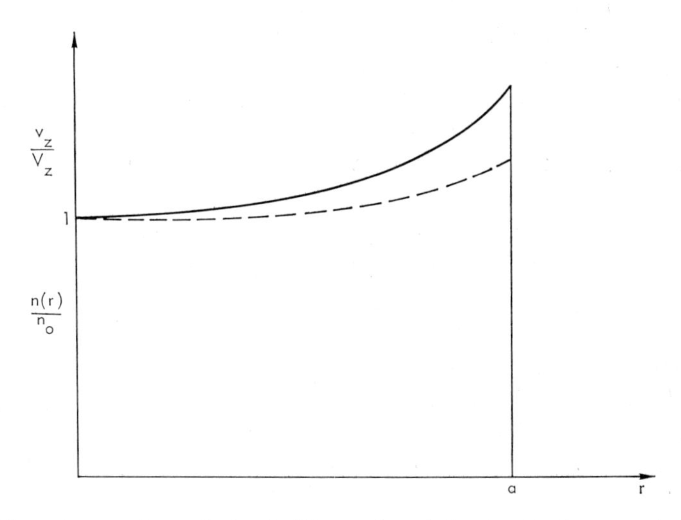

Fig. 4. Plot of streaming velocity (solid curve) and density (broken curve) for $f < 1$, showing the hollowed appearance of the beam in the model of Hammer and Rostoker (3).

The input parameters in this model are γ_0, β_z, and f. How much freedom have we in choosing them? To answer this, we first observe that eq. 1-59 has no solution if $f = 0$. For, in that case, we should have

$$\frac{1}{\gamma_0{}^2(1 - \beta_z{}^2)} = I_0{}^2\left(\frac{a}{L}\right) > 1$$

or

$$\gamma_0{}^2 < \frac{1}{1 - \beta_z{}^2}$$

But $\gamma_0{}^2 = [1 - \beta_z{}^2 - \langle v_\perp{}^2 \rangle / c^2]^{-1}$, where $\langle v_\perp{}^2 \rangle$ is the mean square transverse velocity at $r = 0$; this implies

$$\gamma_0{}^2 > \frac{1}{1 - \beta_z{}^2} \tag{1-66}$$

If $0 < f \leq 1$, eq. 1-59 does have a solution. For $f \ll 1$, the value of a required to satisfy eq. 1-59 is very large. This reflects the enhanced repulsive tendency of the unneutralized beam electrons. But no matter how small f is chosen, if a is sufficiently large, the beam will carry enough current to cohere through self-pinching.

In beam experiments, the background ions appear as a result of ionization by the primary electrons. Each ionizing collision produces an ion and a slow secondary electron. The secondary electrons blow off because of electrostatic repulsion, until enough ions are present to neutralize the beam ($f = 1$). Thereafter, the secondary electrons stay around, so that it is difficult to conceive of a real beam with an excess of positive over negative charge. Nevertheless, it is interesting to observe that the present model is capable of describing such a beam.

For $f > 1$, $\sqrt{1 - f}$ is imaginary, and we must take $I_0(r\sqrt{1 - f}/L) = J_0(r\sqrt{f - 1}/L)$. The solution 1-57 to 1-63 thus rewritten is valid, provided that

$$[\beta_z{}^2 \gamma_0{}^2 I_0{}^2(r/L) + 1]^{1/2} \leq \gamma_0 J_0(r\sqrt{f - 1}/L)$$

When equality holds, this expression defines the boundary radius a, as before. Clearly $a\sqrt{f - 1}/L > x_{01}$, where x_{01} is the first zero of $J_0(x)$.

Since every individual electron is limited to $r \leq a$, v_r goes to zero somewhere near the edge of this region. Consequently, we expect $\langle v_z(r) \rangle$ to have a maximum at $r = a$, and this is indeed so. The condition that this maximum be less than c, from eq. 1-65 is $\beta_z I_0(a/L) < I_0(a\sqrt{1 - f}/L)$. Substituting for $I_0(a\sqrt{1 - f}/L)$ from eq. 1-59, we see that this becomes

$$\beta_z{}^2 I_0{}^2(a/L) < \beta_z{}^2 I_0{}^2(a/L) + 1/\gamma^2$$

which is always satisfied. Thus, whenever a is defined (i.e., for any $f > 0$), $v_z < c$ throughout the beam.

By similar methods we can show that $0 < \langle v_\perp{}^2 \rangle / c^2 < 1$ for all radii. No new restrictions are supplied in this fashion. Thus the only limits on the input parameters are

$$\gamma_0{}^2 > \frac{1}{1 - \beta_z{}^2}$$

and

$$f > 0$$

The total current carried by the beam is found by integrating $j(r) = -nev_z$. The result is

$$I = -\frac{mc^3}{2e}\gamma_0\beta_z\frac{a}{L}I_1\left(\frac{a}{L}\right) \tag{1-67}$$

It is interesting to compare this with the Alfvén-Lawson limit (3). Since the beam velocity is not uniform across the beam, the Alfvén definition of a critical current,

$$I_A = \frac{mc^3}{e}\gamma_0\beta_z$$

is inappropriate. However, as noted previously, I_A is just the current for which an electron gyroradius in the maximum B_θ equals one-half the beam radius. Substituting eqs. 1-63 to 1-65 in

$$\left[\frac{\gamma mc}{e}\frac{v_z}{B_\theta}\right]_{r=a} = \frac{a}{2}$$

we find

$$\frac{a}{2L}I_1\left(\frac{a}{L}\right) = I_0\left(\frac{a}{L}\right) \tag{1-68}$$

For a beam of radius $a = a_c$ determined by eq. 1-68, the "critical" current I_c flows. Equation 1-67 gives a value

$$I_c = \frac{-mc^3}{e}\beta_z\gamma_0 I_0\left(\frac{a_c}{L}\right) \tag{1-69}$$

for this current. The ratio of eq. 1-69 to eq. 1-67 is

$$\frac{I}{I_c} = \frac{a}{2L}\frac{I_1(a/L)}{I_0(a_c/L)}$$

Thus, for $a \gg L, a_c$, it is possible to have an equilibrium with $I \gg I_c$. It is clear that electrons in such a beam are in no real danger of being turned around by B_θ because they traverse such a small portion of their orbits in the region of maximum field.

The number of electrons per unit length of beam is

$$N = 2\pi\int_0^a r\,dr n(r) = \frac{mc^2}{2e^2}\gamma_0\frac{a}{\sqrt{1-f}\,L}I_1\left(\frac{a\sqrt{1-f}}{L}\right) \tag{1-70}$$

If $f = 1$, eq. 1-70 becomes $N = \pi n_0 a^2$. If $f > 1$,

$$N = mc^2\gamma_0\,a/(L\sqrt{f-1}\,2e^2)\,J_1(a\sqrt{f-1}/L)$$

From eq. 1-70 we find v/γ by evaluating γ at $r = a$:

$$v/\gamma = \tfrac{1}{2}a/(L\sqrt{1-f})\, I_1(a\sqrt{1-f}/L)/I_0(a\sqrt{1-f}/L)$$

Thus large values of a/L imply large v/γ, as well as high currents.

For most experimental situations it is reasonable to take $f = 1$, as Hammer and Rostoker (3) show. They also note that, if $\beta_\perp c$ is the transverse velocity of an electron at $r = 0$, $\beta_\perp^2 \gg \beta_z^2$, when $I \gg I_A$. There is considerable transverse motion, but the current may still exceed I_A. This is the same as Section I-A, where the choice of f_0 assumes that individual electron motion is primarily transverse.

The case $I \ll I_A$ (low density, $a \ll L$) yields a uniform beam of radius a. This solution was found before by Mjolsness (5) and nonrelativistically by Longmire (6). Then n, γ, and $\langle v_z \rangle$ are constant to radius a and

$$B_\theta = \frac{2\pi v_z ne}{c} r$$

As we mentioned in the Introduction, this leads to a maximum current

$$I \simeq \frac{m\gamma v_\perp^2 c^2}{4ev_z}.$$

So, an arbitrarily large current may be carried by low-velocity beams. (The energy content of such a beam is, of course, quite low, so it is not of great interest.)

Hammer and Rostoker treat particle orbits for their equilibrium in some detail. They show that their hollowed-out model has a large fraction of its energy deposited in particle motion, rather than in the electric and magnetic fields, and that it is much more efficient in this regard than the uniform beam. Also, enclosure of their model beam in a conducting cylinder inhibits the pinching instability because energy must be added to the fields if the beam contracts; this tendency becomes stronger, the higher the current.

Returning to the general solution of eqs. 1-54 and 1-55, we note that the coefficients D_1, D_2 can take on nonzero values. The analysis just carried out applies as well to this more general model, with considerably more complicated algebra. The only important new physics involved is the necessity of imposing a boundary current and line charge in some cases, similar to Section I-A. This arises because B_θ and E_r diverge at the axis as $1/r$; the boundary current and charge are necessary to push the beam out from the axis or pull it in closer than the natural equilibrium found for $D_1 = D_2 = 0$.

The general solution might conceivably be of interest in describing a particular experimental setup. For purposes of illustration, we restrict ourselves to discussion of a simple example.

Let $C_1 = 0 = C_2$ and $D_1 = 1 = D_2 K_0(b\sqrt{1-f}/L)$, where $K_0(b/L) = 1$. If now we take $f \to 1$, we find

$$\phi(r) = 0$$

$$A_z(r) = \frac{mcV_z\gamma_0}{e}\left[K_0\left(\frac{a}{L}\right) - 1 - \left(\frac{a}{L}\right)K_1\left(\frac{a}{L}\right)\ln\frac{r}{a}\right], \qquad r < a$$

$$= \frac{mcV_z\gamma_0}{e}\left[K_0\left(\frac{r}{L}\right) - 1\right] \qquad r \geq a$$

where as before we write $E_0 = mc^2\gamma_0$, $\bar{P}_z = mV_z\gamma_0$. The first expression for A_z is found by smoothly connecting the vacuum solution to that for $r \geq a$, and a is defined by equating the left-and right-hand sides of eq. 1-48:

$$\gamma_0{}^2\beta_z{}^2K_0{}^2(a/L) + 1 = \gamma_0{}^2 \tag{1-71}$$

Assume $b > a$. Then A_z vanishes at $r = b$; hence $v_z = V_z$ there. Again writing $\beta_z = V_z/c$, we have $\gamma_0{}^2 > (1 - \beta_z{}^2)^{-1}$; this is so because $K_0(a/L) > 1$ (K_0 decreases monotonically).

The beam has associated with it a magnetic field

$$B_\theta = \frac{mcV_z\gamma_0}{eL}K_1\left(\frac{a}{L}\right)\frac{a}{r}, \qquad r < a$$

$$= \frac{mcV_z\gamma_0}{eL}K_1\left(\frac{r}{L}\right), \qquad r \geq a \tag{1-72}$$

Of course, $E_r = 0$. Also, for $r < a$, $n(r) = 0 = v_z(r)$. For $r \geq a$, $n(r) = n_0$, $\gamma(r) = \gamma_0$, and

$$v_z = V_z K_0(r/L) \tag{1-73}$$

(see Figure 5). Thus $\beta_z K_0(a/L) < 1$ should hold; from eq. 1-71 we see that it does. The current density is $j(r) = -n_0 e v_z(r \geq a)$; integration yields a total current

$$I = -\frac{mc^3}{2e}\beta_z\gamma_0\frac{a}{L}K_1\left(\frac{a}{L}\right)$$

The total number of electrons is infinite, because $n(r)$ is constant for $r \geq a$. However, we can cut off the beam at some $r = r_{\max} \gg a$ without affecting its properties much, since at large r by eq 1-73 most of the electron motion is in the transverse direction. Then

$$N \simeq n_0 \pi r_{\max}^2$$

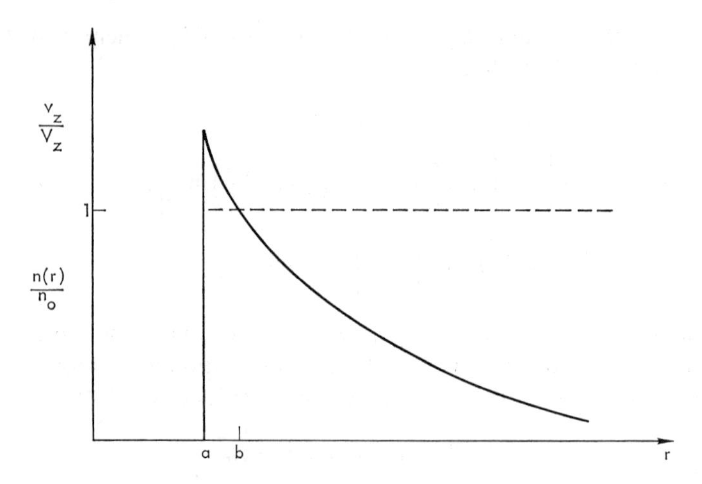

Fig. 5. Plot of streaming velocity (solid curve) and density (broken curve) for $f = 1$ showing the effect of the repulsive hard-core current on the axis in the double-delta-function beam model.

Since the electrons with $r \gg a$ contribute little to the total current, the correct formula for the Alfvén-Lawson current in this model is

$$I_A = \frac{mc^3}{e} \bar{\beta}_z \gamma_0$$

where $\bar{\beta}_z < \beta_z$. Consequently, the ratio of I to I_A is

$$I/I_A = (\bar{\beta}_z/\beta_z)(a/L)K_1(a/L)$$

This can again exceed one.

Returning to eq. 1-72, we note that B_θ diverges for small r. To make this consistent, we must postulate a current I_0 on the axis, with

$$I_0 = \frac{mc^3}{2e} \beta_z \gamma_0 \left(\frac{a}{L}\right) K_1 \left(\frac{a}{L}\right)$$

This is equal to the total beam current but oppositely directed. Thus I_0 exerts a repulsive force that constrains the beam electrons to keep away from the axis by at least a distance a.

The equilibria of this section and the generalized Bennett model of Section I-A are probably the most extensively investigated theoretical models thus far advanced. The Hammer-Rostoker model has the advantage of being monoenergetic and confined to a finite radius; it is particularly appropriate to problems involving experiments in which the beam nearly fills its cylindrical channel, and to time scales such that an originally mono-energetic beam has not had time to randomize the particle energy.

2. Semi-relativistic Uniform Beam

We can repeat the derivation of the preceding model in a clearer form by making some simplifying assumptions (25). The resulting solution corresponds to the low ν/γ limit of the exact solution. It is convenient to begin with the equations of motion for a single electron:

$$\frac{d}{dt}\left(m\gamma\frac{dx}{dt}\right) = -e\frac{\partial U}{\partial x}$$

$$\frac{d}{dt}\left(m\gamma\frac{dy}{dt}\right) = -e\frac{\partial U}{\partial y} \tag{1-74}$$

$$\frac{d}{dt}\left(m\gamma\frac{dz}{dt}\right) = \frac{e}{c}\frac{d}{dt}A_z$$

where $U = -e(\phi - \beta_z A_z)$. We assume $\gamma = \gamma_0$ for all electrons, so that $d\gamma/dt = 0$, and combine the first two equations of the set 1-74 to get

$$W = \tfrac{1}{2}m\gamma_0 {v_\perp}^2 + U = \text{const} \tag{1-75}$$

where ${v_\perp}^2 = {v_x}^2 + {v_y}^2$. The third equation on integration yields

$$P_z = m\gamma_0\frac{dz}{dt} - \frac{e}{c}A_z = \text{const}$$

We take $m\gamma_0\, dz/dt \gg e/c\, A_z$, so that v_z is an approximate constant of motion. Then we can choose a distribution function f_0 in the form

$$f_0 = \overline{K}\delta(W - W_0)\delta(v_z - V_z) \tag{1-76}$$

Here f_0 is normalized by $\int d^3v f_0 = n(r)$. The absence of a cutoff at $|\mathbf{v}| = c$ appears as an additional approximation.

Taking moments, we find that $v_z = V_z$ and $n(r) = 2\pi\overline{K}/mr_0$ out to some limit $r = a$. The beam radius a is determined by setting $v_\perp = 0$ in eq. 1-75; if $r > a$, there is no value of $\mathbf{v}_\perp$ for which the argument of the first delta function in eq. 1-76 vanishes, so $n(r) = 0$.

The equations for ϕ and A_z may be combined by means of the technique that led to the introduction of ψ in Section I-A. Thus, for $r < a$, we have

$$\frac{1}{r}\frac{d}{dr}\left(r\frac{d\phi}{dr}\right) = 4\pi ne(1 - f) = \frac{8\pi^2\overline{K}e(1 - f)}{m\gamma_0}$$

and

$$\frac{1}{r}\frac{d}{dr}\left(r\frac{dA_z}{dr}\right) = -\frac{4\pi j}{c} = \frac{8\pi^2 \bar{K} e\beta_z}{m\gamma_0}$$

These combine to give

$$\frac{1}{r}\frac{d}{dr}\left(r\frac{dU}{dr}\right) = \frac{8\pi^2 \bar{K} e^2}{m\gamma_0}[f - 1 + \beta_z^2]$$

for $r \leq a$, while for $r > a$

$$\frac{1}{r}\frac{d}{dr}\left(r\frac{dU}{dr}\right) = 0$$

The solution is

$$\begin{aligned} U &= W_0(r^2/a^2), \qquad r \leq a \\ &= W_0[1 + 2\ln(r/a)], \qquad r > a \end{aligned}$$

where a is now determined:

$$a^{-2} = \frac{2\pi^2 e^2 \bar{K}}{m\gamma_0 W_0}[f - 1 + \beta_z^2]$$

It follows that

$$\begin{aligned} B_\theta &= -\frac{4\pi^2 \bar{K} e\beta_z}{m\gamma_0} r, \qquad r \leq a \\ &= -\frac{4\pi^2 \bar{K} e\beta_z}{m\gamma_0}\frac{a^2}{r} \qquad r > a \end{aligned}$$

and $E_r = B_\theta(1 - f)/\beta_z$.

The approximation $m\gamma_0 V_z \gg e/c\, A_z$ implies

$$\gamma_0(f - 1 + \beta_z^2) \gg W_0/mc^2 \tag{1-77}$$

The approximation $W = \text{const}$ implies

$$\gamma_0 v_\perp^2/c^2 \ll 2$$

or

$$\gamma_0 \gg W_0/mc^2 \tag{1-78}$$

Clearly, eq. 1-77 is a more exacting condition that eq. 1-78.

Thus the electrons are moving almost entirely in the z-direction. Their orbits are only slightly perturbed by the self-field. Consequently, it is not surprising that the ratio of the beam current I to the Alfvén current,

$$\frac{I}{I_A} = -\frac{W_0 V_z}{e(f - 1 + \beta_z^2)} \Big/ \frac{mc^3 \beta_z \gamma_0}{e} = \frac{-W_0}{mc^2 \gamma_0}$$

is small, by eq. 1-78. All of the formulas that describe this model follow from the results of the previous section in the limit $a/L \ll 1$. The model has the virtues of simplicity and (since the orbits are sinusoidal) is easy to use in stability calculations.

II. Azimuthal Current Models

The Astron is an experiment fusion device, the subject of intensive study at the Lawrence Radiation Laboratory, Livermore, California. Its characteristic feature is a rotating cylinder of relativistic electrons confined in a mirror field. It differs from other confinement systems in that internal currents contain the heated plasma. The solenoid of relativistic electrons, gyrating in an external B_z, reverses the external field and produces closed magnetic field lines over a small volume. In this region plasma confinement may be possible, since a hydromagnetically stable, minimum-average-B configuration results. The trapped plasma is also heated by the relativistic electrons through collisions and perhaps the two-stream instability. Historically, equilibrium calculations have focused on the relativistic electrons and the depth of the magnetic wells they may provide. Models truly representative of the experiments apparently demand numerical solution, beyond some point. We shall mostly treat analytic models, with a brief indication, at the end, of the direction that recent numerical work has taken.

The current is now in the θ direction, and the appropriate component of Maxwell's equation is

$$\frac{d}{dr}\left[\frac{1}{r}\frac{d}{dr}(rA_\theta)\right] = -\frac{4\pi}{c} j_\theta(r) \tag{2-1}$$

It is natural to attempt to generalize the model discussed under Section I to azimuthal currents by taking

$$f_0 = \overline{K} \exp\left[-\frac{a}{c} H + \beta P_\theta\right]$$

The effect of this is to replace α by βr everywhere. The restriction $a > |\alpha|$ becomes $a > |\beta| r$, which of course fails for sufficiently large r. It is not permissible to legislate that f_0 vanish for r greater than some critical value, since f_0 must not be an explicit function of r, which is not a constant of motion. There appears to be no way around this difficulty. The azimuthal form of Section I-A-2 fails for the same reason.

A. *Azimuthally Weighted Energy Shell*

We attempt to adapt the model of Section I-A-2-b to the case of azimuthal current. Instead of eq. 1-35 we write

$$f_0 = \overline{K}\delta(H - E_0)e^{\beta P\theta} \tag{2-2}$$

Here $f = 1$ and $\phi = 0$. The equation for $j_\theta(r)$ is identical with that of Section I-A-2-b after we make the substitutions

$$A_z \to A_\theta$$
$$\alpha \to \beta r$$

Thus

$$j_\theta(r) = -2\pi e\overline{K} \exp\left(\frac{-e}{c}\,\beta r A_\theta\right)$$
$$\times \left\{\frac{mc\gamma}{\beta r}\left[e^{\beta rmc\gamma} + e^{-\beta rmc\gamma}\right] - \frac{1}{(\beta r)^2}\left[e^{\beta rmc\gamma} - e^{-\beta rmc\gamma}\right]\right\}$$

When this is inserted in the right-hand side of eq. 2-1, the resulting differential equation cannot be solved, even in the limit $mc\beta r\gamma \gg 1$. However, in the opposite limit it becomes identical with eq. 1-3 under the transformations.

$$\Phi = \frac{e}{c}\,\beta r A_\theta$$

and

$$x = (r/\lambda)^2$$

where

$$\lambda^{-4} = \frac{4\pi^2}{3}\,\overline{K}\left(\frac{e\beta}{c}\right)^2\left(\frac{E_0}{c}\right)^3$$

Recourse to eq. I-4 once again supplies explicit expressions for vector potential, field, and current profile:

$$rA_\theta = -\frac{c}{\beta e}\ln\left\{\frac{\cosh^2\left[\eta\lambda^{-2}(r^2 - r_0{}^2)\right]}{2\eta^2}\right\}$$
$$B_z = -\frac{4c\eta}{\lambda^2\beta e}\tanh\left[\eta\left(\frac{r^2 - r_0{}^2}{\lambda^2}\right)\right]$$
$$j_\theta(r) = \frac{8\pi}{3}\,\eta^2\overline{K}e\left(\frac{E_0}{c}\right)^3\beta r\,\mathrm{sech}^2\left[\eta\,\frac{r^2 - r_0{}^2}{\lambda^2}\right]$$

Since $\overline{K}$ and η appear in measurable quantities only in the product $\overline{K}\eta^2$, it is clear that without loss of generality we can set $\eta = 1$. The solution breaks down for $r \geq (mc\beta\gamma)^{-1}$; there the exact form of j_θ must be used. But if $\lambda \ll r_0 \ll (mc\beta\gamma)^{-1}$, $j_\theta(r)$ is negligibly small for such values of r and the error may be disregarded.

B. Heckrotte-Neil Model

An alternate approach (27) is to choose a desired form for j_θ and invert the calculation to find the distribution function. Here P_θ and P are constants of the motion, so that any function of these is a legitimate f_0. Heckrotte and Neil chose

$$f_0(P, P_\theta) = \phi(P)\delta(P_\theta - P_0)$$

with $\phi(P)$ to be determined. They normalized so that $N(z)$, the total number of particles per unit length along z, is given by

$$N(z) = \int \phi(P)\, dP_z\, d\theta$$

They treated a thin E-layer of radius R for which the current is given by

$$j_\theta(z) = \frac{-e}{2\pi R} \int \frac{P_\theta}{m} \phi(P)\, dP_z\, d\theta$$

where m is the particle mass in the relativistic limit (approximated as P/c) and

$$P_\theta = \frac{P_0}{R} + \frac{e}{c}(A_0 + A_\theta)$$

is the kinetic momentum. Since A_0 is the vector potential of the uniform external field B_0, $A_0 = -B_0 R/2$. Changing variables,

$$j_\theta(z) = -\frac{2epc}{R} \int_{P_\theta}^{P_m} \frac{\phi(P)\, dP}{\sqrt{P^2 - P_\theta{}^2}} \tag{2-3}$$

where P_m is the maximum allowed value of P and P_θ is a function of z through its dependence on $A_\theta(z)$. All electrons at a particular point z have the same P_θ, regardless of their total momentum P. An electron oscillates back and forth along z with an amplitude determined by P. It will only reach the end of the E-layer at $z = \pm l/2$ if its $P = P_m$. We can use this to find a simple form for the angular canonical momentum, P_0.

Take an electron with $P = P_m$. At $z = \pm l/2$, $A = 0$ and

$$P_\theta\left(\pm \frac{l}{2}\right) = \frac{P_0}{R^2} + \frac{e}{c} RA_0 = P_m$$

Now the greatest momentum an electron can have in the field B_0 is

$$P_m = -\frac{eB_0 R}{c}$$

So we find

$$P_0 = -\frac{eB_0 R^2}{2c}$$

The angular momentum of any electron is given by

$$P_\theta = P_m + \frac{eA_\theta}{c}$$

Now let us pick some simple form for j_θ, say,

$$j_\theta = j_0 \cos kz$$

It is easy to show that

$$A_\theta = \frac{2\pi R}{c} j_\theta$$

and so

$$P_\theta = P_m\left(1 - \frac{2\pi j_0}{cB_0}\cos kz\right)$$

Now we are ready to return to the integral equation for $\phi(P)$, eq. 2-3. We find

$$\int_{P_\theta}^{P_m} \frac{\phi(P)\,dP}{\sqrt{P^2 - P_\theta^{\,2}}} = \frac{c}{4\pi e^2}\left(\frac{P_m - P_\theta}{P_\theta}\right)$$

This is a standard form of integral equation (26) which may be written as

$$G(P_\theta) = \int_1^{P_m/P_\theta} \frac{\phi(P_\theta v)\,dv}{\sqrt{v^2 - 1}}$$

The solution is

$$\phi(P) = -\frac{2P}{\pi}\int_1^{P_m/P} \frac{G'(Pw)\,dw}{\sqrt{w^2 - 1}}$$

Therefore

$$\phi(P) = \frac{c\sqrt{P_m^{\,2} - P^2}}{2\pi^2 e^2 P}$$

We include this treatment primarily to show that the task of working "backward" from j_θ to f_0 can be done for simple choices of j_θ. This program becomes much harder when the E-layer is taken with a finite thickness in r, however. The reader is referred to the paper of Heckrotte and Neil for a full discussion of the thin E-layer case, as well as some treatment of Astron instabilities.

C. *Uniform Cold Azimuthal Beam*

$$f_0 = \bar{K}\delta(H - E_0)\delta(P_\theta - \bar{P}_\theta) \tag{2-4}$$

This is the version of the model in I-B-1 appropriate to an azimuthal current. Spreads in E-layer energy or angular momentum can stabilize many instabilities that occur in the Astron. This equilibrium, which lacks such spreads, may not be appropriate to the detailed study of the E-layer. Both $n(r)$ and $j_\theta(r)$ vanish, except where

$$\left[\left(\frac{\bar{P}_\theta}{r} + \frac{e}{c}A_\theta\right)^2 + m^2c^2\right]^{1/2} c - e\phi < E_0$$

when this inequality is satisfied,

$$j_\theta(r) = -\frac{2\pi\bar{K}e}{r}\left(\frac{\bar{P}_\theta}{r} + \frac{e}{c}A_\theta\right)$$

and

$$n(r) = 2\pi\bar{K}\frac{1}{r}\frac{E_0 + e\phi}{c^2}$$

The equation for A_θ is

$$\frac{d}{dr}\left[\frac{1}{r}\frac{d}{dr}(rA_\theta)\right] = \frac{8\pi^2 e\bar{K}}{r^2 c}\bar{P}_\theta\left(1 + \frac{erA_\theta}{c\bar{P}_\theta}\right) \tag{2-5}$$

while ϕ satisfies

$$\frac{1}{r}\frac{d}{dr}\left(r\frac{d\phi}{dr}\right) = \frac{8\pi^2 e\bar{K}}{rc^2}E_0(1 - f)\left(1 + \frac{e\phi}{E_0}\right) \tag{2-6}$$

We can solve eq. 2-5 by letting $\rho = r/L$ and

$$\chi = \frac{erA_\theta}{c\bar{P}_\theta} + 1$$

where

$$\frac{1}{L} = \frac{8\pi^2 e^2\bar{K}}{c^2}$$

so that

$$\frac{d^2\chi}{dp^2} - \frac{1}{\rho}\frac{d\chi}{dp} - \frac{1}{\rho}\chi = 0 \tag{2-7}$$

Equation 2-7 can be related to that for modified Bessel functions, with

$$\chi = \rho[C_1 I_2(2\rho^{1/2}) + C_2 K_2(2\rho^{1/2})]$$

where C_1 and C_2 are arbitrary constants. Similarly, by writing $\chi = (e\phi/E_0) + 1$, $\rho = r/L'$, where $L' = L(1 - f)^{-1/2}$, we find

$$\frac{d^2\chi}{d\rho^2} + \frac{1}{\rho}\frac{d\chi}{d\rho} - \frac{1}{\rho}\chi = 0$$

the solution of which is

$$\chi = D_1 I_0(2\rho^{1/2}) + D_2 K_0(2\rho^{1/2})$$

A discussion similar to that of I-B-1 applies, with the result that we can obtain a solution that for $r < r_{\max}$ has the form

$$j_\theta(r) = -\frac{2\pi \overline{K} e \overline{P}_\theta}{r^2}\frac{r}{L} I_2\left[2\left(\frac{r}{L}\right)^{1/2}\right]$$

$$n(r) = \frac{2\pi \overline{K} E_0}{rc^2} I_0\left[2\left(\frac{r}{L'}\right)^{1/2}\right]$$

and zero otherwise, while

$$B_z(r) = \frac{1}{r}\frac{d}{dr}(rA_\theta) = \frac{c\overline{P}_\theta}{eL^2}\left(\frac{L}{r}\right)^{1/2} I_1\left[2\left(\frac{r}{L}\right)\right]^{1/2}, \qquad r \leq r_{\max}$$

$$= \frac{c\overline{P}_\theta}{eL^2}\left(\frac{L}{r_{\max}}\right)^{1/2} I_1\left[2\left(\frac{r_{\max}}{L}\right)^{1/2}\right], \qquad r > r_{\max}$$

Here, $r_{\max}$ is determined by

$$\left\{\left[\frac{\overline{P}_\theta}{L} I_2[2(r_{\max}/L)^{1/2}]\right]^2 + m^2c^2\right\}^{1/2} c = E_0 I_0\left[\frac{2r_{\max}^{1/2}}{L'^{1/2}}\right]$$

which always has a solution, provided only that $E_0 > mc^2$.

If we allow $C_2 \neq 0$, it is possible to construct a rotating shell, where $j_\theta(r)$ vanishes except for $r_{\min} < r < r_{\max}$. If $(r_{\max} - r_{\min})/L \ll 1$, then $n(r)$ and $j_\theta(r)$ are constant to a good approximation.

This model was considered by Marx (23) who derived it starting from a nonrelativistic Hamiltonian (taking $f = 1$):

$$H = \frac{p^2}{2m}$$

This modification does not produce any change in the field equation.

For applications to Astron stability, it is useful to look at solutions with z-dependence as well. The field equation is now eq. 2-5 with an additional term:

$$\frac{\partial}{\partial r}\left[\frac{1}{r}\frac{\partial}{\partial r}(rA_\theta)\right]+\frac{\partial^2 A}{\partial z^2}=\frac{8\pi^2 e\bar{K}}{r^2 c}\bar{P}_\theta\left(1+\frac{erA_\theta}{c\bar{P}_\theta}\right)$$

or

$$\frac{\partial^2\Phi}{\partial\rho^2}-\frac{1}{\rho}\cdot\frac{\partial\Phi}{\partial\rho}+\frac{\partial^2\Phi}{\partial\zeta^2}=\frac{1}{\rho}\Phi$$

where $\zeta = z/L$ and $\Phi = (e/c)\,rA_\theta + \bar{P}_\theta$. Writing $\Phi = \psi(\rho)\cos\kappa\zeta$, we find

$$\frac{\partial^2\psi}{\partial\rho^2}-\frac{1}{\rho}\frac{\partial\psi}{\partial\rho}-\left(\kappa^2+\frac{1}{\rho}\right)\psi=0$$

The solution can be written in the form

$$\psi = e^{-\kappa\rho}F\left(\frac{1-\kappa}{2\kappa}, -1; 2\kappa\rho\right)$$

where $F(a, c; \xi)$ is any solution of Kummer's equation,

$$\xi F'(\xi)+(c-\xi)F'(\xi)-aF(\xi)=0$$

We take as our basic solutions ϕ_1 and ϕ_2, which are defined in standard notation (29) by

$$e^{\kappa\rho}\phi_1(\kappa\rho)=y_2=\xi^2\,\Phi\left(\frac{1-\kappa}{2\kappa},3;\,2\kappa\rho\right)$$

$$e^{\kappa\rho}\phi_2(\kappa\rho)=y_5=\psi\left(\frac{1-\kappa}{2\kappa},-1,\,2\kappa\rho\right)$$

The region in which $j_\theta \neq 0$ is defined by

$$\left(\frac{\psi}{r}\right)^2=\left(\frac{E_0}{c}\right)^2-m^2c^2$$

For fixed z, j_θ is localized between a minimum and maximum radius. This zone looks as shown in Figure 6*a*.

$$\begin{aligned}j_\theta(r,z)&=\frac{1}{Lr^2}\,\psi(r,z)\\&=\frac{e^{\kappa\rho}}{L^3}\frac{1}{\rho}\left[C_1\phi_1(\kappa\rho)+C_2\,\phi_2(\kappa\rho)\right]\cdot\cos\kappa z\end{aligned}$$

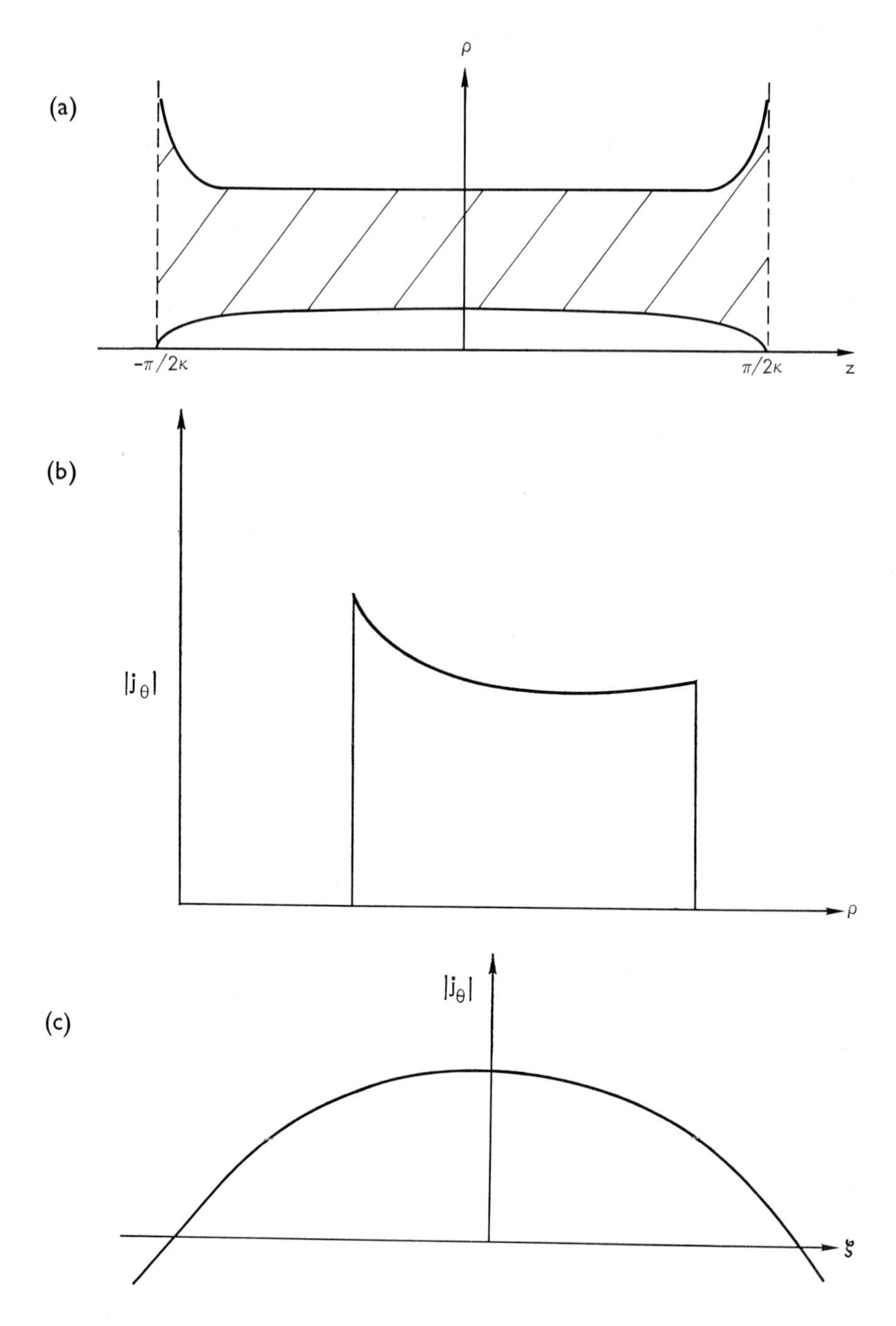

Fig. 6. (*a*) Region inside which *E*-layer is localized in azimuthal form of double-delta function model with z dependence; (*b*) radial dependence of current profile at $z = 0$. (*c*) current vs z for typical value of r.

Since $\psi =$ const on this perimeter, j_θ is infinite at $\rho = 0$, $\zeta = \pm \pi/2\kappa$ and zero at $\rho = \infty$, $\zeta = \pm \pi/2\kappa$. As a function of ρ for fixed ζ, $j_\theta(r)$ looks like Figure 6*b*, while as a function of ζ, for fixed ρ, it looks like Figure 6*c*.

D. *Marx Model* (*Relativistic Rigid Rotor*)[23]

$$f_0 = \overline{K}(p^2 + m^2c^2)^{1/2} \exp\left[-\frac{p^2}{2\chi^2} + \frac{m\omega_0 P_\theta}{\chi^2}\right] \tag{2-8}$$

Complete neutralization is assumed. Except in the ultrarelativistic (or classical) limit, $n(r)$ cannot be found in closed form, but $j_\theta(r)$ is

$$j_\theta(r) = -\overline{K}ecm\omega_0 r(2\pi\chi^2)^{3/2} \exp\left[\frac{e}{c}\frac{m\omega_0 r}{\chi^2}A_\theta + \frac{1}{2}\frac{(m\omega_0 r)^2}{\chi^2}\right]$$

Writing

$$\Phi = -\frac{e}{c}A_\theta \frac{m\omega_0 r}{\chi^2} - \frac{1}{2}\left(\frac{m\omega_0 r}{\chi}\right)^2,$$

$$\chi = r^2\lambda$$

where

$$\lambda = \tfrac{1}{2}(\overline{K}ecm\omega_0)^{1/2}(2\pi\chi^2)^{3/4}$$

we obtain eq. I-3. The solution of eq. 2-1 is

$$rA_\theta = -\frac{c\chi^2}{em\omega_0}\left[\frac{\omega_0^2 m^2 r^2}{2\chi^2} - \ln 2 - B + 2\ln\cosh\left(e^{B/2}\frac{r^2 - r_0^2}{\delta^2}\right)\right]$$

where B and r_0 are arbitrary constants and δ is the geometric mean of the Debye length and the average gyroradius

$$\delta^{-4} = 8\pi\omega_0^2 e^2 m^2 \overline{K}(2\pi\chi^2)^{1/2}/c$$

Since $\overline{K}$ multiplies $e^{B/2}$ wherever they occur, we can take $B = 0$ without loss of generality. Then

$$j_\theta(r) = -2(2\pi\chi^2)^{3/2}\,\overline{K}e\omega_0 mcr\,\mathrm{sech}^2\,\frac{r^2 - r_0^2}{2\delta^2}$$

$$B_z = \frac{c\chi^2}{em\omega_0}\left(\frac{m^2\omega_0^2}{\chi^2} + \frac{2}{\delta_2}\tanh\frac{r^2 - r_0^2}{2\delta^2}\right)$$

The physical significance of the various parameters has been discussed by Marx, who applied the model to the question of the stability of the Astron machine. One difficulty of this model is that a finite number of electrons are counterrotating. This can be seen in eq. 2-8, where changing the sign of P_θ leads to a nonzero particle density. In the actual Astron experiment all particles are injected in the same rotational sense, so that their P_θ is always positive.

E. *Marder-Weitzner Numerical Model*

To obtain a realistic Astron model with both r and z dependence, it is necessary to undertake numerical calculations, since the nonlinear partial differential equation concerned is analytically intractable. A commonly used distribution has the "rigid rotor" form

$$f_0 = n_0 \left(\frac{m}{KT}\right)^{3/2} g_0\left(\frac{\frac{1}{2}mv^2 + \omega(mrv_0 + q\psi)}{KT}\right)$$

where $\psi(r, z) = rA_\theta(r, z)$; n_0, ω, and T are electron density on the axis, mean rotation frequency, and temperature. The precise form of g_0 is not crucial; any positive monotone function of ψ will give essentially the same behavior. For convenience we take g_0 to be a nonrelativistic Maxwellian and define a length

$$L = \left(\frac{KT}{n_0 \omega^2 q^2}\right)^{1/4}$$

Then with dimensionless coordinates

$$\rho = r/L, \qquad y = \sqrt{2}\,\frac{z}{L}$$

Maxwell's equation becomes (*cf.* eq. I-3 and the preceding section)

$$\frac{\partial^2 \Psi}{\partial x^2} + \frac{1}{x}\frac{\partial^2 \Psi}{\partial y^2} = e^{-\Psi}$$

where

$$\Psi = \left(\omega q \psi - \frac{m\omega^2 r^2}{2}\right) / KT$$

This equation must be solved numerically. We look for a solution in a rectangular box, with $\Psi = 0$ on three boundaries; on the fourth side ($y = 0$), we take $\partial\Psi/\partial y = 0$. This gives a half-section of an E-layer. These conditions do not impose a unique solution. Because of problems related to bifurcation, all solutions are not obtainable by standard numerical methods. Marder and Weitzner (30) have developed an iteration scheme that converges to all solutions. Some of their results are shown in Figure 7. Figure 7(a) shows a possible E-layer with well depth $\Psi = -0.38$. A much deeper solution, with the same homogeneous boundary data, is shown in Figure 7(b). These E-layers differ

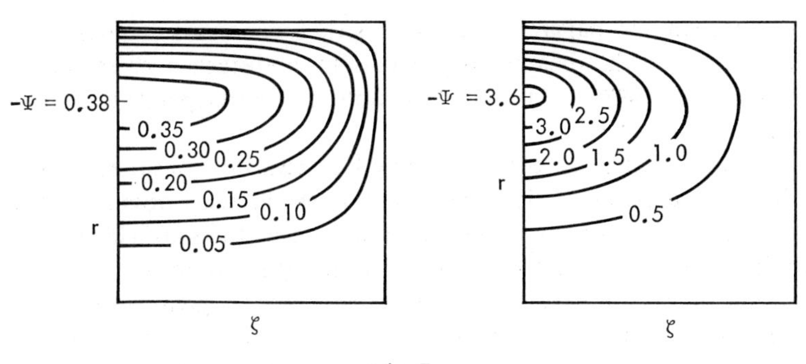

Fig. 7.

only in the initial guess made in the iteration technique. Since the density behaves like $e^{-\Psi}$, the deeper solution is almost 30 times as dense as the shallow one. The object of the Astron program is to produce just such a deep magnetic well in which a thermonuclear plasma may be trapped.

III. Conclusion

The usefulness of any study of relativistic beam equilibria is limited by the fact that steady states are difficult to achieve in the laboratory. Thus far, experiments last at most a few hundred nanoseconds and beams travel down chambers only occasionally longer than a few meters. This is much like firing a cannon into a phone booth, the transients are everything.

Nevertheless, a knowledge of the range of possible equilibria available to a beam with given injection parameters is useful. Papers on plasma instabilities nearly always begin with an assumption about the equilibrium of the plasma. Without this knowledge instability analysis is made much more difficult. The equilibrium determines the unperturbed particle orbits, and these orbits are used to integrate the Vlasov equation. (Other approaches to instability theory are used but are less rigorous.) Thus the conventional wisdom of this field demands an equilibrium for a starting point.

Even so, there are obvious defects in the equilibria we have presented here. We have made very naïve assumptions about the processes by which the background gas is ionized. We have neglected the presence of walls or conducting boundaries which must be present in many experiments. The anode phenomena have been legislated away by taking the beams to be very long, with no external electric fields present. The entire question of just how a relativistic beam propagates through a gas, ionizing as it goes, has been ignored; it may be that beams propagate in fits and starts over an appreciable range of gas pressures, so that the assumption of z-independent equilibria is unjustified.

However much these equilibria may lack realism, considerable stability work has been done in the field. There are, after all, many obvious sources of instability in rapidly streaming beams: pinching by the self-field, two-stream instability between the relativistic electrons and the cold ions, and transverse instabilities such as the firehose. These have yielded a great forest of papers that we cannot summarize here. The best introductory works are the papers by Budker (10), Finkelstein and Sturrock (12), and two papers by Bludman, Watson, and Rosenbluth (31). More recent work is contained in Ref. 32–41.

The models we have discussed exhaust the list of exactly soluble relativistic beam equilibria known to us. They clearly show that the Alfvén limit on current may be exceeded by hollowing out the density profile. By proper shaping, an arbitrarly large current can be made to flow. Also, a wide range of experimental limitations may be spanned by choosing the appropriate model; e.g., beams with fixed energy or linear momentum can be represented several ways. In any case, physical conditions of considerable complexity may be represented by these mathematical models, and it is hoped that the range of such models will grow as the study of relativistic beams goes forward.

References

1. W. H. Bennett, *Phys. Rev.*, **45**, 890 (1934).
2. H. Alfvén, *Phys. Rev.*, **55**, 425 (1939).
3. D. Hammer and N. Rostoker, *Phys. Fluids* **13**, 1831 (1970).
4. S. E. Graybill, J. Uglam, and S. V. Nablo, *Bull. Am. Phys. Soc.*, **13**, 56 (1968).
5. G. Yonas and P. Spence, *Proc. Tenth Symposium on Electron, Ion, and Laser Beam Technology*, L. Marton, ed., San Francisco Press, 1969.
6. M. L. Andrews, H. E. Davitian, H. H. Fleischmann, D. A. Hammer, and J. A. Nation, to be published.
7. J. Rander, B. Ecker, G. Yonas, and D. J. Drickey, *Phys. Rev. Letters*, **24**, 2321 (1966).
8. S. Putnam, *Phys. Rev. Letters*, **25**, 1129 (1970).
9. J. G. Linhart, *Proc. Roy. Soc.*, (London) **A249**, 318 (1959).
10. G. I. Budker, *Sov. J. Atomic Energy*, **5**, 673 (1956).
11. J. D. Lawson, *J. Electronics*, **3**, 587 (1957) and **5**, 146 (1958); *Plasma Physics*, **1**, 31 (1959).
12. D. Finkelstein and P. A. Sturrock, "Stability of Relativistic Self-Focusing Streams," Ch. 8, in *Plasma Physics*, J. E. Drummond, ed., McGraw-Hill, New York, 1961.
13. P. B. Spence, B. Ecker, and G. Yonas, unpublished work.
14. W. F. Ames, *Nonlinear Partial Differential Equations in Engineering*, Academic Press, New York, 1967.

15. G. Walker, *Proc. Roy. Soc.* **91**, 410 (1915).
16. D. Pfirsch, *Z. Naturforsch.* **17A**, 861 (1962).
17. O. Buneman, "The Bennett Pinch," Ch. 7, in *Plasma Physics*, J. E. Drummond, ed., McGraw-Hill, New York, 1961.
18. J. Benford, private communication.
19. H. P. Furth, *Plasma Physics*, AEA, Vienna, 1965, p. 411.
20. G. Benford, D. L. Book, and R. N. Sudan, *Phys. Fluids*, **13**, 2621 (1970).
21. P. W. Spence, B. Ecker, and G. Yonas, *Bull. Am. Phys. Soc.*, **14**, 1007 (1969). See also Physics International Report PFIR-106-2.
22. T. G. Roberts and W. H. Bennett, *Plasma Physics*, **10**, 381 (1968).
23. K. D. Marx, *Phys. Fluids*, **11**, 357 (1968).
24. R. N. Sudan, private communication.
25. R. C. Mjolsness, *Phys. Fluids*, **6**, 1730 (1963).
26. C. L. Longmire, *Elementary Plasma Physics*, Interscience, New York, 1963, p. 105.
27. W. Heckrotte and V. K. Neil, Report UCRL-6942, Lawrence Radiation Laboratory, Livermore, California, 1962.
28. Paul I. Richards, *Manual of Mathematical Physics*, Pergamon Press, 378 (1959).
29. A. Erdelyi, ed., *Higher Transcendental Functions*, Vol. 1, Bateman Manuscript Project, McGraw-Hill, New York, 1953, p. 253.
30. M. B. Marder and H. Weitzner, to be published.
31. K. M. Watson, S. A. Bludman and M. N. Rosenbluth, *Phys. Fluids*, **3**, 741–757 (1960).
32. M. N. Rosenbluth, *Phys. Fluids*, **3**, 932 (1960).
33. G. Benford, D. L. Book, N. C. Christofilos, T. K. Fowler, V. K. Neil, and L. D. Pearlstein, *Plasma Physics and Controlled Nuclear Fusion Research*, Novosibirsk, USSR, 1967, Paper CN24-F10.
34. S. V. Yadavalli, *Z. Physik*, **196**, 255 (1966).
35. S. Weinberg, *J. Math. Phys.*, **5**, 1371 (1964); also *J. Math. Phys.*, **8**, 614 (1967).
36. A. J. R. Prentice, *Plasma Physics*, **9**, 433 (1967).
37. A. M. Sessler, *Proc. Fifth International Conference on High Energy Accelerators*, CNEN, 1966.
38. V. V. Zheleznyakov and E. V. Surorov, *Sov. Phys.—JETP*, **27**, 335 (1968).
39. G. Dorman, *J. Plasma Physics*, **2**, 557 (1968); also *J. Appl. Phys.*, **37**, 2321 (1966).
40. P. C. W. Fung, *Plasma Physics*, **11**, 285 (1969); also *Can. J. Phys.*, **47**, 161 (1969).
41. A. S. Roberts, Jr., *Plasma Physics*, **8**, 53 (1966).

Plasma Collective Modes Involving Geometry and Velocity Spaces*

B. COPPI

Massachusetts Institute of Technology
Cambridge, Massachusetts

Introduction

A dominant characteristic of plasma dynamics is that of involving long-range forces of interaction between particles. As a consequence, the macroscopic geometry of a given plasma configuration plays an essential role in determining the types of collective effect that can be excited in it and are

* This work was supported in part by the U.S. Atomic Energy Commission, Contract AT(30-1)-3980.

connected with a microscopic, nonfluid description of the system. This is markedly different from the case of a gas of short-range interacting particles, where the geometry of the system does not affect, generally, the microscopic types of mode that it can sustain (1,2). So a plasma will behave in an inherently different way, depending on whether, for instance, it is two-dimensional or three-dimensional.

We define as microscopic modes for a classical system of many particles those that involve knowledge of the one-particle distribution function $f(\mathbf{r}, \mathbf{v}, t)$ and as macroscopic those described by the ordinary moments of f, i.e., density, flow velocity, and pressure. A plasma can spontaneously give rise to modes of both types when it is not in thermal equilibrium and there is a source of energy that can be released in it. Therefore it can be viewed as a system decaying from a given initial state that we shall call equilibrium (3). The effects of the various kinds of modes excited in this decay can be to destroy the initial plasma configuration (4); in this case they are properly called instabilities; to set it into a coherent oscillation (5); to set it into a turbulent state (6,7); or, more generally, to make it exhibit transport properties, e.g., particle diffusion, electrical resistivity (8,9), etc., that cannot be described by a theory considering only particle-particle interactions. This is usually referred to as *anomalous* plasma transport.

The relevant collective modes are generally described by nonlinear equations; a common process is to separate the overall problem into the following:

1. An equilibrium problem aimed at describing an initial state that is time-independent or slowly varying.
2. A linear (stability) problem in which the properties and the time evolution of small perturbations from the equilibrium are analyzed by properly linearizing the relevant equations.
3. A dynamic nonlinear problem in which the perturbations mentioned above, when found growing in time, are followed through the regime in which the linearized approximation is no longer valid. Thus it becomes part of this last problem to identify the final state of decay of the system, evaluating the oscillations that can be set up in it, the anomalous transport, etc.

In particular, any mode obtained from the linearized approximation, and having a time dependence of the form exp $[i\omega t]$ with Im $\omega < 0$, is commonly called an "instability." This is irrespective of what the final evolution of this mode will be and, in most cases, is quite different from that of a real instability which leads to a destruction of the initial configuration such as the

Rayleigh-Taylor instability of a heavy fluid supported by a light one under the effect of gravity (1,2,4).

The solution of the linearized problem is therefore of significant importance in order to have a physical insight and an approximate evaluation of the collective effects that can be excited in a given plasma configuration. This significance is further reinforced by the fact that in many important cases (5,10) the final state "remembers" the features of the modes described by the linear approximation and it is possible to interpret experiments on the basis of it.

I. Two-dimensional Initial State

In the past three years considerable progress has been made in the analysis and understanding of the collective effects (11,12) taking place in a two-dimensional plasma equilibrium configuration (Figure 1) in which the kinetic pressure nkT is much smaller than the magnetic pressure.

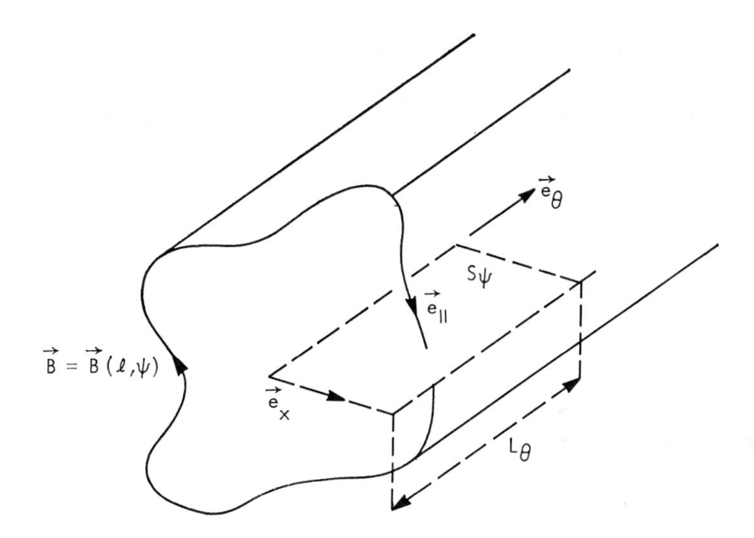

Fig. 1. Two-dimensional magnetic configuration.

Work in this direction has been stimulated by the flourishing of multipole (13) experiments (Figure 2) characterized by having current conductors, inside a low-pressure plasma, which generate a two-dimensional magnetic

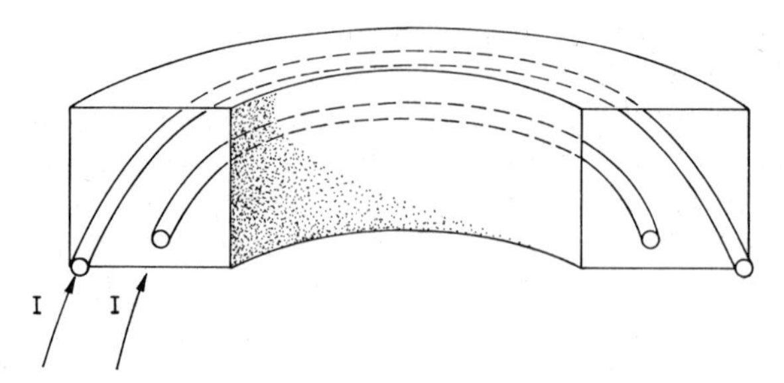

Fig. 2. Schematics of a quadrupole experiment.

configuration of the type indicated in Figure 1. These types of experiment, by having to deal with spatially varying magnetic field and magnetic curvature, showed the appearance of modes that could not be accounted for by the previously known theory. A second stimulus in this direction had come from the fact that a theoretical understanding of one-dimensional configurations (Figure 3) had essentially been reached and an interpretation of relevant experiments of plasmas in a constant magnetic field had become possible (5).

As we shall see, progress in the new direction will involve understanding of new collective effects that include consideration of the particle velocity space and of the macroscopic geometry of the system at the same time.

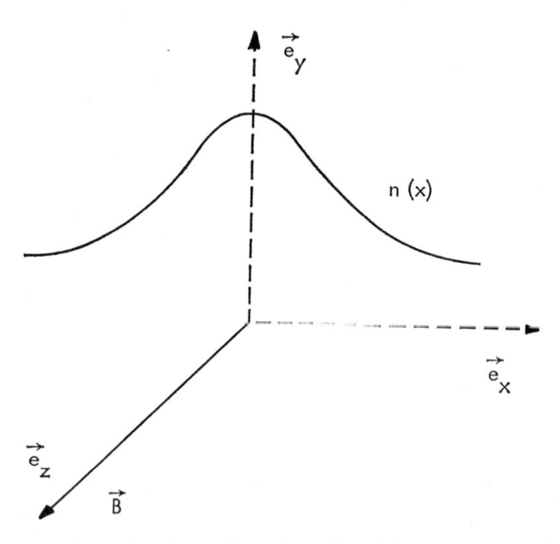

Fig. 3. One-dimensional configuration of an inhomogeneous plasma.

Mathematically, one will have to deal with integral equations that are linear in the eigenfunction (the electric potential), for the linearized approximation mentioned before, and highly nonlinear in the complex eigenvalue (the frequency).

Because of the $\beta \ll 1$ assumption, where $\beta = 8\pi nT/B^2$, n is the number of particles/cm^3, T is the plasma temperature in energy units (erg), and B is the magnetic field. The two-dimensional field $\mathbf{B}$ is of vacuum type, to lowest order in β; we can therefore introduce a magnetic potential χ, so that $\mathbf{B} = \nabla\chi$ and $dl = d\chi/B$, l being a measure of the distance along a given line of force. Therefore, referring to the directions

$$\mathbf{e}_\theta \text{ (of symmetry)}, \quad \mathbf{e}_\parallel \equiv \frac{\mathbf{B}}{B}, \quad \mathbf{e}_x \equiv \mathbf{e}_\theta \times \mathbf{e}_\parallel$$

we shall use the set of coordinates (14)

$$\mathbf{r}_\theta, l, r_x \quad \text{or} \quad r_\theta, \chi, \psi$$

depending on their convenience. Here ψL_θ measures the magnetic flux through a surface perpendicular to the lines of force (Figure 1) and having length L_θ along $\mathbf{e}_\theta$, so that $d\psi = B\, dr_x$.

The particle collision, mean free path is assumed to be longer than all macroscopic dimensions of the system, so that we can use the Vlasov equation,

$$\frac{\partial f}{\partial t} + \mathbf{v} \cdot \nabla f + \frac{e}{m}\left(\mathscr{E} + \frac{\mathbf{v}}{c} \times \mathbf{B}\right) \cdot \nabla_v f = 0 \tag{1-1}$$

where $\mathbf{v}$ is the velocity, e the charge, and m the mass for the considered particles. The initial equilibrium state is given by eq. 1-1 with $\partial/\partial t = 0$, so that f is a function of the invariants E and p_θ of the particle motion, and we shall choose

$$f_0 = \left(\frac{m}{\pi T}\right)^{3/2} N(p_\theta) \exp\left(-\frac{mE}{2T}\right) \tag{1-2}$$

a distribution with constant isotropic temperature T and whose spatial dependence is given through the function $N(p_\theta)$. Here $E \equiv v_\perp{}^2 + v_\parallel{}^2$ represents the particle energy and p_θ the $\mathbf{e}_\theta$ component of the particle canonical momentum, $mv_\theta + (e/c)A_\theta$, A being the magnetic vector potential and assuming (14) that $\mathscr{E}_0 = 0$.

In a configuration of the type indicated in Figure 1, all particle orbits are periodic. The particles whose motion is sampling the entire length of a given magnetic field line will be called "circulating," while the particles sampling only a portion of this length will be called "trapped." Due to the presence of a density gradient, electrons and ions have a diamagnetic flow velocity v_d in opposite directions along $\mathbf{e}_\theta$, given by

$$v_{dj} = -\frac{cT_j}{e_j B n_0}\frac{dn}{dr_x} = -\frac{cT_j}{e_j n_0}\frac{dn_0}{d\psi} \tag{1-3}$$

where $j = i$ or e indicates the species, and n_0 the equilibrium density. Clearly, v_{dj} is a constant along a given line of force. On the other hand, the magnetic field curvature gives rise to a guiding-center drift along $\mathbf{e}_\theta$, given by

$$v_{Dj} = c\frac{m_j}{e_j}\frac{E - \frac{1}{2}\mu B}{B^2}\frac{\partial B}{\partial r_x} = c\frac{m_j}{e_j B}\left(E - \frac{1}{2}\mu B\right)\frac{\partial B}{\partial\psi} \tag{1-4}$$

where $\mu = v_\perp{}^2/B$ is the magnetic moment. Since we shall consider processes whose characteristic frequency

$$\omega < \Omega_i$$

Ω_i being the ion gyrofrequency, μ will be conserved in all cases. In particular, the velocity space integral will be transformed in one on μ and E after integrating over the phase angle of the perpendicular velocity, so that

$$\int d^3\mathbf{v} = \frac{1}{2}\pi \iint dE\, d\mu \frac{B}{|v_\parallel|}$$

with a convention that contributions from positive and negative values of $v_\parallel = (E - \mu B)^{1/2}$ are to be added.

For order of magnitude evaluations we introduce the density gradient scale distance, r_n, the local radius of magnetic curvature, R_c, and the connection length, G, so that

$$\frac{1}{r_n} \approx \left|\frac{1}{n_0}\frac{dn_0}{dr_x}\right|, \qquad \frac{1}{R_c} \approx \left|\frac{1}{B}\frac{\partial B}{\partial r_x}\right|, \qquad \frac{1}{L} \approx \left|\frac{1}{B}\frac{\partial B}{\partial l}\right| \tag{1-5}$$

In particular, referring to Figure 1, L will represent the order of magnitude distance between a minimum and the successive maximum of the B field along the same line of force. Now, if $v_{\text{th}j} = \sqrt{2T_j/m_j}$ is the thermal velocity and $a_j = v_{\text{th}j}/\Omega_j$ the average gyroradius, we have for instance

$$v_{di} \approx \frac{1}{2}\frac{a_i v_{\text{th}i}}{r_n} \qquad \text{and} \qquad v_{Di} \approx a_i \frac{v_{\text{th}i}}{R_c}$$

so that $v_{di}/v_{Di} \approx R_c/2r_n$.

In order to represent a realistic experimental situation we assume from here on that

$$a_j < r_n < L \sim R_c \tag{1-6}$$

in the asymptotic sense, so that $v_{Di}/v_{di} \ll 1$, and in the following we shall always expand in this ratio.

We also consider the velocity space average of a quantity $X(l, \mathbf{v})$,

$$\langle X \rangle \equiv \frac{1}{n_0} \int d^3\mathbf{v} f_0 X \tag{1-7}$$

the space average over a tube of flux,

$$[X]_{\text{av}} \equiv \frac{\oint (dl/B)X}{\oint dl/B} \tag{1-8}$$

and the average over a particle orbit time,

$$X^{(0)}(E, \mu) \equiv \frac{\oint (dl/v_{\|})X}{\oint dl/v_{\|}} \tag{1-9}$$

In particular, if $Y = Y(l)$, independent of $\mathbf{v}$, then

$$Y^{(0)}(E, \mu) = Y^{(0)}(\lambda)$$

where $\lambda \equiv \mu/E$ represents the pitch angle. We also define

$$\tau(E, \mu) \equiv \oint \frac{dl}{v_{\|}} \qquad \text{and} \qquad \omega_b(E, \mu) \equiv \frac{2\pi}{\tau}$$

respectively, as the orbit period and the "bounce frequency." An evident order of magnitude is then

$$\langle \omega_{bj} \rangle \sim \frac{v_{\text{th}j}}{L}$$

In the following we shall make frequent interchange of the order of integration on E, μ, and l, showing for instance that

$$[\langle X \rangle]_{\text{av}} = \frac{\pi}{2 \oint dl/B} \iint dE\, d\mu\, |\tau(E, \mu)|\, X^{(0)}(E, \mu) \tag{1-10}$$

and

$$[\langle X^{(0)} \rangle]_{\text{av}} = [\langle X \rangle]_{\text{av}} \tag{1-11}$$

In addition we notice that, if $Y = Y(l)$,

$$[Y^2(l)]_{\text{av}} - \langle (Y^{(0)})^2 \rangle]_{\text{av}} > 0 \tag{1-12}$$

from the Schwartz inequality. In fact, if we decompose $Y(l)$ in harmonics of the orbit periodicity, we have

$$Y(l) = \sum Y^{(n)}(E, \mu) \exp\left(in2\pi \frac{\hat{t}}{\tau}\right) \tag{1-13}$$

where

$$\hat{t} \equiv \int^{l} \frac{dl'}{v_{\|}}$$

is a parametric variable replacing l and being related to it by the equation of motion, and

$$Y^{(n)}(E, \mu) = \frac{1}{\tau} \int_0^{\tau} Y[l(\hat{t})] \exp\left(-in2\pi \frac{\hat{t}}{\tau}\right) d\hat{t} \tag{1-14}$$

so that the left-hand side of eq. 1-12 becomes

$$\sum_n [\langle | Y^{(n)}(E, \mu)|^2 \rangle]_{\text{av}}$$

evidently positive definite, where $n \neq 0$.

Considering specifically the magnetic-curvature drift, we shall analyze cases in which the local-curvature drift is less than the one averaged over a line of force, i.e.,

$$\langle v_{Di} \rangle < [\langle v_{Di} \rangle]_{\text{av}}$$

This condition is relevant to realistic configurations, such as quadrupoles, having a "shallow magnetic well," in the sense that the average radius of magnetic curvature determining the stability of hydromagnetic modes is much larger than r_n. In fact, as will become apparent later, the sign of $v_{di} v_{Di}$ is important in determining the stability of a number of modes such as those resulting from the expansion of the plasma across the lines of force, when magnetic field and particle density decrease in the same direction, i.e., $v_{di} v_{Di} < 0$.

Finally, it is useful to recall that, in terms of the set of coordinates introduced earlier,

$$\bar{v}_{Di} \equiv [\langle v_{Di} \rangle]_{\text{av}} = 2c \frac{T_i}{e_i} \frac{(\partial/\partial\psi) \oint \partial\chi/B^2}{\oint d\chi/B^2} = \frac{2c}{e_i} T_i \frac{(\partial/\partial\psi) \oint dl/B}{\oint dl/B} \tag{1-15}$$

$$\begin{aligned} v_{Di}^{(0)} &\equiv \frac{\oint (dl/v_{\|}) v_{Di}}{\oint dl/v_{\|}} = \frac{cm_i}{e_i \oint dl/v_{\|}} \oint d\chi \frac{\partial}{\partial\psi} \left[\frac{(2E - 2\mu B)^{1/2}}{B}\right] \\ &= -\frac{\sqrt{2}\, cm_i}{e_i \oint dl/v_{\|}} \frac{\partial}{\partial\psi} \oint dl v_{\|} \equiv -\frac{cm_i}{e_i \oint dl/v_{\|}} \frac{\partial J(E, \mu)}{\partial\psi} \\ &= -\frac{cm_i E}{e_i} \frac{(\partial/\partial\psi) \oint dl(1 - \lambda B)^{1/2}}{\oint dl/(1 - \lambda B)^{1/2}} \end{aligned} \tag{1-16}$$

The former quantity, whose sign depends on the gradient, with respect to ψ, of $\oint dl/B$, determines the stability of hydromagnetic modes of the flute type (see the next section) and gives a measure of the so-called magnetic well depth (15). The latter quantity, which is due to the gradient of the action integral (12) $J(E, \mu) = \oint v_{\|} \, dl$, is different for different particles of given E and μ, and will play an important role in some of the kinetic type of mode we shall consider later.

II. Mode Classification

Having considered a $\beta \ll 1$ configuration, it is intuitively clear (and can be proven) that perturbations from the equilibrium are more likely to grow if they do not vary the magnetic field so that the magnetic energy of the system is not increased. Hence, since the equilibrium electric field is zero, we consider a small electrostatic perturbation such that $\mathscr{E} = -\nabla\Phi$. We shall also limit our attention to normal-mode solutions of the form

$$\Phi = \phi(l, r_x) \exp [i(\omega t + kr_\theta)] \tag{2-1}$$

Then a first-mode classification will be given by the magnitude of ω, so that we have:

1. Hydromagnetic modes, independent of the magnitude of L in the sense that, for the limit $L/r_n \to \infty$,

$$\langle \omega_{bi} \rangle < \langle \omega_{be} \rangle < \omega \tag{2-2}$$

2. Fast kinetic modes with

$$\langle \omega_{bi} \rangle < \omega < \langle \omega_{be} \rangle \tag{2-3}$$

These modes are adiabatically J-conserving for the electrons but not for the ions.

3. Slow kinetic modes with

$$\omega < \langle \omega_{bi} \rangle < \langle \omega_{be} \rangle \tag{2-4}$$

These modes are adiabatically J-conserving for both ions and electrons.

The driving mechanism for an unstable mode of these types is related to the energy—free energy—that can be released. So we have:

1. Driftlike modes, requiring only a density gradient and growing at the expense of the average particle kinetic energy along the magnetic field.
2. Curvature-driven modes if they require a density gradient and local magnetic curvature of proper sign, so that expansion energy is released when particle density and magnetic field decrease along the same direction.

A further distinction comes from the topology of the analyzed modes. We restrict ourselves to considering localized profiles in the $\mathbf{e}_x$ direction, around the point $r_x = r^0$ and over a scale distance less than r_n. The wavelength in the $\mathbf{e}_\theta$ direction is assumed to be sufficiently short so that $k \gg |d\Phi/dr_x|\Phi^{-1}$ and we can ignore from here on the dependence of Φ on r_x. Therefore, $\phi(l, r_x) \approx \phi(l, r^0) \equiv \phi(l)$.

Considering the profile along l and defining $\phi = \bar{\phi} + \tilde{\phi}$, where $\bar{\phi} \equiv [\phi]_{av}$, we then have

1. Wiggly (ballooning-type) modes, when $\bar{\phi} = 0$ or $\tilde{\phi} \gg \bar{\phi}$.
2. Flutelike modes, when $\bar{\phi} \gg \tilde{\phi}$.

Finally, depending on whether a small portion of velocity space is involved in the growth of a given mode, we find

1. Resonant modes, with small growth rates and depending on wave-particle resonances (Landau-damping-like).
2. Nonresonant (fluidlike) modes with growth rates larger or of the order of the oscillation frequency which involve the entire velocity space.

III. Linearized Vlasov Equation for Low-Frequency Modes

This equation is clearly

$$\frac{\partial f_\varepsilon}{\partial t} + \mathbf{v} \cdot \nabla f_\varepsilon + \frac{e}{m}(\mathbf{v} \times \mathbf{B}) \cdot \nabla_v f_\varepsilon = +\frac{e}{m} \nabla\Phi \cdot \nabla_v f_0 \tag{3-1}$$

If f_ε is the perturbed distribution function. In particular f_ε is obtained by integration along unperturbed particle orbits

$$f_\varepsilon = \frac{e}{m} \int^t \nabla\Phi \cdot \nabla_v f_0 \, dt' \tag{3-2}$$

In addition, if we consider the Larmor radius as much smaller than the transverse wavelengths of Φ, we can carry out the integral in eq. 3-2 along guiding-center orbits that are evidently given by

$$\frac{dl}{dt} = v_{\|}$$

and

$$\frac{dr_\theta}{dt} = v_D$$

as defined by eq. 1-4. So we have, omitting the index j,

$$\nabla_v f_0 = 2\mathbf{v}\frac{\partial f_0}{\partial E} + \mathbf{e}_\theta m \frac{\partial f_0}{\partial p_\theta}$$

and, recalling that $\mathbf{v} \cdot \nabla\Phi = d\Phi/dt - \partial\Phi/\partial t$,

$$\hat{f} = 2e\left\{\frac{\partial f_0}{\partial(mE)}\Phi - i\left[\omega\frac{\partial f_0}{\partial(mE)} - k\frac{\partial f_0}{\partial p_\theta}\right]\int^t \Phi(\mathbf{r}(t'))\,dt'\right\}$$

where $\hat{f}$ stands for the perturbed guiding-center distribution. Then, making use of the form of Φ and of f_0, we obtain (16,17):

$$\hat{f}_j = -\frac{2e_j}{T_j}f_{0j}\left\{\Phi - i(\omega - \omega d_j)\int_{-\infty}^{t} dt'\phi(l(t'))\exp\left[i\omega t' + ik\int^{t'} v_{Dj}(l(\hat{t}))\,d\hat{t}\right]\right\} \tag{3-3}$$

where $\omega d_j = kv_{dj}$, v_{dj} being the diamagnetic velocity defined by eq. 1-3. Equation 3-3 can easily be extended to the case in which the ion Larmor radius is not negligible in comparison with the transverse wavelength and, as we consider frequencies $\omega < \Omega_j$, only the average of the distribution function over the gyration angle $f_1 = (\frac{1}{2}\pi)\int_0^{2\pi} f_\varepsilon\, d\varphi$ is of interest. Then we obtain (18)

$$f_{1j} = -\frac{2e_j}{T_j}f_{0j}\left\{\Phi - i(\omega - \omega d_j)J_0\left(\frac{kv_\perp}{\Omega_j}\right)\int^t dt'\phi(l')J_0\left(\frac{kv_\perp}{\Omega_j'}\right)\right. \tag{3-4}$$
$$\left.\cdot\exp\left[i\omega t' + ik\int^{t'} d\hat{t}\, v_{Di}(\hat{l})\right]\right\},$$

where $l' \equiv l(t')$, $\Omega' = \Omega_j'(l')$ and $\hat{l} = l(\hat{t})$, by recalling that, within the considered limit, the particle trajectory perpendicular to the field can be written as $\mathbf{r}_\perp(t) = \Omega\left(\frac{\mathbf{B}}{B} \times \mathbf{V}\right) + \int \mathbf{V}_D\, dt''$, and by taking the φ average of $\exp[i\mathbf{k} \cdot \mathbf{r}_\perp(t') - i\mathbf{k} \cdot \mathbf{r}(t)]$. The average electric field felt by a particle is in fact

$$\Phi_1 = \frac{1}{2\pi}\int_0^{2\pi} d\varphi\, \Phi(\mathbf{r}_g + \mathbf{a}(\varphi)) = \frac{1}{2\pi}\int_0^{2\pi} d\varphi\, \phi(l)\exp\left(i\frac{kv_\perp}{\Omega}\cos\varphi\right)$$
$$= \phi(l)J_0\left(\frac{kv_\perp}{\Omega}\right)$$

The same factor J_0 transforms the part of the guiding-center distribution that is not due to change in the particle energy, such as the first term inside brackets of eq. 3-4, into a particle distribution function. When $f_0 = N(p_\theta) F(E, \mu)$ we have (19)

$$f_1 = \frac{2e}{m}\left\{\Phi\left[\frac{\partial f_0}{\partial E} + \frac{1}{B}\frac{\partial f_0}{\partial \mu}(1 - J_0^2)\right] - i\left(\omega\frac{\partial f_0}{\partial E} - \frac{kc}{2e}\frac{\partial f_0}{\partial \psi}\right)\right. \tag{3-5}$$

$$\left.\cdot J_0(l)\int^t dt' \exp\left[i\omega t' + ik\int^{t'} v_D(l'')\, dt''\right]\phi(l')J_0(l')\right\}$$

The justification of the term in $\partial f_0/\partial \mu$ comes from the fact that, to the first order in Φ, the correct adiabatic invariant is:

$$\mu_\varepsilon = \frac{v_\perp{}^2}{B} + \frac{2e}{Bm}(1 - J_0{}^2)\Phi \tag{3-6}$$

A derivation of this, based on considering the expression for Φ_1, is given in Ref. 20.

The dispersion equation is obtained in our case by coupling the solution of eqs. 3-3 or 3-4 with Poisson's equation

$$-\nabla^2\phi = 4\pi e(n_{i1} - n_{e1}) \tag{3-7}$$

n_{j1} being the perturbed density. If we assume, as we do from here on, that $k \gg |\partial\phi/\partial l|/\phi$, we have

$$n_{ik} = n_{ek} + \frac{k^2}{4\pi e}\phi$$

and we shall neglect the last term whenever finite Debye-length effects are not important.

IV. Fast (Electron) Modes

We consider at first modes (11) with $\langle\omega_{bi}\rangle < \omega < \langle\omega_{be}\rangle$ and completely neglect the electron Larmor radius, so that f_{e1} is given by eq. 3-3. Thus, to solve for the integral in t', we write $v_{De}(\hat{l}) = v_{De}^{(0)} + \tilde{v}_{De}$, recalling the definition eq. 1-16, and expand

$$\phi(l')\exp\left[ik\int^{t'} d\hat{t}\, \tilde{v}_{De}(\hat{l})\right] = \sum_n A_e^{(n)}(E, \mu)\exp\left[i(n\omega_b t' + kr_\theta)\right] \tag{4-1}$$

as indicated in eq. 1-13. Then we have, for $f_{ek} = f_{e1}\exp[-i(\omega t + ikr_\theta)]$,

$$f_{ek} = \frac{e}{T_e} f_{0e}\left\{\phi - (\omega - \omega d_e)\left[\frac{A_e^{(0)}(E, \mu)}{\omega + \omega_{De}^{(0)}} + \sum_{n\neq 0}\frac{A_e^{(n)}\exp(in\omega_b \hat{t})}{\omega + \omega_{De}^{(0)} + n\omega_b}\right]\right\} \tag{4-2}$$

where $\omega_{De}^{(0)} = kv_{De}^{(0)}$ and $\omega_b \hat{t} = 2\pi(\int^l dl'/v_\parallel)/(\oint dl/v_\parallel)$. Clearly, the last term inside the square brackets is of higher order in $\omega/\langle\omega_{be}\rangle$ with respect to the first one if $A^{(n)} \lesssim A^{(0)}$. In addition, if we assume $\omega_{De} \lesssim \omega$, as we shall do from here on, we have $A^{(0)} \approx \Phi^{(0)} = [\oint dl\phi(l)/v_\parallel]/\oint dl/v_\parallel$ to lowest order in $\omega/\langle\omega_{be}\rangle$. The expansion of ϕ in harmonics of the orbit periodicity is, in fact, indicated by $\phi(l) = \sum \Phi^{(n)}(E, \mu) \exp[in\omega_b \hat{t}]$. The corresponding perturbed density is

$$n_{ek} = \frac{e}{T_e} n_{e0}\left\{\phi - (\omega - \omega_{de})\left\langle \sum_n \frac{A^{(n)} \exp(in\omega_b \hat{t})}{\omega + \omega_{De}^{(0)} + n\omega_b}\right\rangle\right\} \tag{4-3}$$

Now we consider the ion species and if we define

$$f_{ik} = f_{i1} \exp[-i(\omega t + kr_\theta)]$$

and recall that $\omega > \langle\omega_{bi}\rangle$, the integration over t' can be performed by writing

$$f_{ik} = -\frac{e}{T_e} f_{0i} \tag{4-4}$$

$$\left\{\phi - \frac{\omega - \omega_{di}}{\omega + \omega_{Di}} J_0\left(\frac{kv_\perp}{\Omega_i}\right) \int_{-\infty}^{0} dt_*\left(\frac{\partial}{\partial t_*} \exp[i(\omega + \omega_{Di})t_*]\right) A(l(\hat{t} + t_*))\right\}$$

where

$$A(l_*) \equiv \phi(l_*) J_0(l_*) \exp\left[ik \int_0^{t_*} \{v_{Di}[l(t'_*)] - v_{Di}(l)\}\, dt'_*\right]$$

$$l_* \equiv l(\hat{t} + t_*) \qquad \text{and} \qquad \hat{t} \equiv \int^l dl/B$$

So, if we perform two integrations by parts, the integral in eq. 4-4 becomes

$$A(l) + \frac{i}{\omega + \omega_{Di}} v_\parallel \frac{\partial A}{\partial l} - \left(\frac{1}{\omega + \omega_{Di}}\right)^2 \frac{\partial^2 A}{\partial \hat{t}^2} \tag{4-5}$$

Now, if we evaluate the perturbed ion density n_{ik}, the term odd in $v_\parallel$ appearing in eq. 4-5 does not contribute, so that finally

$$n_{ik} = -\frac{e}{T_i} n_{i0}\left\{\phi(l) - \left\langle\frac{\omega - \omega_{di}}{\omega + \omega_{Di}} J_0\left[J_0\,\phi(l) - \left(\frac{1}{\omega + \omega_{Di}}\right)^2 \frac{\partial^2 A}{\partial \hat{t}^2}\right]\right\rangle\right\} \tag{4-6}$$

Notice that, in deriving both n_{ik} and n_{ek}, we have kept higher order terms in the respective expansions, in view of the fact that the final dispersion relation for the different modes will lead to a cancellation of the lowest order terms. Then we assume from here on that

$$\omega_{De} \sim \omega_{Di} < \omega$$

consistently with the fact that often $\omega \sim \omega_{de} \sim \omega_{di}$ and $r_n \ll R_c$ and that, consequently, particle-wave (Landau) resonances involving $\omega = \omega_{Dj}$ would affect only very few particles in the tail of the distribution. Therefore, we can write the final dispersion relation as

$$\begin{aligned}\phi(l)\left[1-\left\langle\frac{\omega-\omega_{di}}{\omega+\omega_{Di}}J_0{}^2\left(\frac{kv_\perp}{\Omega_i}\right)\right\rangle\right]\frac{T_e}{T_i} \\ +\frac{T_e}{T_i}\left(1-\frac{\omega_{di}}{\omega}\right)\left\langle\frac{1}{\omega^2}J_0(l)\frac{\partial^2}{\partial\hat{t}^2}J_0(l)\phi(l)\right\rangle+\phi(l) \\ +(\omega-\omega_{de})\left\langle\sum_n\frac{A^{(n)}\exp(in\omega_b\hat{t})}{\omega+\omega_{De}^{(0)}+n\omega_b}\right\rangle+\frac{k^2T_e}{4\pi e^2n_0}\phi(l)=0\end{aligned} \tag{4-7}$$

If we assume that

$$\lambda_{De}{}^2<a_i{}^2\frac{T_e}{T_i}$$

we can neglect for the modes now considered the last term in eq. 4-7. There will be a slow mode, presented later, for which this term will play an important role. Now, before looking for explicit solutions of eq. 4-7, it is useful to consider the quadratic form resulting from taking its inner product by $\phi^*(l)$ over the volume of a tube of flux,

$$\iiint dV\,\phi^*\propto\int\frac{dl}{B}\phi^*$$

So we have

$$\begin{aligned}\int\frac{dl}{B}|\phi|^2\left\{1+\frac{T_e}{T_i}\left[1-\left\langle\frac{\omega-\omega_{di}}{\omega+\omega_{Di}}J_0{}^2\right\rangle\right]\right\} \\ -\left(1-\frac{\omega_{di}}{\omega}\right)\frac{T_e}{T_i\omega^2}\frac{\pi}{2}\iint dE\,d\mu\,f_{io}\int_0^\tau d\hat{t}\left|\frac{\partial}{\partial\hat{t}}J_0\phi\right|^2 \\ -(\omega-\omega_{de})\frac{\pi}{2}\iint dE\,d\mu|\tau|f_{eo}\left\{\frac{|\Phi^{(0)}|^2}{\omega+\omega_{De}^{(0)}}+\sum_{n\neq0}\frac{|\Phi^{(n)}|^2}{\omega+n\omega_b}\right\}=0\end{aligned} \tag{4-8}$$

and we shall use this equation as the basis for the order of magnitude estimates leading to the various modes.

V. Resonant Electron Ballooning Mode

We consider at first the relatively long wavelength limit, where $J_0\simeq 1-\frac{1}{4}k^2v_\perp{}^2/\Omega_i{}^2$. The importance of this limit is due to the fact that modes with wavelengths comparable with the macroscopic dimension of the system are thought to cause the largest particle transport. Then eq. 4-8 reduces, in lowest order, to

$$\oint \frac{dl}{B} \left\{ \left(1 - \frac{\omega_{de}}{\omega}\right)(|\phi|^2 - \langle|\Phi^{(p)}|^2\rangle) + \left\langle \frac{\omega_{De}^{(0)}(\omega - \omega_{de})}{\omega(\omega + \omega_{De}^{(0)})} |\Phi^{(0)}|^2 - \frac{\omega_{De}(\omega - \omega_{di})}{\omega(\omega + \omega_{Di})} |\phi|^2 \right. + \left(1 - \frac{\omega_{di}}{\omega}\right) \frac{T_e}{m_i} \left(\frac{k^2}{\Omega_i^2} |\phi|^2 - \frac{1}{\omega^2} \left|\frac{\partial \phi}{\partial l}\right|^2 \right) \right\} = 0 \quad (5\text{-}1)$$

Clearly, the first term is the lowest order one. So, we have two choices (11,21):

1. $\omega = \omega_{de} + \delta\omega$ with $\delta\omega < \omega$.
2. $\phi = \phi^{(0)} + \delta\phi$, so that, for $\phi \equiv \bar{\phi} + \tilde{\phi}$, where $\bar{\phi} \equiv [\phi(l)]_{\text{av}} = \text{const}$, $\tilde{\phi} < \bar{\phi}$. The inequalities here are intended in an asymptotic sense.

In this section we shall consider the first case (11,21) and in addition assume $\Phi^{(0)} = 0$. If we have a configuration in which the magnetic-field profile is symmetric around each minimum of it, the present choice corresponds to taking $\phi(l)$ antisymmetric around each point of minimum. In fact,

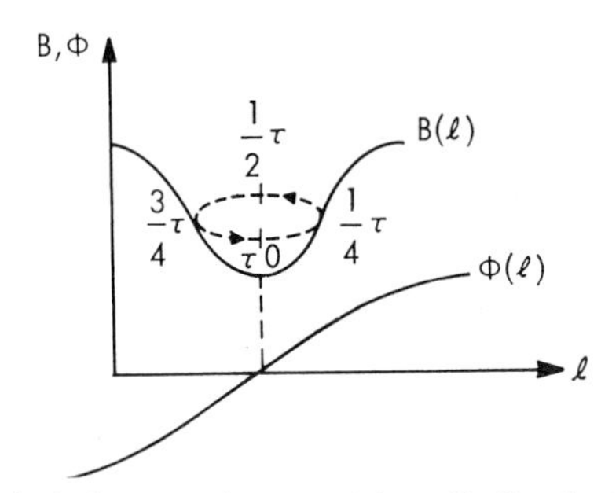

Fig. 4. Typical electrostatic-potential profile for electron modes.

referring to Figure 4, if we compute the integral $\int_0^{\tau} \phi(l(\hat{t}))\, d\hat{t}$ by splitting it into

$$\int_0^{1/4\tau} + \int_{1/4\tau}^{1/2\tau} + \int_{1/2\tau}^{3/4\tau} + \int_{3/4\tau}^{\tau}$$

we recognize that, if $\phi(l)$ is antisymmetric as specified, these integrals cancel each other. We see that, by making this choice, we have directly eliminated a direct contribution of trapped particles. Then eq. 5-1 reduces to

$$\frac{\partial\omega}{\omega_{de}} \oint \frac{dl}{B} |\phi|^2 - \oint \frac{dl}{B} |\phi|^2 \left(1 + \frac{T_i}{T_e}\right)\left(\frac{\langle\omega_{De}\rangle}{\omega_{de}} - b_{(e)}\right) - \frac{T_e + T_i}{\omega_{de}{}^2 m_i} \oint \frac{dl}{B} \left|\frac{\partial\phi}{\partial l}\right|^2 = 0 \quad (5\text{-}2)$$

where $b_{(e)} \equiv k^2 T_e/m_i \Omega_i^2$, and can be used as a variational principle in $\delta\omega$. We anticipate that, if $\delta\omega/\omega_{de} < 0$, i.e., $\omega < \omega_{de}$, the resulting mode becomes unstable when the effects of wave-particle resonance are taken into account. Therefore, we see that the last term, representing sound propagation along the lines of force, is always stabilizing, while the previous term related with the ion inertia across the lines of force (i.e., polarization and finite Larmor-radius drift) is destabilizing. Clearly, the curvature term in $\langle\omega_{De}\rangle/\omega_{de}$ contributes to instability only where it is negative. So, we expect the present mode to have its maximum around the point at which

$$\mathscr{K}(k) \equiv \frac{k_\perp{}^2 T_e}{m_i \Omega_i{}^2} - \frac{\langle v_{De}\rangle}{v_{de}} \equiv b_{(e)} - \frac{d \ln B}{d \ln n}$$

is maximum. Actually, we can explicitly solve the dispersion equation for normal modes that are strongly localized around this point $l = l_0$. We then have the differential equation (11,21),

$$\left\{\frac{T_i + T_e}{m_i \omega_{de}{}^2} \frac{d^2}{dl^2} - \frac{\delta\delta\omega}{\omega_{de}} - \frac{l^2}{2}\left|\frac{d^2\mathscr{K}}{dl^2}\right|_{l=l_0}\left(1 + \frac{T_i}{T_e}\right)\right\}\phi(l) = 0 \tag{5-3}$$

where $\delta\delta\omega = \delta\omega - (1 + T_i/T_e)\mathscr{K}(l_0)\omega_{de}$. If we define

$$\frac{\mathscr{K}}{\mathscr{L}^2} \equiv \left|\frac{d^2\mathscr{K}}{dl^2}\right|_{l=l_0}$$

we see that the eigensolutions of eq. 5-3 are of the form

$$\phi_\nu(l) = H_\nu\left(\frac{l}{\Delta}\right) \exp\left(-\frac{l^2}{2\Delta^2}\right)$$

where the index ν is odd and

$$\Delta \approx \frac{(\mathscr{L} r_n)^{1/2}}{(b_{(e)}\mathscr{K}_0)^{1/4}} \tag{5-4}$$

Now, for the validity of the present result, we require $\Delta^2 < \mathscr{L}^2$, so that we can give a stability criterion, for their nonexistence, of the form

$$\mathscr{L} < \frac{r_n}{(\mathscr{K}_0 b_{(e)})^{1/2}} \tag{5-5}$$

that is clearly independent of the ion temperature. Another, based on invoking the effect of ion Landau damping and a different breakdown of the used approximation, is so that

$$L < r_n \left(\frac{T_i}{T_e b_{(e)}} \right)^{1/2} \tag{5-6}$$

where numerical factors have been neglected. A comparison of these results with the types of fluctuation that have been experimentally observed will be discussed in a later section; the present treatment can be easily extended to cover the regime in which $b \equiv k^2 T_i / m_i \Omega_i^2 > 1$. In particular, a nonlocalized solution of eq. 5-2 has been analyzed in Ref. 22, for a specific magnetic curvature profile.

VI. Particle-wave Resonances

As indicated, we can find the growth rate for the mode considered earlier only by considering the resonance of the wave with the particles having bounce frequency (11,18):

$$\omega_b(E, \mu) = \omega \tag{6-1}$$

Now we recall that, in a configuration with straight, constant, magnetic-field lines (Figure 3), the corresponding resonance involves particles with $v_\parallel = \omega / k_\parallel$, $k_\parallel$ being the wave number along the magnetic field. Two consequences of this are: (1) the resonant portion of velocity space is represented by a plane, since $\mathbf{v}_\perp$ can take any value, and (2) in order to have a sufficient number of particles involved in the resonance, $k_\parallel$ must be finite, so that $\omega / k_\parallel$ is not much larger than the thermal velocity.

In this respect a configuration with varying B field differs in that a mode with ϕ constant along the B field, i.e., $\omega / k_\parallel = \infty$, can have a finite number of particles contributing to the resonance given by eq. 6-1. In fact, this circumstance will appear in the successive sections. The second difference comes from the fact that the resonance represented by eq. 6-1 in the limit, where $\omega < \langle \omega_b \rangle \sim v_{\text{th}} / L$, generally involves a smaller amount of phase space than in the corresponding case of constant magnetic field. In fact, ω_b can be written as

$$\omega_b = \frac{2\pi}{L_0(\lambda)} \sqrt{E}$$

where

$$L_0(\lambda) \equiv \oint \frac{dl}{\sqrt{1 - \lambda B}}$$

and $\lambda \equiv \mu/E$, so that the resonant portion of phase space is generally represented by a small spherical surface of radius $\sqrt{E}$. So, while in the case of constant magnetic field, the particle-wave resonance contribution is of the order $\omega/k_{\parallel} v_{\text{th}}$; in the present case it is smaller, typically of order $(\omega/\langle\omega_b\rangle)^2$, where, for a comparison, $k_{\parallel}$ is $\sim 1/L$.

For the modes considered in the previous two sections, only the electrons with low bounce frequency will resonate with the wave. These are particles that are just barely trapped or just barely circulating, or particles with very low energies. Then, returning to eq. 4-8, the higher-order terms in $\omega/\langle\omega_{be}\rangle$ neglected in deriving eq. 5-1, are

$$\left(1 - \frac{\omega_{de}}{\omega}\right) \iint dE\, d\mu |\tau| f_{eo} \sum_{n \neq 0} |\Phi^{(n)}|^2 \left[\frac{\omega}{n\omega_b}\left(1 - \frac{\omega}{n\omega_b}\right) + i\pi|\omega|\,\delta(\omega + n\omega_b)\right] \tag{6-2}$$

Notice that the first-order term in ω/ω_b cancels and the lowest order contribution (23) of the nonresonant terms is of the order $\omega^2/\langle\omega_b\rangle^2$.

Now we consider the resonant terms

$$\begin{aligned} i\pi|\omega| \iint dE\, d\mu |\tau| f_{eo} |\Phi^{(n)}|^2\, \delta(\omega - n\omega_b) \qquad\qquad (6\text{-}3)\\ = i\pi \int d\lambda\, L_0(\lambda) \int dE\, f_0(E) E^{1/2} |\Phi^{(n)}(E, \lambda)|^2 E_{(n)}\, \delta(E - E_{(n)}) \end{aligned}$$

where

$$\tau = L_0(\lambda)/\sqrt{E}$$

$$\delta(\omega - n\omega_b) = E_{(n)}\, \delta(E - E_{(n)})|\omega|$$

and

$$E_{(n)} = \omega^2 L_0{}^2(\lambda)/(2\pi n)^2$$

So the expression 6-2 becomes

$$\begin{aligned} \left(1 - \frac{\omega_{de}}{\omega}\right) \sum_{n \neq 0} \Bigg\{ \frac{\omega^2}{\hat{\omega}_{be}{}^2 n^2} \left(\frac{m_e}{T_e \pi^3}\right)^{1/2} \int dE \frac{\exp(-m_e E/2T_e)}{E^{1/2}} \int d\lambda \frac{L_0{}^3(\lambda)}{L^2} |\Phi^{(n)}(E, \lambda)|^2 \\ - \frac{i}{\pi} \frac{|\omega^3|}{\hat{\omega}_{be}{}^3 n^3} \int d\lambda \frac{L_0{}^4(\lambda)}{L^3} \exp\left[-\frac{m_e}{2T_e}\left(\frac{\omega L_0(\lambda)}{2\pi n}\right)^2\right] |\Phi^{(n)}(E_{(n)}, \lambda)|^2 \Bigg\} \end{aligned} \tag{6-4}$$

where $\lambda = \mu/E$ and $\hat{\omega}_{be} \equiv (2\pi/L)\sqrt{2T_e/m_e}$, and shows that, in the considered limit, the particle-wave contributes to the order $\omega^3/\langle\omega_{be}{}^3\rangle$.

Referring again to a configuration with constant magnetic field (24), we recall that evaluation of Landau damping is done through the integral

$$W\left(\frac{\omega}{k_{\parallel} v_{\text{th}}}\right) = \frac{1}{2} \int_{-\infty}^{+\infty} \frac{\partial f/\partial v}{\omega/k_{\parallel} v_{\text{th}} + v}\, dv \tag{6-5}$$

where $v \equiv v_{\|}/v_{\text{th}}$. Now, if $f = (1/\sqrt{\pi}) \exp(-v^2)$, in the limit $\omega < k_{\|} v_{\text{th}}$, we have

$$W \approx -1 + i\sqrt{\pi} \frac{\omega}{k_{\|} v_{\text{th}}}$$

If instead we have a distribution function (24) with a "plateau" around $v = 0$ such as $f \propto \exp(-v^4)$, we have

$$W \propto -\alpha_1 - \alpha_2 \frac{\omega^2}{k_{\|}{}^2 v_{\text{th}}{}^2} + i\alpha_3 \frac{\omega^3}{k_{\|}{}^3 v_{\text{th}}{}^3} \tag{6-6}$$

for $\omega < k_{\|} v_{\text{th}}$. As said previously, the appearance in eqs. 6-4 and 6-6 of a resonant term of order $\omega^3/\hat{\omega}_{be}{}^3$ is related with the contribution of the non-divergent nonresonant term to the order $(\omega/\hat{\omega}_{be})^2$.

Finally returning to the mode considered in the earlier section, we can deduce its growth rate from eqs. 4-8 and 5-2:

$$\frac{\gamma}{\omega_{de}} \oint \frac{dl}{B} |\phi|^2 = -(\delta\omega) \frac{\pi}{2} \iint dE\, d\mu f_0(E) |\tau| \sum_n |A^{(n)}|^2\, \delta(\omega + n\omega_b) \tag{6-7}$$

where $\gamma = -\text{Im}(\omega)$ is the growth rate. Therefore, instability occurs if $\delta\omega/\omega_{de} < 0$, as anticipated.

VII. Quasifluid Electron Drift Mode

We define as quasifluid a mode that can be unstable, i.e., self-excited, independent of particle-wave resonance effects, which involves the entire velocity space but, on the other hand, is not given by the classical MHD fluid approximation.

The mode we consider here (24) is fast in the sense that $\hat{\omega}_{bi} < \omega < \hat{\omega}_{be}$ and comes from having the term in $(\omega/\hat{\omega}_{be})^2$ play a role to the lowest order inside the dispersion relation. If we refer to eq. 4-8, this clearly implies taking $\Phi^{(0)} = 0$, e.g., $\Phi(l)$ antisymmetric around each point of minimum magnetic field as indicated in Section V. Then eq. 4-8 reduces to

$$\oint \frac{dl}{B} |\phi|^2 \left\{1 + \frac{T_e}{T_i}\left[1 - \left\langle \frac{\omega - \omega_{di}}{\omega + \omega_{Di}} J_0{}^2 \right\rangle\right]\right\} + \left(1 - \frac{\omega_{de}}{\omega}\right) \frac{\omega^2}{\hat{\omega}_{be}{}^2} \frac{\pi}{2} \iint dE\, d\mu f |\tau^3| \sum_{n \neq 0} \left| \frac{2\pi\hat{\omega}_{be} \Phi^{(n)}}{n} \right|^2 = 0 \tag{7-1}$$

and, to make the last term relevant to the lowest order, we are led to take $\omega < \omega_{de}$, $\omega_{Di} < \omega < \omega_{di}$ and $J_0^2 \ll 1$, so that $kv_{\text{th}i}/\Omega_i > 1$. Therefore, the significant dispersion relation is

$$\left(1 + \frac{T_e}{T_i}\right)\left(\oint \frac{dl}{B} |\phi|^2\right) - \frac{\omega_{de}}{\omega}\left(\frac{1}{(2\pi b_0)^{1/2}} \oint \frac{dl}{B_0} |\phi|^2\right) \tag{7-2}$$

$$- \frac{\omega_{de}\omega}{\hat{\omega}_{be}^2}\left(\frac{\pi}{2}\iint dE\, d\mu\, f|\tau|^3 \sum \left|\frac{2\pi\hat{\omega}_{be}\Phi^{(n)}}{n}\right|^2\right) = 0$$

$$\frac{\omega\omega_{de}}{\hat{\omega}_{be}^2} + \frac{\omega_{de}}{\omega}\frac{\Upsilon}{(2\pi b_0)^{1/2}} - \left(1 + \frac{T_e}{T_i}\right)\Xi = 0 \tag{7-3}$$

where

$$\Xi \cdot (2\pi\hat{\omega}_{be})^2 \frac{\pi}{2}\iint dE\, d\mu f_{0e}|\tau^3| \sum_{n\neq 0} |\Phi^{(n)}|^2/n^2$$

$$\equiv \int dl|\phi|^2/B = \frac{\pi}{2}\iint dE\, d\mu f_{0e}|\tau| \sum_{n\neq 0} |\Phi^{(n)}|^2$$

$$\Upsilon \equiv \Xi \oint dl|\phi|^2 \Big/ \left(B_0 \oint dl|\phi|^2/B\right)$$

$$b_0 = k^2 T_i/m_i\Omega_0^2, \quad \Omega_0 \equiv eB_0/m_i$$

and B_0 is a typical value of the magnetic field introduced here to show that Υ is proportional to $\oint dl|\phi|^2$. In fact, $\langle J_0^2\rangle = I_0(b)\exp(-b)$, where

$$b \equiv \frac{k^2 T_i^2}{m_i \Omega_i^2} \tag{7-4}$$

and for $b > 1$, $I_0(b)\exp(-b) \approx (2\pi b)^{-1/2}$. The solution is clearly

$$\frac{\omega}{\hat{\omega}_{be}} = \frac{1}{2}\left(1 + \frac{T_e}{T_i}\right)\frac{\hat{\omega}_{be}}{\omega_{de}}\Xi \pm \left[\frac{1}{4}\left(1 + \frac{T_e}{T_i}\right)^2 \frac{\hat{\omega}_{be}^2}{\omega_{de}^2}\Xi^2 - \frac{\Upsilon}{(2\pi b_0)^{1/2}}\right]^{1/2} \tag{7-5}$$

So if

$$\omega_{de} > \frac{1}{2}\hat{\omega}_{be}\left(1 + \frac{T_e}{T_i}\right)\frac{\Xi}{\Upsilon^{1/2}}(2\pi b_0)^{1/4} \tag{7-6}$$

a fluidlike, macroscopic, instability occurs. Notice that the root with the negative sign that becomes unstable is a proper continuation of the drift ballooning mode treated in Section V. The root with the positive sign corresponds to a new mode. The importance of this comes out by considering the regime in which condition 7-6 is not satisfied. Then the roots of eq. 7-5 are both stable and the requirement $\omega < \omega_{de}$ implies

$$\omega_{de} > \hat{\omega}_{be}\left[\left(1 + \frac{T_e}{T_i}\right)\Xi\right]^{1/2} \tag{7-7}$$

which is less restrictive than eq. 7-6. In this case if, by extension of what is done in the case in which $\phi(l) \approx \exp(ik_{\parallel} l)$, we define a wave energy $U_k = \omega_k(\partial\varepsilon/\partial\omega_k)$ where ε is the effective dielectric constant,

$$\varepsilon \propto \left(1 + \frac{T_e}{T_i}\right)\Xi - \frac{\omega_{de}}{\omega}\frac{\Upsilon}{(2\pi b_0)^{1/2}} - \frac{\omega\omega_{de}}{\hat{\omega}_{be}^{\,2}}$$

and find

$$U_k \propto \mp 2\frac{\omega_{de}}{\hat{\omega}_{be}}\left[\frac{1}{4}\left(1 + \frac{T_e}{T_i}\right)^2 \frac{\hat{\omega}_{be}^{\,2}}{\omega_{de}^{\,2}}\Xi^2 - \frac{\Upsilon}{(2\pi b_0)^{1/2}}\right]^{1/2} \tag{7-8}$$

Therefore, the two modes have opposite energy sign. So, we can argue that by taking into account the particle-wave resonance effects considered in Section VI, one of these roots remains unstable, although it now involves a limited region of phase space. In addition, by extension of the theory of nonlinear interaction between negative and positive energy waves (25,26), we expect that the nonlinear effects will contribute to enhance the instability. So, when condition 7-7 is satisfied, we expect a considerable anomalous particle diffusion by the onset of these modes. In fact, either they start as a fluid-type instability when condition 7-6 is also satisfied or, although they may start as a resonant, weak instability, the nonlinear interactions will tend to enhance the growth rate and the resulting anomalous transport.

To evaluate the coefficients Υ and Ξ, one has to know the eigenfunction $\phi(l)$, in general. However, we can see that, in the special cases in which the last term on the left-hand side of eq. 7-2 can be neglected, a variational principle for the eigenvalue ω^2 is obtained and can be used to evaluate more specifically the instability growth rate.

Finally, we notice that we have classified these modes as of drift type, since their instability does not involve the effects of magnetic curvature.

VIII. Resonant Electron Flute Mode (11)

We refer back to the quadratic form (eq. 5-1) and to the discussion following it. Then we consider the second of the two consistent asymptotic limits that are possible, i.e., $\bar{\phi} > \tilde{\phi}$, where $\bar{\phi} + \tilde{\phi} \equiv \phi$ and $\bar{\phi} \equiv [\phi]_{\mathrm{av}}$. Then the quadratic form reduces to

$$\oint \frac{dl}{B}\left\{\left[\left(1-\frac{\omega_{di}}{\omega}\right)b_{(e)}-\frac{\omega_{de}\langle\omega_{De}\rangle}{\omega^2}\left(1+\frac{T_i}{T_e}\right)\right.\right. \tag{8-1}$$

$$\left.-\left(1-\frac{\omega_{di}}{\omega}\right)\left\langle\frac{\omega_{De}{}^2}{\omega^2}\right\rangle\frac{T_i}{T_e}-\left(1-\frac{\omega_{de}}{\omega}\right)\left\langle\left(\frac{\omega_{De}^{(0)}}{\omega}\right)^2\right\rangle\right]_{\text{av}}\bar{\phi}^2$$

$$+\left(1-\frac{\omega_{de}}{\omega}\right)(|\tilde{\phi}|^2-\langle|\tilde{\Phi}^{(0)}|^2\rangle)-\frac{\tilde{T}_e}{m_i\omega^2}\left|\frac{\partial\tilde{\phi}}{\partial l}\right|^2$$

$$\left.+\left\langle\frac{\omega_{De}^{(0)}}{\omega}\left(1-\frac{\omega_{de}}{\omega}\right)\bar{\phi}(\tilde{\Phi}^{(0)}+\tilde{\Phi}^{(0)*})-\frac{\omega_{De}}{\omega}\left(1-\frac{\omega_{di}}{\omega}\right)\bar{\phi}(\tilde{\phi}+\tilde{\phi}^*)\right\rangle\right\}=0$$

Here we have taken into account terms of the order $(\omega_D/\omega)^2$ and $\omega_D\tilde{\phi}/\omega\bar{\phi}$, so that we may analyze the case in which $[\langle\omega_{De}\rangle]_{\text{av}} < \{[\langle\omega_{De}{}^2\rangle]_{\text{av}}\}^{1/2}$ and these terms become important, as will be seen later. In the present section we do not consider such a case; therefore eq. 8-1 reduces to

$$\left(1-\frac{\omega_{di}}{\omega}\right)\bar{b}_{(e)}-\frac{\omega_{de}\,\overline{\omega}_{De}}{\omega^2}\left(1+\frac{T_i}{T_e}\right) \tag{8-2}$$

$$+\left(1-\frac{\omega_{de}}{\omega}\right)\left[\left|\frac{\tilde{\phi}}{\bar{\phi}}\right|^2-\left\langle\left|\frac{\tilde{\Phi}^{(0)}}{\bar{\phi}}\right|^2\right\rangle\right]_{\text{av}}-\frac{T_e}{m_i\omega^2}\left[\left|\frac{\partial\tilde{\phi}}{\partial l}\frac{1}{\bar{\phi}}\right|^2\right]_{\text{av}}=0$$

To proceed further, we need an evaluation of $\tilde{\phi}/\bar{\phi}$. Therefore, we consider the periodic part over $\oint dl/B$ of the relevant lowest order dispersion relation,

$$\phi(l)-\left\langle\Phi^{(0)}\left(1-\frac{\omega_{De}^{(0)}}{\omega}\right)\right\rangle \tag{8-3}$$

$$+\frac{\omega-\omega_{di}}{\omega-\omega_{de}}\left[b_{(e)}-\frac{\langle\omega_{De}\rangle}{\omega}+\frac{T_e}{m_i\omega^2}B\frac{\partial}{\partial l}\frac{1}{B}\frac{\partial}{\partial l}\right]\phi(l)=0$$

and we then have

$$\left(1+\frac{T_e}{m_i\omega^2}B\frac{\partial}{\partial l}\frac{1}{b}\frac{\partial}{\partial l}\right)\tilde{\phi}(l)-\langle\tilde{\Phi}^{(0)}\rangle=-\bar{\phi}\left\{\frac{\langle\tilde{\omega}_{De}^{(0)}\rangle}{\omega}+\frac{\omega-\omega_{di}}{\omega-\omega_{de}}\left[b_{(e)}-\frac{\langle\tilde{\omega}_{De}\rangle}{\omega}\right]\right\} \tag{8-4}$$

where $\tilde{b}_{(e)} \equiv b_{(e)} - \bar{b}_{(e)}$ and $\tilde{\omega}_{De} - \omega_{De} - \overline{\omega}_{De}$. So, if $\omega \neq \omega_{de}$,

$$\frac{\tilde{\phi}}{\bar{\phi}} \approx \frac{\omega_{de}}{\omega}$$

and in eq. 8-2 we can consistently neglect all terms in $|\tilde{\phi}/\bar{\phi}|^2$. Then the real part of the frequency is given by

$$\omega^0=\frac{1}{2}\omega_{di}\pm\left[\frac{1}{4}\omega_{di}{}^2+\frac{\omega_{di}\,\overline{\omega}_{Di}}{\bar{b}}\left(1+\frac{T_e}{T_i}\right)\right]^{1/2} \tag{8-5}$$

Since we are interested in modes that are stable within the MHD approximation, we consider here lines of force on which $\omega_{di}\bar{\omega}_{Di} > 0$. The growth rate $\gamma = -\text{Im}(\omega)$ is obtained by perturbation from eq. 8-2 to which the particle-wave resonant term,

$$-\frac{i\pi^2}{2}(\omega - \omega_{de})\iint dE\, d\mu |\tau| f_{e0} \sum_{n\neq 0} |A^{(n)}|^2\, \delta(\omega - n\omega_b) \tag{8-6}$$

is to be added. Therefore, we have

$$\gamma \frac{\bar{b}_{(e)}}{\omega_0{}^2}(2\omega_0 - \omega_{di}) = \frac{i\pi^2}{2}(\omega_{de} - \omega_0) \iint dE\, d\mu |\tau| f_{e0} \sum_{n\neq 0} |A^{(n)}|^2\, \delta(\omega - n\omega_b) \tag{8-7}$$

so that for instability we require $0 < \omega_0/\omega_{de} < 1$, which in turn gives the condition

$$\frac{\bar{\omega}_{Di}}{\omega_{di}} > \bar{b}\,\frac{T_e}{T_i}$$

The relevant wave has phase velocity in the direction of the electron diamagnetic velocity and has been labeled a "gravitational" wave in Ref. 11, recalling the simulation that had been made of the average favorable magnetic curvature by an upward effective gravity.

We also recall that

$$\frac{\bar{\omega}_{Di}}{\omega_{di}} \equiv 2h\,\frac{r_n}{R_c}$$

where $0 < h < 1$, is usually referred to as the average magnetic-well depth with regard to the fact that the average magnetic curvature when $\bar{\omega}_{Di}/\omega_{di} > 0$ contributes toward stability of hydromagnetic types of mode.

IX. Quasifluid Electron Flute Mode (27)

We now consider the practically interesting case in which the magnetic well is sufficiently shallow so that second-order terms in ω_{De}/ω, ω_{Di}/ω and in $(\tilde{\phi}/\bar{\phi})^2$ become important. The condition on h will be specified later. In this case it is necessary to produce a solution for $\tilde{\phi}$. So, it becomes simpler to use, instead of the quadratic form eq. 8-1, the equivalent equation that comes from taking the $\oint dl/B$ average of eq. 8-3

$$\bar{\phi}\left\{\left(1 - \frac{\omega_{di}}{\omega}\right)\frac{T_e}{T_i}\left(\bar{b} - \frac{\bar{\omega}_{Di}{}^2}{\omega^2}\right) - \frac{\omega_{de}\bar{\omega}_{De}}{\omega^2}\left(1 + \frac{T_i}{T_e}\right) - \left(1 - \frac{\omega_{de}}{\omega}\right)\frac{\hat{\omega}_{De}{}^2}{\omega^2}\right\} \tag{9-1}$$
$$+\left(1 - \frac{\omega_{de}}{\omega}\right)\left[\left\langle\frac{\tilde{\omega}_{De}^{(0)}}{\omega}\tilde{\phi}^{(0)}\right\rangle + \frac{\omega - \omega_{di}}{\omega - \omega_{de}}\frac{T_e}{T_i}\left(\tilde{b} + \frac{\langle\tilde{\omega}_{Di}\rangle}{\omega}\right)\tilde{\phi}(l)\right]_{\text{av}} = 0$$

Here again we have assumed $\tilde{\phi} < \bar{\phi}$ on the basis of eq. 8-4. In addition, we have defined $\overline{\omega}_{Di}{}^2 \equiv [\langle \omega_{Di}{}^2 \rangle]_{\rm av}$ and $\hat{\omega}_{De}{}^2 \equiv [\langle \omega_{De}^{(0)2} \rangle]_{\rm av}$, and have recalled that $\overline{\omega}_{Di} \equiv [(\omega_{Di})]_{\rm av} = -(T_i/T_e)[\langle \omega_{De}^{(0)2} \rangle]_{\rm av}$, due to an inversion of the order of integration on $d^3\mathbf{v}$ and dl/B. Following the same inversion procedure, we notice that

$$\oint \frac{dl}{B} \int d^3\mathbf{v} f_{e0}\, \omega_{De}^{(0)} \tilde{\Phi}^{(0)} = \frac{\pi}{2} \oint dl \iint \frac{dE\, d\mu}{v_{\|}} f_{e0}\, \tilde{\omega}_{De}^{(0)} \frac{\oint (dl/v_{\|}) \tilde{\phi}(l)}{\oint dl/v_{\|}}$$

$$= \frac{\pi}{2} \oint \frac{dl}{B} \tilde{\phi}(l) \iint \frac{dE\, d\mu}{v_{\|}} B f_{e0}\, \tilde{\omega}_{De}^{(0)}$$

so that

$$[\langle \tilde{\omega}_{De}^{(0)} \tilde{\Phi}^{(0)} \rangle]_{\rm av} = [\langle \tilde{\omega}_{De}^{(0)} \rangle \tilde{\phi}(l)]_{\rm av} \tag{9-2}$$

Therefore the last term of eq. 9-1 becomes

$$\left(1 - \frac{\omega_{de}}{\omega}\right) \left[\left(\frac{\langle \tilde{\omega}_{De}^{(0)} \rangle}{\omega} + \frac{\omega - \omega_{di}}{\omega - \omega_{de}} \frac{T_e}{T_i} \left(\tilde{b} + \frac{\langle \tilde{\omega}_{Di} \rangle}{\omega} \right) \right) \tilde{\phi}(l) \right]_{\rm av} \tag{9-3}$$

and replacing in it the left-hand side of eq. 8-4 we finally obtain

$$\begin{aligned} \bar{b}\left(1 - \frac{\omega_{di}}{\omega}\right) \frac{T_e}{T_i} - \frac{1}{\omega^2} \left[\omega_{de} \overline{\omega}_{De} + \hat{\omega}_{De}{}^2 + \frac{T_i}{T_e} (\omega_{de} \overline{\omega}_{De} + \overline{\omega_{De}{}^2}) \right] \\ + \frac{\omega_{de}}{\omega^3} \left(\hat{\omega}_{De}{}^2 - \frac{T_i^2}{T_e^2} \overline{\omega_{De}{}^2} \right) - \left(1 - \frac{\omega_{de}}{\omega}\right) \left[\frac{\tilde{\phi}^2}{\bar{\phi}^2} - \frac{\langle \tilde{\Phi}^{(0)2} \rangle}{\bar{\phi}^2} \right]_{\rm av} \\ + \frac{T_e}{m_i \omega^2} \left[\left(\frac{1}{\phi} \frac{\partial \tilde{\phi}}{\partial l} \right)^2 \right]_{\rm av} = 0 \end{aligned} \tag{9-4}$$

This equation is equivalent to eq. 8-1 which has the property of having real coefficients.

Reconsidering the inhomogeneous integro-differential equation 8-4, we notice that, if $T_e \sim T_i$, the second-derivative term can be neglected on account of the assumption $\omega < \langle \omega_{bi} \rangle$. On the other hand, after evaluating various alternatives, we can see that the most interesting asymptotic limit is found for $T_e > T_i$, $\omega \sim \omega_{di} < \omega_{de}$, $\tilde{b} > \tilde{\omega}_{De}/\omega$. This case is also of practical interest, since many significant experiments (see Section XVI) on two-dimensional configurations involve plasmas with $T_e > T_i$. Then the right-hand side of eq. 8-4 reduces to $-\bar{\phi}\langle \tilde{\omega}_{De}^{(0)2} \rangle/\omega$ to the lowest order and, if $\omega\omega_{de} > T_e/M_i L^2$, the second-derivative term in eq. 8-4 is again negligible and $\tilde{\phi}/\bar{\phi} \sim \omega_{De}/\omega \ll 1$. In this case the quadratic terms in $(\tilde{\phi}/\bar{\phi})^2$ give a contribution of the same order as $\omega_{de}\hat{\omega}_{De}^2/\omega^2$, and we can see that $\omega \sim [\omega_{de} T_e/M_i R_c^2]^{1/3}$. Therefore, if $L^2 \sim R_c^2$, the fact that $\omega_{de} > \omega$ implies also $\omega_{de} > T_e/M_i L^2\omega$ and we can consistently take

$$\tilde{\phi}(l) - \langle\tilde{\Phi}^{(0)}\rangle = -\bar{\phi}\,\frac{\langle\tilde{\omega}_{De}^{(0)}\rangle}{\omega} \tag{9-5}$$

as the lowest order form of eq. 8-4.

Now, it is important to observe that the solution of eq. 9-5 can be formally written as

$$\tilde{\phi}(l) = \frac{\hat{\omega}_{De}}{\omega}\,G(l)\bar{\phi} \tag{9-6}$$

where $G(l)$ is a finite real function of l depending on the geometry of the system. Then eq. 9-4 reduces simply to

$$\bar{b}_{(e)}\left(1 + \frac{\omega_{de}}{\omega}\frac{T_i}{T_e}\right) - \frac{\omega_{de}\bar{\omega}_{De}}{\omega^2} + \frac{\omega_{de}\hat{\omega}_{De}^2}{\omega^3}(1 + \Gamma_0) = 0 \tag{9-7}$$

where $\Gamma_0 \equiv [G^2(l) - \langle G^{(0)2}\rangle]_{av}$, recalling that eq. 1-12 is a positive definite quantity from the Schwartz inequality, and $\bar{b}_{(e)} \equiv T_e\bar{b}/T_i$. Here we have clearly taken $T_e/T_i > (\bar{\omega}_{De}^2/\hat{\omega}_{De}^2)^{1/2}$.

Notice that eq. 9-7 exhibits a third root in addition to the two given by eq. 8-2 coming from the splitting of the electron gravitational mode discussed earlier into two modes, both with phase velocity in the direction of the electron diamagnetic velocity. One of these becomes unstable independently of wave-particle resonance effects, when

$$(3W + I^2)^{3/2} < (\tfrac{3}{2}^3 + I^3) + \tfrac{9}{2}WI \tag{9-8}$$

where

$$I \equiv \frac{T_i}{T_e}\left(\frac{\omega_{de}}{\hat{\omega}_{De}}\right)^{2/3}\left(\frac{\bar{b}_{(e)}}{1 + \Gamma_0}\right)^{1/3} = \frac{T_i}{T_e}\left(\frac{R_c}{2r_n}\right)^{2/3}\left(\frac{\bar{b}_{(e)}}{1 + \Gamma_0}\right)^{1/3}$$

and

$$W \equiv \bar{\omega}_{De}\left[\frac{\omega_{de}}{\hat{\omega}_{De}^4\bar{b}_{(e)}(1 + \Gamma_0)^2}\right]^{1/2} = h\left[\frac{R_c}{2r_n\bar{b}_{(e)}(1 + \Gamma_0)^2}\right]^{1/3}$$

Condition 9-8 is obtained by rewriting the dispersion relation 9-7 as

$$F(\xi) = 1 + \frac{I}{\xi} - \frac{W}{\xi^2} + \frac{1}{\xi^3} = 0$$

and imposing $F(\xi_0) > 0$, where ξ_0 is the positive root of $dF/d\xi = 0$ and $\xi \equiv \omega\{\bar{b}_{(e)}/[\omega_{de}\hat{\omega}_{De}(1+\Gamma_0)]\}^{1/3}$. In particular, in the very low limit of T_i/T_e, where the terms in I are negligible, the instability condition reduces to

$$h < 3(1+\Gamma_0)^{2/3}\left[\frac{r_n}{2R_c}\,\bar{b}_{(e)}\right]^{1/3} \tag{9-9}$$

In the same limit, when this condition is well satisfied, the relevant mode is overstable with frequency of oscillation of the same order as the growth rate:

$$\gamma \sim \left(\omega_{de}\frac{T_e}{M_i R_c^2}\right)^{1/3} \tag{9-10}$$

Notice that the assumed limit $\omega < \omega_{de}$ entails

$$b_{(e)} > \left(\frac{r_n}{R_c}\right)^2 \tag{9-11}$$

In particular, conditions 9-9 and 9-11 are compatible and realistically satisfied in interesting experimental situations such as in stellarator experiments, where the magnetic well is shallow and typically $h \sim r_n/R_c$. When the magnetic well is sufficiently deep, so that condition 9-8 is not met, we have, as shown in the previous section, a weak instability due to particle-wave resonance effects.

Finally, we recall that the lower limit of T_i/T_e, above which the fluidlike instability discussed above occurs, is a function of the geometry and can be evaluated as follows. We assume $L > R_c$, so that it is consistent to take $b \sim \bar{\omega}_{Di}{}^2/\omega^2$ and consider the asymptotic limit $T_i \sim T_e$, $\omega < \omega_{de} \sim \omega_{di}$. Then eq. 8-4 to reduces $\tilde{\phi} - \langle\tilde{\Phi}^{(0)}\rangle = -(\bar{\phi}/\omega)\{\langle\tilde{\omega}_{De}^{(0)}\rangle + T_i\langle\tilde{\omega}_{De}\rangle/T_e\}$, so that formally $\tilde{\phi}\bar{\phi} = (\hat{\omega}_{De}/\omega)\{G(l) + (T_i/T_e)H(l)\}$ and the dispersion relation becomes

$$\bar{b} - \frac{\bar{\omega}_{De}}{\omega}\left(1+\frac{T_i}{T_e}\right) + \frac{\hat{\omega}_{De}{}^2}{\omega^2}\left\{1 - \frac{T_i^2}{T_e^2}\left[\left(G+\frac{T_i}{T_e}H\right)^2 - \left\langle\left(G^{(0)}+\frac{T_i}{T_e}H^{(0)}\right)^2\right\rangle\right]_{\text{av}}\right.$$
$$\left. - \left(\frac{T_i}{T_e\hat{\omega}_{De}}\right)^2[\langle\omega_{De}{}^2\rangle - \langle(\omega_{De}^{(0)})^2\rangle]_{\text{av}}\right\} = 0$$

where the quantities inside square brackets are positive for the Schwartz inequality. Thus the condition for marginal stability can easily be obtained and the stabilization that is occurring by increasing T_i/T_e can be attributed to a finite ion gyroradius effect.

X. Slow (Ion) Modes

This class of modes has very low frequency, $\omega < \langle \omega_{bi} \rangle < \langle \omega_{be} \rangle$, and therefore is $J = \oint v_{\parallel}\, dl$ conserving for both ions and electrons. In particular, the perturbed electron density is

$$n_{ek} = \frac{e}{T_e} n_{e0}\left\{\phi - (\omega - \omega_{de})\left\langle \frac{A_e^{(0)}}{\omega + \omega_{De}^{(0)}} \right\rangle\right\} \tag{10-1}$$

and the ion density

$$n_{ik} = -\frac{e}{T_e} n_{i0}\left\{\phi - (\omega - \omega_{di})\left\langle J_0\left(\frac{kv_\perp}{\Omega}\right) \sum_n \frac{A_i^{(n)} \exp(in\omega_b \hat{t})}{\omega + \omega_{Di}^{(0)} + n\omega_b} \right\rangle\right\} \tag{10-2}$$

Then, proceeding as in Section IV, we can derive a quadratic form by taking the $\oint (dl/B)\phi^*$ average of Poisson's equation,

$$k^2\phi = 4\pi e(n_{ik} - n_{ek})$$

So we obtain

$$\left[\left(k^2\lambda_{De}{}^2 + 1 + \frac{T_e}{T_i}\right)|\phi|^2 - (\omega - \omega_{de})\left\langle \frac{|A_e^{(0)}|^2}{\omega + \omega_{De}^{(0)}} \right\rangle \right. \tag{10-3}$$

$$\left. - (\omega - \omega_{di}) \frac{T_e}{T_i} \left\langle \frac{|A_i^{(0)}|^2}{\omega + \omega_{Di}^{(0)}} \right\rangle \right]_{\text{av}}$$

$$+ \left(1 - \frac{\omega_{di}}{\omega}\right) \frac{T_e}{T_i} \frac{\pi}{2} \sum_{n \neq 0} \left\{ \frac{\omega^2}{\hat{\omega}_{bi}{}^2 n^2} \int \frac{dE}{E^{1/2}} \exp\left(-\frac{m_i}{2T_i} E\right) \int d\lambda \frac{L_0{}^3(\lambda)}{L^2} |A_i^{(n)}(E, \lambda)|^2 \right.$$

$$\left. - \frac{i}{\pi} \frac{|\omega^3|}{(n\hat{\omega}_{bi})^3} \int d\lambda \frac{L_0{}^4(\lambda)}{L^3} \exp\left[-\frac{m_i}{T_i}\left(\frac{\omega L(\lambda)}{2\pi n}\right)^2\right] |A_i^{(n)}(\lambda)|^2 \right\} = 0$$

where higher-order terms in $\omega/\hat{\omega}_{bi}$ have been included as indicated by eq. 6-4, and $\lambda_{De}{}^2 \equiv T_e/(4\pi n e^2)$.

XI. Slow Flute Modes Driven by Local Magnetic Curvature

We consider now modes with perpendicular wavelengths larger than the ion gyroradius, so that $ka_i < 1$. This implies that $\langle \omega_{Di}{}^2 \rangle / \langle \omega_{bi} \rangle^2 \sim (ka_i)^2 L^2/R_c{}^2 < 1$. Therefore,

$$|A_i^{(0)}|^2 \approx \left|\left\{\phi(\hat{t})\left[1 - \frac{1}{4}\frac{k^2 v_\perp{}^2}{\Omega_i{}^2}(\hat{t}) + ik \int^{\hat{t}} (v_{Di}(t') - v_{Di}^{(0)})\right]\right\}^{(0)}\right|^2$$

and

$$|A_e^{(0)}|^2 \approx |\Phi^{(0)}|^2$$

Neglecting at first the higher order terms in $\omega/\hat{\omega}_{bi}$, we find that eq. 10-3 reduces to

$$\Bigg[\left(1+\frac{T_e}{T_i}\right)(|\phi|^2 - \langle|\Phi^{(0)}|^2\rangle) - \left(1+\frac{T_e}{T_i}\right)\frac{T_e}{T_i}\left\langle\frac{\omega_{Di}^{(0)}(\omega_{di}+\omega_{Di}^{(0)})}{\omega^2+\omega\omega_{Di}^{(0)}(1-T_e/T_i)-\omega_{Di}^{(0)2}T_e/T_i}|\Phi^{(0)}|^2\right\rangle + (\omega-\omega_{di})\frac{T_e}{T_i}\left\langle\frac{|\Phi^{(0)}|^2 - |A_i^{(0)}|^2}{\omega+\omega_{Di}^{(0)}}\right\rangle\Bigg]_{av} = 0 \tag{11-1}$$

Now, if we take the limit $\omega < \omega_{Di}$, it is easy to see that a sufficient condition for stability is $\omega_{di}\omega_{Di}^{(0)} > 0$, implying

$$\frac{\partial J}{\partial \psi} < 0 \tag{11-2}$$

i.e., maximum J for all cases in which $\partial n_0/\partial\psi < 0$.

Condition 11-2 implies that the average curvature drift of each particle is favorable to stability and is quite strict. In the present section we consider configurations or magnetic surfaces on which eq. 11-2 is not satisfied and the magnetic well is not too deep, in the sense that we shall define. Then it is convenient again to separate ϕ into $\bar{\phi} + \tilde{\phi}$ as in Section IX and take the average and the periodic part, over $\oint dl/B$, of the dispersion relation,

$$\phi - \langle\Phi^{(0)}\rangle - \frac{T_e}{T_i}\left\langle\frac{\omega_{Di}^{(0)}(\omega_{di}+\omega_{Di}^{(0)})}{\omega^2+\omega\omega_{Di}^{(0)}(1-T_e/T_i)-\omega_{Di}^{(0)2}T_e/T_i}\Phi^{(0)}\right\rangle = 0 \tag{11-3}$$

Here we have neglected the term of order $(ka_i)^2$, corresponding to the last one in eq. 11-1 on the assumption that

$$\frac{\omega_{Di}{}^2}{\omega^2} \sim (ka_i)^2\frac{\hat{\omega}_{bi}{}^2}{\omega^2}\frac{L^2}{R_c{}^2} > (ka_i)^2 \tag{11-4}$$

So, to the second order in $\omega_{Di}{}^2/\omega^2$, we have

$$\phi(l) - \langle\Phi^{(0)}\rangle = \frac{T_e}{T_i}\left\langle\left[\frac{\omega_{di}\omega_{Di}^{(0)}}{\omega^2} + \frac{\omega_{Di}^{(0)2}}{\omega^2}\left(1-\frac{\omega_{di}}{\omega}+\frac{T_e}{T_i}\frac{\omega_{di}}{\omega}\right)\right]\Phi^{(0)}\right\rangle = 0 \tag{11-5}$$

and, after averaging,

$$\left[\bar{\omega}_{Di}\frac{\omega_{di}}{\omega^2} + \frac{[\langle\omega_{Di}^{(0)2}\rangle]_{av}}{\omega^2}\left(1-\frac{\omega_{di}}{\omega}+\frac{T_e}{T_i}\frac{\omega_{di}}{\omega}\right)\right]\bar{\phi} + \frac{\omega_{di}}{\omega^2}[\langle\omega_{Di}^{(0)}\tilde{\Phi}^{(0)}\rangle]_{av} = 0 \tag{11-6}$$

and

$$\tilde{\phi}(l) - \langle\tilde{\Phi}^{(0)}\rangle = \frac{T_e}{T_i}\frac{\omega_{di}}{\omega^2}[\langle\tilde{\omega}_{Di}^{(0)} - \overline{\omega}_{Di}\rangle\overline{\phi} + \langle\omega_{Di}^{(0)}\tilde{\Phi}^{(0)} - [\langle\omega_{Di}^{(0)}\tilde{\Phi}^{(0)}\rangle]_{\text{av}}\rangle] \tag{11-7}$$

We see that typically $\tilde{\phi} \sim (\omega_{de}\omega_{Di}/\omega^2)\overline{\phi}$, so, in order to proceed and consider the case $\tilde{\phi}/\overline{\phi} < 1$, we assume

$$\frac{\omega_{de}\,\omega_{Di}}{\omega^2} < 1 \tag{11-8}$$

and to make the term in $\tilde{\phi}$ relevant within eq. 11-6:

$$\overline{\omega}_{Di} < \hat{\omega}_{Di} \tag{11-9}$$

roughly corresponding to the average magnetic curvatures being smaller than the typical local one. Here $\overline{\omega}_{Di} \equiv [\langle\omega_{Di}\rangle]_{\text{av}}$ and $\hat{\omega}_{Di}{}^2 \equiv [\langle\omega_{Di}^{(0)2}\rangle]_{\text{av}}$. In these conditions eq. 11-7 reduces to

$$\tilde{\phi}(l) - \langle\tilde{\Phi}^{(0)}\rangle = \frac{T_e}{T_i}\frac{\omega_{di}}{\omega^2}\langle\omega_{Di}^{(0)}\rangle\overline{\phi} \tag{11-10}$$

to the lowest order; its solution can be formally expressed as

$$\tilde{\phi} = \frac{T_e}{T_i}\frac{\omega_{di}\,\hat{\omega}_{Di}}{\omega^2}G(l)\overline{\phi} \tag{11-11}$$

where $G(l)$ is a real finite function of l that depends on the detailed geometry of the system. On the other hand, considering the last term of eq. 11-6 and recalling how eq. 9-2 was proved, we have

$$[\langle\omega_{Di}^{(0)}\tilde{\Phi}^{(0)}\rangle]_{\text{av}} = [\phi(l)\langle\omega_{Di}^{(0)}\rangle]_{\text{av}} \tag{11-12}$$

and

$$\overline{\phi}[\tilde{\phi}(l)\langle\omega_{Di}^{(0)}\rangle]_{\text{av}} = \frac{T_i}{T_e}\frac{\omega^2}{\omega_{di}}[\tilde{\phi}^2(l) - \langle\tilde{\Phi}^{(0)2}\rangle]_{\text{av}} \tag{11-13}$$

Finally, we replace $\tilde{\phi}$ by its formal expression 11-11 and obtain from eq. 11-6 the dispersion relation

$$\varepsilon_0 \equiv -\frac{\omega_{di}\overline{\omega}_{Di}}{\omega^2} + \frac{\omega_{di}{}^2\hat{\omega}_{Di}{}^2}{\omega^4}\frac{T_e}{T_i}\Gamma - \frac{\hat{\omega}_{Di}{}^2}{\omega^2}\left(1 - \frac{\omega_{di}}{\omega} + \frac{T_e}{T_i}\frac{\omega_{di}}{\omega}\right) = 0 \tag{11-14}$$

Here $\Gamma \equiv [G(l)^2 - \langle (G^{(0)})^2 \rangle]_{av}$ is positive definite for the Schwarz inequality and finite for B having finite variation along l, and ε_0 is the dielectric constant. The finite solutions of eq. 11-14 are

$$\omega_0 = \frac{1}{2}\,\omega_{di}\,\frac{[1 - T_e/T_i \pm D]}{1 + w}$$

where

$$w \equiv \bar{\omega}_{Di}\,\frac{\omega_{di}}{\hat{\omega}_{Di}^{\,2}} \qquad \text{and} \qquad D^2 \equiv \left(1 - \frac{T_e}{T_i}\right)^2 - 4\Gamma\left(\frac{T_e}{T_i}\right)(1 + w)$$

In particular, for order of magnitude estimates we can write

$$w = \frac{hR_c}{2r_n}$$

where $h \equiv R_c[1/R_c]_{av}$. Therefore, we obtain the following results.

1. If $\Gamma(w + 1) > (1 - T_e/T_i)^2 T_i/4T_e$, an algebraically unstable mode is obtained, independent of wave-particle resonance effects. Notice that the real part of the frequency gives a phase velocity in the direction of the electron diamagnetic velocity for $T_e > T_i$ and one in the ion direction for $T_i > T_e$.

In regions of fairly deep well depth, this mode grows with the rate

$$\gamma \approx \hat{\omega}_{Di}\left(\frac{\Gamma\omega_{di}\,T_e}{\langle\hat{\omega}_{Di}\rangle T_i}\right)^{1/2}$$

i.e., decreasing as the well depth increases. Notice that the condition $\tilde{\phi}/\bar{\phi} \approx \omega_{*i}\omega_{Di}\,T_e\,G/\omega^2 T_i < 1$ is verified for $\bar{\omega}_{Di} < G\hat{\omega}_{Di}$, i.e.,

$$h < G$$

in configurations (see Section XVI) with a relatively deep magnetic well.

2. If $(T_i/4T_e)(1 - T_e/T_i)^2 > (w + 1)\Gamma$, a condition that can be realized in regions of shallow magnetic well and for relatively large values of T_e/T_i or T_i/T_e, eq. 11-14 gives two purely oscillating modes. One can see that they have phase velocity in the direction of the ion diamagnetic velocity or in the contrary one, depending on whether $T_i/T_e > 1$ or $T_e/T_i > 1$.

It is important to notice that the two modes are of opposite energy, i.e., positive and negative, so that we expect, adding to the present treatment the wave-particle resonance terms given in Section X, one of the two modes to become unstable. In fact, the wave energy U_k is proportional to

$$\frac{\omega\,\partial\varepsilon_0}{\partial\omega} \propto \frac{\omega}{\omega_{di}}\left[1 - \frac{T_e}{T_i} - 2(w + 1)\frac{\omega}{\omega_{di}}\right] \propto \mp\left(1 - \frac{T_e}{T_i}\right) - D$$

and the dielectric constant, including the effects of resonance between the wave and the ions with orbit frequency $\omega_b(E, \mu) = \omega_0$ (see Section X) is

$$\varepsilon_1 = \varepsilon_0 - i(\omega - \omega_{di}) \frac{\Delta}{|\omega|}$$

where Δ is positive definite and

$$\Delta \sim \frac{\omega^3}{\hat{\omega}_{bi}^3} \frac{|A^{(1)}|^2}{|\bar{\phi}|^2} \frac{1}{1 + T_e/T_i}$$

Now

$$|A^{(1)}|^2/\bar{\phi}^2 \sim (\tilde{\phi}/\bar{\phi})^2 + (\hat{\omega}_{Di}/\hat{\omega}_{bi})^2 \sim \frac{\hat{\omega}_{Di}^2}{\omega^2}\left(\frac{\omega_{de}^2}{\omega^2} + \frac{\omega^2}{\hat{\omega}_{bi}^2}\right) \sim \left(\frac{\hat{\omega}_{Di}\,\omega_{de}}{\omega^2}\right)^2$$

so that

$$\Delta \sim \frac{(\omega_{de}\,\hat{\omega}_{Di}\,G)^2}{|\omega|\,\hat{\omega}_{bi}^3(1 + T_e/T_i)} \tag{11-15}$$

In this case the dispersion relation can be solved by perturbation, obtaining $\omega = \omega_0 - i\gamma$ and

$$\gamma = \left[D \mp \left(2w + 1 + \frac{T_e}{T_i}\right)\right] \omega_0^4 \frac{\Delta}{[2\omega_{Di}^2|\omega^0|(1 + w)]}$$

From all of this it is reasonable to expect the observation of flutelike fluctuations in configurations where the magnetic well depth is shallow over a sufficiently large region of ψ and when T_i/T_e or T_e/T_i are sufficiently larger than one.

3. The significance of the two large roots of eq. 11-14 can be understood by assuming that $R_c^2 > L^2$, so that $(ka_i)^2 \sim \omega_{Di}^2/\omega^2$. Then, reconsidering the equations of Section X to include finite ion gyroradius, we arrive at the dispersion relation,

$$\varepsilon_2 \equiv b\left(1 - \frac{\omega_{di}}{\omega}\right) - \frac{\omega_{Di}^2}{\omega^2}\left(1 - \frac{\omega_{di}}{\omega} + \frac{T_e}{T_i}\frac{\omega_{di}}{\omega}\right) - \frac{\omega_{di}\,\overline{\omega}_{Di}}{\omega^2} - \frac{\omega_{di}^2\hat{\omega}_{Di}^2}{\omega^4}\frac{T_e}{T_i}\Gamma - i(\omega - \omega_{di})\frac{\Delta}{|\omega|} = 0, \tag{11-16}$$

where $b \approx k_\perp^2 a_i^2$. Now, we can see that the two mentioned roots are essentially the hydromagnetic flute modes that come from balancing the second against the fourth term in the expression for ε_2.

Comparing the flute modes treated in this section with those treated in Section IX, we notice that:

1. The slow modes ($\omega < \hat{\omega}_{bi}$) tend to be of resonant type, i.e., not algebraically unstable, in conditions of extreme temperature ratio (high T_i/T_e or high T_e/T_i) and in regions or configurations with shallow magnetic well. Therefore, a fluidlike instability appears only where the magnetic well becomes sufficiently deep, or when $T_i = T_e$.
2. The fast modes ($\omega > \hat{\omega}_{be}$) on the contrary become fluidlike unstable for shallow magnetic well and relatively large T_e/T_i. When one of these conditions does not exist, then the relevant instability is of resonant type.

Finally we notice that, if one proceeds gradually from regions where the condition $\bar{\omega}_{Di} < \hat{\omega}_{Di}$ is satisfied to regions where it is not, the solution $\phi(l)$ of interest becomes more and more localized, so that the present modes are expected to proceed to those considered in Section XV. This corresponds to going from regions in which there is a relatively large number of particles with unfavorable average curvature drift $v_D^{(0)}$ to regions with relatively few particles of this type (see Section XVI).

XII. Low-Density, Ion Flute Mode (29)

Now, if we consider the low-density regime, where finite Debye-length effects are important, we can obtain another kind of unstable modes of flute type, provided that

$$\frac{T_e}{T_i}\frac{\omega_{di}\hat{\omega}_{Di}}{\omega^2}G < 1 \tag{12-1}$$

as follows from eq. 11-11. The relevant dispersion relation then becomes

$$\varepsilon_3 \equiv \frac{T_i}{T_i + T_e}\left[k^2\lambda_{Di}{}^2 + \bar{b}\left(1 - \frac{\omega_{di}}{\omega}\right)\right] - \frac{\hat{\omega}_{Di}{}^2}{\omega^2}\left(1 - \frac{\omega_{di}}{\omega} + \frac{T_e}{T_i}\frac{\omega_{di}}{\omega}\right) \tag{12-2}$$

$$\frac{\omega_{di}\bar{\omega}_{Di}}{\omega^2} - \frac{\omega_{di}{}^2\omega_{Di}{}^2}{\omega^4}\frac{T_e}{T_i}\Gamma - i(\omega - \omega_{di})\frac{\Delta}{|\omega|} = 0$$

where $\lambda_{Di}{}^2 = T_i/4\pi ne^2$.

We notice that, if $T_e/T_i \ll 1$, the interesting mode has ω independent of T_e; therefore from eq. 11-11 it follows that $\tilde{\phi}$ is negligible with respect to $\bar{\phi}$. So, the present mode becomes really a flute. Then the dispersion relation reduces to

$$k^2\lambda_{Di}{}^2 - \frac{\omega_{di}\bar{\omega}_{Di}}{\omega^2} + \left(1 - \frac{\omega_{di}}{\omega}\right)\left(\bar{b} - \frac{\hat{\omega}_{Di}{}^2}{\omega^2} - i\Delta\frac{\omega}{|\omega|}\right) = 0 \tag{12-3}$$

where Δ is now independent of $\tilde{\phi}$, since $|A^{(1)}|^2 \sim |\bar{\phi}\hat{\omega}_{Di}/\omega|^2$.

Therefore, from eq. 10-3, we have

$$\Delta \sim \frac{|\omega| \hat{\omega}_{Di}^{\ 2}}{\hat{\omega}_{bi}^{\ 3}} \tag{12-4}$$

If the magnetic well is sufficiently shallow so that $\overline{\omega}_{Di}/\hat{\omega}_i < \hat{\omega}_{Di}/\omega$, then we can have a residual instability with $\omega = \omega_{di} + \delta\omega$ and

$$\frac{\delta\omega}{\omega_{di}} = \left(k^2 \lambda_{Di}^{\ 2} - \frac{\overline{\omega}_{Di}}{\omega_{di}}\right) \frac{1}{\hat{\omega}_{Di}^{\ 2}/\omega_{di}^{\ 2} + i\Delta} \tag{12-5}$$

in the proper asymptotic limit. In this case instability occurs if (29):

$$k^2 \lambda_{Di}^{\ 2} > \frac{2hr_n}{R_c} \tag{12-6}$$

On the other hand, a fluidlike instability can also be obtained from eq. 12-3, which, in the limit $L/R_c \sim 1$, reduces to

$$1 - \frac{\overline{\omega}_{Di}\,\omega_{di} + \hat{\omega}_{Di}^{\ 2}}{k^2 \lambda_{Di}^{\ 2} \omega^2} + \frac{\omega_{di}\,\hat{\omega}_{Di}^{\ 2}}{k^2 \lambda_{Di}^{\ 2} \omega^3} = 0 \tag{12-7}$$

Then the instability condition is

$$k\lambda_{Di} > \frac{2}{3^{3/2}} \left[h \left(\frac{R_c}{2r_n}\right)^{1/3} + \left(\frac{2r_n}{R_c}\right)^{2/3} \right]^{3/2} \tag{12-8}$$

and can be compared to eq. 12-6 for the resonant type of instability (29,30).

XIII. Slow Fluidlike Drift Modes (14,23,24)

Now we consider modes that are not of flute type in the sense that $\tilde{\phi} \gtrsim \bar{\phi}$. As we shall see, these modes can become unstable even in the absence of magnetic curvature effects and differ from those considered in Section VII because they essentially rely on $\Phi^{(0)} \neq 0$. For this we refer to the quadratic form eq. 10-3, take the limit $\omega_{Di} < \omega$, and recall the derivation of eq. 11-1. Thus we have, for $k^2 \lambda_{De}^{\ 2} < 1$ and $k^2 a_i^{\ 2} < 1$,

$$\Bigg[\left(1 + \frac{T_e}{T_i}\right)(|\phi|^2 - \langle|\Phi^{(0)}|^2\rangle) - \left(1 + \frac{T_e}{T_i}\right) \frac{T_e}{T_i} \frac{\omega_{di}}{\omega^2} \langle \omega_{Di}^{(0)} |\Phi^{(0)}|^2 \rangle \tag{13-1}$$

$$+ \left(1 - \frac{\omega_{di}}{\omega}\right) \frac{T_e}{T_i} \langle |\Phi^{(0)}|^2 - |A_i^{(0)}|^2 \rangle \Bigg]_{\text{av}} + \left(1 - \frac{\omega_{di}}{\omega}\right) \frac{T_e}{T_i} \frac{\omega^2}{\hat{\omega}_{bi}^{\ 2}}$$

$$\cdot \sum_{n \neq 0} \iint dE\, d\mu\, f_{i0} |\tau^3| \sum_{n \neq 0} \left| \frac{2\pi}{n} \hat{\omega}_{bi} \Phi^{(n)}(E, \mu) \right|^2 = 0$$

It is then possible consistently to find modes that become unstable when the term in $\omega^2/\hat{\omega}_{bi}{}^2$ becomes important to the lowest order. This leads us to consider the limit $\omega < \omega_{di}$. Then, recalling eq. 1-10 and performing the usual integration, we have

$$\iint dE\, d\mu f_{io} |\tau| \left\{ \left(1 + \frac{T_i}{T_e}\right) \sum_n |\Phi^{(n)}|^2 - \left(1 + \frac{T_e}{T_i}\right) \frac{\omega_{di}\,\omega_{Di}^{(0)}}{\omega^2} |\Phi^{(0)}|^2 \right. \tag{13-2}$$

$$\left. - \frac{\omega_{di}}{\omega} \left(|\Phi^{(0)}|^2 - |A_i^{(0)}|^2 + \frac{\omega^2}{\omega_b{}^2} \sum_{n \neq 0} \left| \frac{\Phi^{(n)}}{n} \right|^2 \right) \right\} = 0$$

where $|\Phi^{(0)}|^2 - |A_i^{(0)}|^2$ is real positive and of order $b|\Phi^{(0)}|^2$. With an obvious identification of the integrals P, Q, R, and S, eq. 13-2 can be rewritten as

$$\left(1 + \frac{T_i}{T_e}\right) P - \frac{\omega_{di}}{\omega} \left\{ \left(1 + \frac{T_e}{T_i}\right) \frac{\hat{\omega}_{Di}}{\omega} Q + \bar{b} R + \frac{\omega^2}{\hat{\omega}_{bi}{}^2} S \right\} = 0 \tag{13-3}$$

Here we have introduced the quantities $\bar{b}$, $\hat{\omega}_{bi}$, and $\hat{\omega}_{Di}$ into the dispersion relation to put the order of magnitude of the various terms into evidence. Notice that it is possible to derive different variational principles in ω, ω^2, or ω^3 from eq. 13-3. We shall use this form, even when it is not variational, to obtain an order of magnitude estimate for ω and for the instability threshold.

We see first of all that, if the magnetic curvature is so small that

$$\bar{b} > \frac{\hat{\omega}_{Di}}{\omega}$$

instability occurs when

$$\omega_{di} > \hat{\omega}_{bi} \left(1 + \frac{T_i}{T_e}\right) \frac{P}{(RSb)^{1/2}} \tag{13-4}$$

The growth rate then is of the order

$$\gamma \sim \bar{b}^{1/2} \hat{\omega}_{bi} \left(\frac{R}{S}\right)^{1/2} \sim k a_i \frac{v_{\text{th}i}}{L} \tag{13-5}$$

The relevant instability can be classified as one of drift type because it does not depend on magnetic curvature and involves transformation of the average ion kinetic energy along the field lines. One can also see from eq. 13-3 that the unstable mode has phase velocity in the direction of the ion diamagnetic velocity. In addition, if one considers the energy of the two modes resulting from eq. 13-3, when $\hat{\omega}_{Di} = 0$, as was done for the modes of Section VII, one can see that they are of opposite energy sign and can argue that their nonlinear interaction will make them unstable, even though eq. 13-4 is not strictly satisfied. Therefore, the least restrictive instability condition can be given as

$$\omega_{di} > \hat{\omega}_{bi}\left(1 + \frac{T_i}{T_e}\right)^{1/2} \tag{13-6}$$

Now if we strictly consider the case in which $L \sim R_c$, the curvature term $\hat{\omega}_{Di}/\omega$ is certainly prevailing over $\bar{b}$ in eq. 13-3. In this case we can reduce eq. 13-3 to the form,

$$1 - \frac{D}{x^2} - x = 0 \tag{13-7}$$

by proper normalization of the variable $\omega \propto x$, where

$$D \equiv \frac{\hat{\omega}_{Di}\,\omega_{di}{}^3}{\hat{\omega}_{bi}{}^4}\,\frac{S^2Q}{(1 + T_i/T_e)^2 P^3} \tag{13-8}$$

Therefore, if $D > 0$, we can see that a fluidlike instability occurs for

$$D > \frac{4}{27} \tag{13-9}$$

which roughly implies, numerical factors apart,

$$\bar{b} > \left(\frac{r_n}{L}\right)^{3/2}\left(\frac{R_c}{L}\right)^{1/2}\left(1 + \frac{T_i}{T_e}\right)$$

Therefore, for $T_i/T_e > 1$, a possible stability condition is

$$\frac{T_i}{T_e} \gtrsim \left(\frac{L}{r_n}\right)^{3/2}\left(\frac{L}{R_c}\right)^{1/2}$$

When condition 13-9 is instead satisfied, the growth rate has the order of magnitude,

$$\gamma \sim \hat{\omega}_{bi}^{2/3}\hat{\omega}_{Di}^{1/3}\left(1 + \frac{T_e}{T_i}\right)^{1/3}\left(\frac{Q}{S}\right)^{1/3} \sim (ka_i)^{1/3}\,\frac{v_{\text{thi}}}{(L^2R_c)^{1/3}}\left(1 + \frac{T_e}{T_i}\right)^{1/3}$$

is equal to that of the frequency of oscillation and is considerable for realistic plasma configurations.

When $0 < D < \frac{4}{27}$, the same mode remains unstable due to the wave-particle resonance term that has been neglected in eq. 13-3 and will later be reconsidered explicitly (Section XV).

Finally, if $D < 0$, another mode becomes fluid unstable, and this belongs to the class of modes we discuss in the next section.

XIV. Curvature Driven Trapped-Particle Modes

These are the modes (12,16,17) obtained from eq. 11-1 in the limit, where the last term, due to the finite ion gyroradius, is negligible and occurs on lines of force on which only a few particles, which spend a relatively long time in regions of unfavorable curvature, have $\omega_{Di}^{(0)}\omega_{di} < 0$, i.e., $\partial J/\partial\psi > 0$. These modes have relatively strong localization along the lines of force, so that $\bar{v}_{Di}$ no longer enters into the picture and clearly $\tilde{\phi}/\bar{\phi}$ is no longer small.

We shall consider for simplicity the case in which resonances involving $\omega - \omega_{Di}^{(0)} = 0$ are not important; then eq. 11-1 becomes a variational form that we can write as

$$\omega^2 = \frac{T_e}{T_i}\omega_{di}\frac{(\pi/2)\iint dE\, d\mu f_{0i}|\tau|\omega_{Di}^{(0)}|\Phi^{(0)}|^2}{\oint dl/B\{|\phi|^2 - \langle|\Phi^{(0)}|^2\rangle\}} \tag{14-1}$$

It is evident that, if there are few particles with $\omega_{Di}^{(0)} < 0$, it is important that the corresponding τ be relatively large so that $\omega^2 < 0$. Therefore, the particles contributing to instability will be barely trapped ones and the barely circulating ones that spend a long time in the region of unfavorable curvature.

In fact, if we take a highly localized trial function, such as $\phi(l) \approx \delta(l - l_0)$, we have

$$\Phi^{(0)}(\lambda) \approx \left\{[1 - \lambda B(l_0)]^{1/2}\oint\frac{dl}{(1-\lambda B)^{1/2}}\right\}^{-1}$$

and the numerator of eq. 14-1 becomes

$$\left(\frac{\pi}{2}\int dE f_{0i}(E)E^{3/2}\right)\int_0^{1/B(l_0)} d\lambda\,\frac{\hat{\omega}_{Di}^{(0)}(\lambda)}{[1-\lambda B(l_0)]\oint dl/(1-\lambda B)^{1/2}}$$

where $\hat{\omega}_{Di}^{(0)}(\lambda) = \omega_{Di}^{(0)}(E, \mu)T_i/m_i E$.

Now, due to the singular factor $1 - B(l_0)$ in the denominator of the λ integral, this is dominated by $\omega_{Di}^{(0)}(\lambda = B(l_0)^{-1})$. Therefore, instability can be found whenever a particle with approximately zero parallel velocity at $B(l_0)$ has an unfavorable curvature drift.

XV. Impurity Particle Modes

When a plasma contains, for instance, two populations of ions with different thermal velocities due to different temperatures or to different masses, then as shown in Ref. 31 new unstable modes can arise. It has been pointed out (3,14,24) that there is a close similarity between these kinds of mode and the trapped-particle modes that we have considered earlier.

To avoid considering at first the effects of magnetic curvature, we refer to the case analyzed in Section XIII. The relevant modes can be simulated by those arising in a configuration with straight magnetic field (Figure 3), where the role of trapped electrons and ions is played by cold electrons and ions with densities n_{eC} and n_{iC} and temperatures T_{eC} and T_{iC}. The normal-mode solutions are now of the form

$$\Phi = \tilde{\phi} \exp (i\omega t + ik_\theta r_\theta + ik_{\|} l) \tag{15-1}$$

and their phase velocity is such that

$$v_{\text{th}iC} < v_{\text{th}eC} < \frac{\omega}{k_{\|}} < v_{\text{th}i} < v_{\text{th}e} \tag{15-2}$$

So, the perturbed densities are, respectively,

$$\tilde{n}_{eC} = \frac{\tilde{\phi}}{\omega} k_\theta \frac{c}{B} \frac{dn_{eC}}{dx} = \frac{e\tilde{\phi}}{T_{eC}} n_{eC} \frac{\omega_{deC}}{\omega} \tag{15-3}$$

with evident definition of ω_{deC}, and

$$\tilde{n}_{iC} = -\frac{e\tilde{\phi}}{T_{iC}} n_{iC} \left[\frac{\omega_{diC}}{\omega} + b_{iC}\left(1 - \frac{\omega_{diC}}{\omega}\right)\right] \tag{15-4}$$

To derive eq. 15-3, we have just taken into account the $\mathscr{E} \times \mathbf{B}$ drift across the lines of force, while from eq. 15-4 we have also added the polarization and finite Larmor radius drift (see Section XVII). On the other hand, for the hot (untrapped) electrons, we have

$$\tilde{n}_e = \frac{e\tilde{\phi} n_e}{T_e} \tag{15-5}$$

and for the corresponding ions,

$$\tilde{n} = -\frac{e\tilde{\phi}}{T_i} n_i \left[1 - \left(1 - \frac{\omega_{di}}{\omega}\right) i\sqrt{\pi} \frac{\omega}{k_{\|} v_{\text{th}i}}\right] \tag{15-6}$$

where the last term in square brackets represents the effect of Landau damping. Now, to simulate the wave-particle type of resonance occurring in a varying magnetic field, we assume that the distribution function of the latter population has a plateau at $v_{\|} = 0$. Therefore, as shown by eq. 6-6, we have

$$\tilde{n}_i = -\frac{e\tilde{\phi}}{T_i} n_i \left[1 + \left(1 - \frac{\omega_{di}}{\omega}\right) \alpha_2 \frac{\omega^2}{k_{\|}{}^2 v_{\text{th}i}^2}\right] \tag{15-7}$$

By imposing the quasineutrality condition in the form $\tilde{n}_i + \tilde{n}_{iC} = \tilde{n}_e + \tilde{n}_{eC}$, with $n_i = n_e$ and $n_{iC} = n_{eC} \equiv n_C$, we arrive at the dispersion relation,

$$\left(1 + \frac{T_i}{T_e}\right) - \frac{\omega_{di}}{\omega}\left(\frac{n_C}{n} b_{iC} + \alpha_2 \frac{\omega^2}{k_{\parallel}{}^2 v_{\text{th}i}^2}\right) \tag{15-8}$$

which reproduces eq. 13-3.

In conclusion we see that a spatially varying magnetic field, besides modifying the process of wave-particle resonance with the consequences indicated in Section VI and later, has the effect that the perturbed particle distribution for frequencies smaller than the typical bounce frequency is not Maxwellian, in the sense that $f_1 \neq -e\Phi f_0/T$ but $f_1 = -(ef_0/T)(\phi - \Phi^{(0)})$. The reason is that there is always a number of sufficiently slow particles such that the wave has a frequency larger than their bounce frequency; the situation is like that of having a cold population of particles such that $k_{\parallel} v_{\text{th cold}} < \omega < k_{\parallel} v_{\text{th main}}$ in the simulating straight magnetic-field configuration that we mentioned earlier.

Finally, one can include the effects of magnetic curvature by adding the proper curvature drift terms (see Section XVII), so that, instead of eqs. 15-3 and 15-4, we have

$$\tilde{n}_{eC} = \frac{e}{T_{eC}} n_{eC}\left[\frac{\omega_{deC}}{\omega} - \frac{\omega_{DeC}}{\omega}\left(1 - \frac{\omega_{deC}}{\omega}\right)\right]\tilde{\phi}$$

and

$$\tilde{n}_{iC} = -\frac{e}{T_{iC}} n_{iC}\left[\frac{\omega_{diC}}{\omega} + \left(b_{iC} - \frac{\omega_{DiC}}{\omega}\right)\left(1 - \frac{\omega_{diC}}{\omega}\right)\right]\tilde{\phi}$$

If, in addition, we take into account the simulated wave-particle resonance effect, eq. 15-6 becomes, after eq. 6-6:

$$\tilde{n}_i = -\frac{e\tilde{\phi}}{T_i} n_i\left[1 + \left(1 - \frac{\omega_{di}}{\omega}\right)\left(\alpha_2 \frac{\omega^2}{k_{\parallel}{}^2 v_{\text{th}i}^2} - i\alpha_3 \frac{\omega^3}{k_{\parallel}{}^3 v_{\text{th}i}^3}\right)\right]$$

The final dispersion relation is

$$\left(\frac{1}{T_i} + \frac{1}{T_e}\right) - \frac{\omega_{DiC}\,\omega_{diC}}{\omega^2}\frac{n_C}{n}\left(\frac{1}{T_{iC}} + \frac{T_{eC}}{T_{iC}{}^2}\right) - \frac{n_C}{n}\frac{b_{iC}}{T_e}\left(1 - \frac{\omega_{diC}}{\omega}\right) + \left(1 - \frac{\omega_{di}}{\omega}\right)\left(\alpha_2 \frac{\omega^2}{k_{\parallel}{}^2 v_{\text{th}i}^2} - i\alpha_3 \frac{\omega^3}{k_{\parallel}{}^3 v_{\text{th}i}^3}\right) = 0 \tag{15-9}$$

and reproduces eq. 13-3 with the addition of the wave-particle resonance term. So, in analogy to eq. 13-7, we can rewrite eq. 15-9, when the term in b_{iC} is negligible, in the form

$$1 - \frac{D}{x^2} - x + i\Delta x^2 = 0$$

for $\omega < \omega_{di}$, where D and x are defined as in eqs. 13-7 and 13-8 and Δ represents the wave-particle resonance term. Then, it is easy to verify that in the presence of this term there is always an unstable mode for any value of D.

XVI. Comparison with Experiments

A comparison of the results previously obtained with exprimental observations is difficult at this stage. On the one hand, what is observed is the end of the nonlinear evolution of collective modes whose features are theoretically known only about their onset. In fact, the problem of the nonlinear effects of the modes described in the previous sections has not been approached yet, in view of the fact that they involve ordinary space and velocity space in an unusually complex fashion. For the same reasons there is not yet a complete picture of all the relevant effects of particle collisions.

On the other hand, most of the experiments that have been carried out in two-dimensional plasma configurations have been primarily aimed at achieving a high confinement time, and there has not yet been sufficient effort devoted to more academic questions such as recognizing why certain modes predicted by theory are observed and why other modes are not observed. The main source of experimental results in this field are the so-called multipole (13) experiments with current conductors inside the plasma, creating the confining magnetic field (Figures 3,5, and 6).

With reference to these experiments we can say that

1. The fast electron modes described in Section V and VII can explain the observation of ballooning modes in the Princeton linear quadrupole with hot-electron plasmas, i.e. $T_e > T_i$. In fact, the observed modes have phase velocity in the direction of the electron diamagnetic velocity, maximum amplitude near the point $l = l_0$ of maximum unfavorable curvature (21,36), an odd profile around the points of minimum magnetic field, so that $\Phi^{(0)} = 0$, and frequency of oscillation

$$\omega \approx \omega_{de} \qquad \text{with} \qquad \hat{\omega}_{bi} < \omega < \hat{\omega}_{be}$$

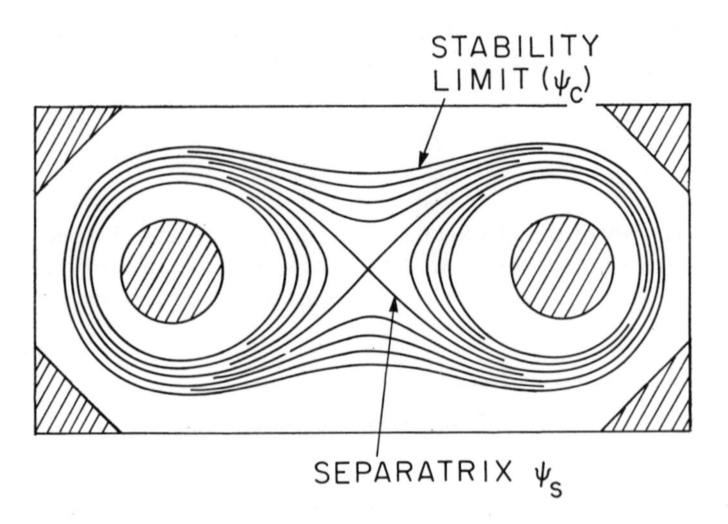

Fig. 5. Magnetic field lines in a quadrupole configuration.

2. The slow ion flute ($\omega < \hat{\omega}_{bi} < \hat{\omega}_{be}$) modes considered in the limit of $T_i > T_e$ as described in Section X, can explain the observation of flute modes in the Gulf General Atomic (GGA) toroidal quadrupole with hot-ion plasmas (28,32), i.e. $T_i > T_e$. In fact, these modes have been detected inside the critical magnetic surface $\psi = \psi_c$, where $[\langle v_{Di} \rangle]_{\text{av}} = (\partial/\partial\psi) \oint dl/B = 0$, separating the (outer) region in which hydromagnetic types of mode are excited from the (inner) region, where modes of this type cannot be excited, since the magnetic curvature is, on the average of phase space, favorable to stability. In addition, the observed phase velocity was in the direction of the ion diamagnetic velocity; the frequency was smaller than $\hat{\omega}_{bi}$; and no modes of this kind appeared to exist within the magnetic surface $\psi = \psi_L$, where $(\partial/\partial\psi) \oint dl$ changes sign.

The last observation can be explained by recalling that, for the existence of these flute modes, it is necessary to have $[\langle \omega_{di} \rangle]_{\text{av}}^2 < [\langle \omega_{Di}^{(0)2} \rangle]_{\text{av}}$, so that a considerable number of particles need have unfavorable drift averaged over the orbit, i.e., $\partial J/\partial\psi > 0$. It has been shown, in fact, by numerical computations of $dJ/d\psi$ that the transition from very few to very many particles with unfavorable drifts occurs roughly at ψ_L.

Note that modes of the same kind have been found to be absent (or of much smaller amplitude) in the octupole configuration (33,34) for equal ranges of plasma parameters (temperatures and density). This can be explained

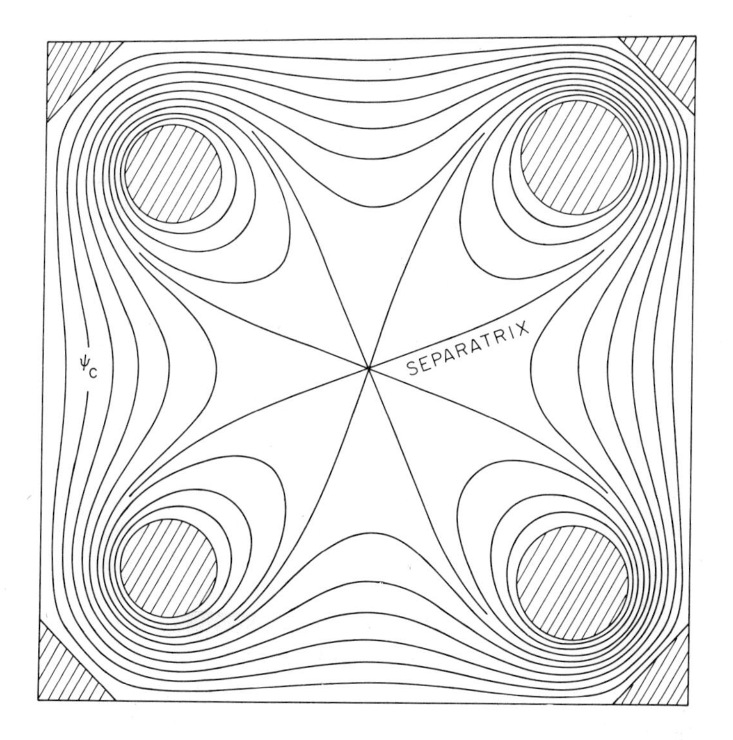

Fig. 6. Magnetic field lines in an octupole configuration.

by referring to eq. 11-14 and noticing that an octupole has a deeper magnetic well (35), so that the inequality

$$(w + 1)\Gamma > \left(1 - \frac{T_e}{T_i}\right)^2 \frac{T_i}{4T_e} \tag{16-1}$$

is better satisfied within $\psi = \psi_c$, and has more flux volume within $\psi = \psi_L$ than a quadrupole. Then the dominant modes are unlikely to be detected as fluctuations and be of flute type. In particular, they may be responsible for the observed anomalous particle losses, from octupole-confined plasmas, that cannot be associated with the very low level of fluctuations appearing inside $\psi = \psi_c$.

More generally, if a mode is expected to be present in a given plasma configuration, where it is likely to give rise to sizable anomalous particle losses, one way to recognize it is by analyzing those regimes or regions of space in which it tends to manifest itself as a fluctuation (28). In fact, in this case the information on frequency, wavelength, and phase velocity is of considerable value in finding a correlation with the theoretical results.

3. Experiments carried out in the Princeton linear quadrupole with hot-electron plasmas ($T_e > T_i$) have also shown the existence of flute modes (36). The observed typical frequency is $\omega < \hat{\omega}_{bi}$ and, different from the modes observed in the GGA quadrupole experiment with $T_i > T_e$, the phase velocity is in the direction of the electron diamagnetic velocity. These facts are all consistent with the conclusions concerning the slow flute modes considered in Section XI in the regime $T_e > T_i$ and with the inequality 16-1 satisfied.

Finally, we recall that preliminary identification of the modes observed in the Culham quadrupole experiment for plasmas with $T_i > T_e$ points toward modes that are not localized in any special region of the lines of force and tend to have phase velocity in the direction of the ion diamagnetic velocity (37).

4. It has been noticed, again in experiments on the GGA toroidal quadrupole with $T_i > T_e$, that, when decreasing the ratio T_i/T_e, a new flute mode, in addition to the one considered at the point 2, tends to appear (38). This mode has a phase velocity in the direction of the electron diamagnetic velocity; in view of the fact that in the experimental situation T_i/T_e is over one, we cannot associate it with the modes discussed in Section XI, but rather with the fast electron flute modes of Section VIII or IX.

5. Experiments carried out on the Princeton minipole and the GGA multipoles have shown that, when the density is smaller than a given amount new flute modes appear (34,39). These have phase velocity in the direction of the ion diamagnetic velocity, frequency $\omega < \hat{\omega}_{bi}$ and are identifiable by the modes treated in Section XII.

6. We now refer to experiments on stellarator-confined plasmas. We notice that the theory developed in the previous sections for configurations with closed lines of force can be applied in the central part of a typical stellarator plasma column, where the magnetic shear is small and the average magnetic curvature is favorable to stability but with shallow magnetic well depth. In fact, the typical order of magnitude of this corresponds to

$$h \sim \frac{r_n}{R_c}$$

Specifically, for experiments in which $T_e \gg T_i$, the conclusions derived in Section IX are of special interest; the flute modes found there can explain the observed fluctuations that are extended over the entire length of the plasma column and are not localized in any special region of the magnetic-field lines (40). In addition, the theoretical growth rates of these modes and the ratio of their wavelengths across the magnetic field to the ion gyroradius are quite adequate to justify the relatively large, particle diffusion coefficients that have been observed in this regime. For the sake of illustration, if we take the diffusion coefficient $\mathscr{D}$ as of the order $\sim \gamma \lambda_\perp^2$, where $\lambda_\perp$ is the typical wavelength transverse to the magnetic field, and $\lambda_\perp \sim r_n$, we have

$$\mathscr{D} \sim \left(\mathscr{D}_B \frac{r_n^4 T_e}{R_c^2 m_i} \right)^{1/3}$$

for $\mathscr{D}_B \sim T_e c/eB$ being the known Bohm diffusion coefficient.

Regimes in which $T_i > T_e$ have been realized through ion cyclotron heating (41). In this case, if we apply to results obtained in Section XI, we again expect the observation of flute modes but with phase velocity in the direction of the ion diamagnetic velocity. A tentative identification of these modes with phase velocity in the direction of the ion diamagnetic velocity has, in fact, been proposed in Ref. 41.

7. Recently, the observation of modes with $\omega < \hat{\omega}_{bi}$, and a phase velocity in the direction of the ion diamagnetic velocity, localized in a region of unfavorable curvature, has been reported for plasmas with $T_i > T_e$ confined in a hard core (Levitron) configuration (42). A precise identification of these modes, which appear to belong to the class of modes considered in Sections XI and XIV, would require a more detailed analysis, taking into account the typical features of a Levitron configuration.

8. In spite of their evident importance, there has not been yet any identification of the slow drift modes considered in Section XIV. One reason for this may be because they have a relatively strong growth rate, of the order of, or larger than, their frequency of oscillation. Thus they may not manifest themselves as fluctuations, but rather as the cause of anomalous transport processes such as the particle diffusion that has been observed to occur even inside regions of multipoles, where the average curvature drift of all particles is favorable (38).

XVII. Appendix: Simplified Treatment of Electron Modes by the Guiding-Center Approximation

There are instances in which a simplified treatment can be adopted for some of the modes discussed above, so that the various physical factors come into a clearer light. We refer here to the electron modes discussed in Section V, characterized by having

$$\hat{\omega}_{bi} < \omega < \hat{\omega}_{be} \qquad \text{and} \qquad \Phi^{(0)} = 0. \tag{17-1}$$

In this case, the electron perturbed density can be obtained from the electron momentum balance equation, where the inertial term is negligible,

$$0 = -\frac{\mathbf{B}}{B} \cdot \nabla(n_{e1} T_e - en\Phi) \tag{17-2}$$

so that

$$n_{e1} = \frac{en}{T_e} \Phi$$

The ion density is instead obtained from the guiding-center conservation equation:

$$i\omega n_{G1} + \nabla \cdot (n_{G1}\mathbf{u}) + \nabla \cdot (n_G \mathbf{u}_{1\perp}) + \nabla \cdot \left(n_G \frac{\mathbf{B}}{B} u_{\parallel 1}\right) = 0 \tag{17-3}$$

where n_G is the guiding-center density and $\mathbf{u}$ their drift velocity. We recall that

$$u_\theta = \frac{2cT_i m_i}{eB} \frac{\partial B}{\partial \psi} = \langle v_{Di} \rangle \tag{17-4}$$

and

$$\mathbf{u}_{1\perp} = \left(1 + \frac{1}{4} a_i^2 \nabla_\perp^2\right) \frac{\mathscr{E}_1 \times \mathbf{B}}{B^2} c + \frac{c}{\Omega_i B} \frac{d\mathscr{E}_1}{dt} \tag{17-5}$$

where we have included, respectively, the curvature drift, the $\mathscr{E} \times \mathbf{B}$ drift, the finite Larmor radius drift, and the polarization drift, with $a_i^2 = 2T_i/(m_i \Omega_i^2)$. The flow velocity along the magnetic field is given by

$$i\omega n m_i u_{\parallel 1} = -\frac{\mathbf{B}}{B} \cdot \nabla(n_{G1} T_i + en\Phi) \tag{17-6}$$

Finally, the guiding-center density is related to the ion density by

$$n_{i1}^2 = [1 + \tfrac{1}{4} a_i^2 \nabla_\perp^2] n_{G1} \tag{17-7}$$

since we are considering the long wavelength approximation where $a_i^2 \nabla_\perp^2 \ll 1$. So, we obtain

$$i\left[\omega + \langle \omega_{Di} \rangle + \frac{T_\perp}{m_i \omega} B \frac{\partial}{\partial} \frac{1}{B} \frac{\partial}{\partial l}\right] n_{Gk} - i\left(1 - \frac{1}{2} b\right) k_\theta c \frac{\partial n}{\partial \psi} \phi - ic \frac{\omega n}{\Omega_i B} k_\theta^2 \phi = 0$$

where $b \equiv \tfrac{1}{2} k_\theta^2 a_1^2$ and $\langle \omega_{Di} \rangle = k_\theta \langle v_{Di} \rangle$.

Now, if we make use of eq. 17-7 and further expand

$$\frac{\langle\omega_{Di}\rangle}{\omega} < 1 \qquad \text{and} \qquad \frac{T_i}{m_i\omega^2} B \frac{\partial}{\partial l}\frac{1}{B}\frac{\partial}{\partial l} < 1$$

we finally obtain

$$n_{ik} = -\frac{en}{T_i}\left\{1 - \left(1 + \frac{\omega_{di}}{\omega}\right)\left[1 - \frac{\langle\omega_{Di}\rangle}{\omega} - b - \frac{T_i}{m_i\omega^2} B \frac{\partial}{\partial l}\frac{1}{B}\frac{\partial}{\partial l}\right]\right\}\phi \quad (17\text{-}8)$$

Therefore, the dispersion equation coming from the quasineutrality condition $n_{ik} = n_{ek}$ is

$$\phi(l) + \frac{\omega - \omega_{di}}{\omega - \omega_{de}}\left\{b\frac{T_e}{T_i} - \frac{\langle\omega_{De}\rangle}{\omega} + \frac{T_e}{m_i\omega^2} B \frac{\partial}{\partial l}\frac{1}{B}\frac{\partial}{\partial l}\right\}\phi(l) = 0 \quad (17\text{-}9)$$

as obtained from eq. 8-3 for $\Phi^{(0)} = 0$. We see that, as known for the ordinary drift wave, the ion temperature does not play an essential role. It just adds a finite Larmor radius correction, resulting in the term ω_{di}. The essential role is instead played by the electron temperature longitudinal to the magnetic field consistently with the fact that the reservoir of energy for the relevant instability is the electron thermal energy parallel to **B**.

Acknowledgments

It is a pleasure to thank the many colleagues and friends with whom the author has shared the understanding or the finding of the results presented in this paper.

Special thanks are due to Miss D. McEntee for her effective collaboration in preparing the manuscript and to Dr. G. Guest for critically reading it.

References

1. H. Lamb, *Hydrodynamics*, 6th ed. rev., New York, 1932.
2. B. Davison, *Neutron Transport Theory*, London, 1958.
3. B. B. Kadomtsev, Review paper given at the 1967 European Conference on Plasma Physics, Royal Institute of Technology, Stockholm. (Unpublished.)
4. M. N. Rosenbluth and C. L. Longmire, *Ann. Phys.* (N.Y.), **1**, 120 (1957).
5. B. Coppi, H. Hendel, F. Perkins, and P. Politzer: *Proc. Conference on Physics of Quiescent Plasmas*, Part 1, Frascati, 1967, p. 201; H. Hendel, B. Coppi, F. Perkins, and P. Politzer: *Phys. Rev. Letters*, **18**, 439 (1967).
6. B. B. Kadomtsev, *Plasma Turbulence*, London, 1965.
7. R. Z. Sagdeev and A. A. Galeev; International Centre for Theoretical Physics Report IC/66/64, Trieste, 1966.
8. B. B. Kadomtsev and O. P. Pogutze: *Sov. Phys.—JETP*, **26**, 1146 (1968).
9. B. Coppi and E. Mazzucato, Princeton University Plasma Physics Report MATT-720, 1969; *Phys. Fluids* (1971).

10. T. H. Dupree, *Phys. Fluids*, **11**, 2680 (1968).
11. B. Coppi, G. Laval, R. Pellat, and M. N. Rosenbluth, International Centre for Theoretical Physics Report IC/66/55, Trieste, 1966; *Plasma Phys.*, **10**, 1 (1968).
12. B. B. Kadomtsev: *Sov. Phys.—JETP Letters*, **4**, 10 (1966).
13. T. Okhawa and W. D. Kerst, *Nuovo Cimento*, **22**, 784 (1961).
14. A. A. Galeev, R. Z. Sagdeev, and V. Wong: *Phys. Fluids*, **10**, 1553 (1967).
15. H. P. Furth, in *Plasma Physics*, Vienna, 1966, p. 391.
16. M. N. Rosenbluth: *Phys. Fluids*, **11**, 869 (1968).
17. B. B. Kadomtsev and O. P. Pogutze, *Sov. Phys.—JETP*, **24**, 1172 (1967).
18. P. Rutherford and E. A. Frieman, *Phys. Fluids*, **11**, 569 (1968).
19. R. J. Hastie, J. B. Taylor, and F. A. Haas, *Ann. Phys.*, (N.Y.), **41**, 302 (1968).
20. J. B. Taylor, *Phys. Fluids*, **10**, 1357 (1967).
21. B. Coppi, M. N. Rosenbluth, and S. Yoshikawa, *Phys. Rev. Letters*, **20**, 190 (1968).
22. J. D. Jukes and K. Rohlena, *Phys. Fluids*, **11**, 891 (1968).
23. P. Rutherford, M. N. Rosenbluth, and B. Coppi, Paper presented at the 1968 Sherwood Meeting on Theoretical Plasma Physics, Berkeley, Calif., 1968.
24. B. Coppi, M. N. Rosenbluth, and P. Rutherford, Princeton Plasma Physics Laboratory MATT-611, 1968; *Phys. Rev. Letters*, **21**, 1055 (1968).
25. V. M. Dikasov, L. I. Rudakov, and D. D. Ryutov, *Sov. Phys.—JETP*, **21**, 608 (1966).
26. B. Coppi, M. N. Rosenbluth, and R. N. Sudan, Princeton University Plasma Physics Report MATT-529, 1967–1969; *Ann. Phys.*, (N.Y.), **55**, 207 (1969).
27. B. Coppi, Princeton University Plasma Physics Report MATT-668, 1969; *Phys. Rev.*, **184**, 130 (1969).
28. B. Coppi, *Phys. Rev. Letters*, **22**, 50 (1969).
29. B. Coppi and S. Ossakow, Princeton Plasma Physics Laboratory MATT-567, 1967; B. Coppi, S. Ossakow, and M. N. Rosenbluth, *Plasma Phys.*, **10**, 571 (1968).
30. C. W. Horton, B. Coppi, and M. N. Rosenbluth, Paper presented at the 1968 Sherwood Meeting on Theoretical Plasma Physics, Berkeley, 1968; P. Rutherford, M. N. Rosenbluth, C. W. Horton, and E. A. Frieman, in *Plasma Physics and Controlled Thermonuclear Fusion Research*, Vol. 1, *IAEA* Vienna, 1969, p. 367.
31. B. Coppi, H. P. Furth, M. N. Rosenbluth, and R. Z. Sagdeev, *Phys. Rev. Letters*, **17**, 377 (1966).
32. T. Okhawa and M. Yoshikawa, *Phys. Rev. Letters*, **19**, 1374 (1968).
33. H. Forsen et al., in *Plasma Physics and Controlled Thermonuclear Fusion Research*, Vol. 1 IAEA, Vienna, 1969, p. 313.
34. T. Okhawa et al., in *Plasma Physics and Controlled Thermonuclear Fusion Research*, Vol. 1 IAEA, Vienna, 1969, p. 683.
35. J. D. Jukes, Gulf General Atomic Report GA-8735, San Diego, Calif., 1968.
36. D. Meade and S. Yoshikawa, Paper presented at the International Symposium on Plasma Fluctuations and Diffusion, Princeton, 1967; Princeton University Plasma Physics Report MATT-Q-25, 1968.
37. T. K. Allen et al., Paper presented at the European Conference on Plasma Physics, Utrecht, 1969.
38. T. Okhawa and M. Yoshikawa, Gulf General Atomic Report GA-9142, San Diego, Calif., 1969, submitted to Plasma Physics.
39. S. Yoshikawa, private communication, 1967.
40. D. J. Grove, private communication, 1969.
41. R. S. Pease, S. Yoshikawa, and H. P. Eubank, *Phys. Fluids*, **10**, 2059 (1967).
42. O. A. Anderson et al., in *Plasma Physics and Controlled Thermonuclear Fusion Research*, Vol. 1, IAEA, Vienna, 1969, p. 443.

Physics of Colloidal Plasmas*

M. S. SODHA AND S. GUHA

Department of Physics,
Indian Institute of Technology,
New Delhi, India

* Work partially supported by ESSA (USA).

Introduction

In recent years, there has been a great deal of interest in the electromagnetic properties of dust suspensions which are commonly known as colloidal plasmas when the free electron density is appreciable. The electron density of such plasmas is controlled by the thermionic or photoelectric emission from the dust particles and the recombination of electrons with them. The investigations in this field have found applications in many areas of science and technology like MHD power generation, ionospheric physics, electromagnetic properties of flames and rocket exhausts, electrostatic precipitation, and a variety of other interesting techniques. The present technology can produce dust with a particle size of $\frac{1}{40}\,\mu$; hence such plasmas are within the realm of modern experimental techniques.

The early experiments by Sugden and Thrush (1951) and Shuler and Weber (1954) showed an electron density in rich hydrocarbon flames which is far in excess of what can be explained by the application of Saha's equation to gaseous species. Further, they have shown that the observed electron density can be explained on the basis of the thermionic emission of electrons from the hot solid carbon particles present in the flames; the existence of such particles has been experimentally demonstrated by various workers (Hofmann and Wilm 1936, Wolfhard and Parker 1949, Parker and Wolfhard 1950, Pepperhoff 1951, Rossler 1953). The assumption that these carbon particles are essentially graphitic in nature is also borne out by the fact that the electron density in rich acetylene-air flames, at temperatures around 2200°K, attains an equilibrium value that can be calculated by assuming the particles to be those of graphite (work function $=4.5$ eV), as has been pointed out by Sugden and Thrush (1951). Thus began the study of ionization of solid particles (i.e., the generation of free electrons by thermionic emission from hot particles), the electromagnetic properties of dust-laden gases, and the various applications of such plasmas.

The colloidal plasmas are also encountered in the exhausts of the rockets that use solid propellants (Fristrom et al. 1962, Smith and Gatz 1963, Friedman et al. 1963). It has long been known that the flight of some rockets using solid propellants is accompanied by electromagnetic effects that interfere with communications and guidance by radio and radar. One of the commonest types of interference is due to free electrons in the neighborhood of the rocket. The electrons causing these effects originate from thermal ionization of atoms and molecules and probably also from the thermionic emission from hot solid particles; particles of alumina and aluminum are present in the combustion of aluminized propellants and should contribute significantly to the electron density.

In the experimental study of the generation of electron rich plasmas by solid propellants, Friedman et al. (1963) have used a caesium nitrate-aluminum system and found that a significant proportion of the product of this system is liquid aluminum oxide in the form of micron-size droplets. The work function of molten aluminum oxide is low enough to contribute significantly to the electron density. Thus these observations have led to the study of the interaction of solid particles with an ionized gas; interest was expressed in the role of these particles in increasing as well as decreasing the electron density.

It was shown experimentally by Soo and Dimick (1963) that the injection of solid particles (iron, alumina) reduces the free electron concentration in a propane flame and it also quenches the glow discharge due to the removal or collection of electrons by these charged particles. Balwanz (1965) has showed that the addition of particles of aluminum to a fuel raises the electron concentration by one to two orders of magnitude. Later Soo and Dimick (1965) demonstrated that particles of high work function increase, while those of low work function decrease, the overall rate of recombination in an arc flame of argon.

The physics of colloidal plasmas is also important in the understanding of the electrostatic precipitators. The gas in the precipitators contains electrons, ions, and charged solid particles. These particles are naturally charged as a result of the method of formation. However, these charges are usually inadequate for effective removal of the particles by the application of an electric field; hence a corona discharge is used to charge the particles.

Another possible industrial application of colloidal plasmas is in the field of MHD power generation. It has been shown that solid particles such as those of ash or soot from combustion or deliberately introduced particles of controlled size of a material having a low work function (i.e., barium oxide, aluminum oxide, and lanthanum hexaboride) contribute favorably to MHD power generation (Sodha and Bendor 1964a, b; Dennery 1964, Halasz et al. 1964, Honma and Fushimi 1966, Hooper et al. 1966, Mori et al. 1966, Honma et al. 1968). The principal advantages of using the colloidal plasmas as the working fluid in MHD generators are the reduction in duct length, low operating temperature throughout the system, and increase in the overall efficiency; the feasibility of this concept and the essential correctness of the theory has been experimentally demonstrated by Waldie and Fells (1967). In electrohydrodynamic generation the desirable opposite effect can be achieved by using dust to carry the positive charges, thus reducing the conductivity of the generator (Lawton 1964, Norris 1966).

In space research the study of microparticles has attracted particular attention, since a large number of solid particles of dimensions larger than atoms or molecules travel in interplanetary space (Davidson and Sandorff

1963, Burbank 1965). Such particles are a hazard to space travel because of the possibility of extensive damage through collisions of the particles with the space vehicles. As a result research has been carried out in laboratories simulating interplanetary particles (Friichtenicht 1962, Vedder 1963, Dingman 1965), in order to gain knowledge of such properties as the impact characteristics of the particles.

James and Vermeulen (1968) have reported that colloidal plasma has been experimentally realized in their laboratory and diagnostic studies are underway.

In the present review we shall present a brief account of the analytical studies of the various methods of the generation of electrons from the solid particles, the application of the statistical mechanics to the ionization of the charged particles (so as to arrive at an expression of equilibrium electron density), and the possible applications of the colloidal plasmas. The relevant experiments are also discussed.

The review has been organized as follows. In Sections I and II we present the equilibrium and nonequilibrium statistical mechanics of the ionization of solid particles, based on the work of Smith (1958, 1959) and Sodha (1963). The rate of generation of electrons from, and recombination with, the charged solid particles and an expression of the electron density have been derived. The results of these sections have been used frequently in the subsequent sections, where the applications of the plasma are discussed. Sections III, IV, and V present a brief review of the work carried out by the Indian Institute of Technology, Delhi (IITD) group on the ionization/recombination of electrons with the negatively charged solid particles (Sodha and Sharma 1967), the dependence of Fermi energy on the size of the particle (Sodha and Dubey 1970), and the field emission from these particles (Sodha and Kaw 1968, Dubey 1970). The importance of these analytical studies has also been discussed.

The analytical studies of the possible applications of the colloidal plasmas are contained in Sections VI to VIII, e.g., the use of high work function particles to reduce the electron density in the rocket exhausts (Sodha et al. 1963, Biasi et al. 1966, 1967, Guha and Arora 1970) and the use of low work function particles in the working gas of MHD generators (Sodha and Bendor 1964a, b; Honma and Fushimi 1966; Hooper et al. 1966, 1967).

Section IX deals with the investigations of the nonequilibrium ionization in wet alkali metal vapors (Smith 1965, Guha 1970); the condensed vapor droplets have been assumed to be spherical metal particles. These vapor droplets are seen in the cooler regions of the MHD duct; if the droplets are negatively charged, the electron density of the working fluid will be modified. Section X is devoted to the study of electrostatic precipitation in which the gas consists of electrons and the charged solid particles. The various mechanisms of charging the particles have also been discussed in this section.

Another method of the generation of the electrons is to irradiate the dust-laden gas to photoelectric light. In Section XI we present the analytical study of this phenomenon (Sodha 1963, Guha and Kaw 1968) and discuss its importance.

We have extended the analysis of gaseous colloidal plasmas to solid-state colloidal plasmas in Section XII. The observed conductivity in additively colored alkali halide crystals is attributed to the thermionic emission of electrons from the colloidal metal particles to the conduction band of the crystals. We present an analysis of this phenomenon (Guha 1968a) and attempt a comparison of our conclusions with the experimental work of Jain and Sootha (1965, 1968). Similarly, we analyze the photoconductivity of the additively colored alkali halide crystals (Guha 1968b).

Section XIII deals with a brief review of the experiments carried out to determine the size, charge, and work function of the solid particles, and the conductivity of the colloidal plasmas.

I. Equilibrium Statistical Mechanics of Ionization of Solid Particles

The first attempt to analyze the problem of thermal ionization of solid particles and to arrive at an expression of equilibrium electron density was made by Sugden and Thrush (1951). They assumed that all the solid particles carry the same positive charge. Einbinder (1957) studied the statistical mechanics of ionization of solid particles and arrived at an expression for free electron density, assuming that electron emission and recombination phenomena were governed by Saha's equation, when the ionization potential was replaced by an appropriate work function. He set up a system of equations for successive ionization (by electron emission) of the particles but did not include the possibility of the formation of negatively charged particles. Arshinov and Musin (1958a, b; 1962) and Smith (1958, 1959) have modified Einbinder's treatment to take into account the formation of negatively as well as positively charged particles. Physically, this is important at high particle densities, where neutral particles may capture free electrons, with the result that the electron density never exceeds what would be found in the case of equilibrium between the electron gas and the bulk solid. The next section is based on the treatment developed by Smith (1958, 1959).

A. *Statistical Mechanics*

The ionization-recombination reaction may be written as

$$A_{z-1} \rightleftharpoons A_z + e \tag{1-1}$$

where A_z represents a particle that has been ionized z times and hence has a charge ze. From the law of mass action

$$\frac{N_z N_e}{N_{z-1}} = K_z(T) \tag{1-2}$$

where $K_z(T)$ is the equilibrium constant for the reaction, N_z and N_e is the density of A_z and free electrons. Applying Saha's ionization equation to this reaction (1-1), one obtains

$$K_z(T) = 2\,\frac{(2\pi mkT)^{3/2}}{h^3}\exp\left(-\frac{\phi_z}{kT}\right) \tag{1-3}$$

where k is the Boltzmann constant, h Planck's constant, m the mass of an electron, T the temperature of the solid particles, and ϕ_z the energy required to remove an electron from a solid particle that is ionized $(z - 1)$ times. In eq. 1-3 it is assumed that the statistical weights of the initial and the final state of the charged particle are equal because the particles are macroscopic in size and differ only in their electrostatic charge.

The effect of the charge of the particle on the work function can be treated from a macroscopic point of view because the size of the particles is much larger than that of an atom, the removal of successive electrons from the lattice to just outside the surface of the particle is governed by the work function ϕ of the solid, and the particles are not noticeably distorted in shape when ionized. Hence it is assumed that the first ionization potential of an uncharged particle ϕ_1 differs from the work function of the bulk solid ϕ by the electrostatic work necessary to remove the electron to infinity from a position just outside the surface of the solid; thus

$$\phi_1 = \phi + \frac{e^2}{2c} \tag{1-4}$$

where c is the electrostatic capacitance of the isolated particle. Einbinder (1957) has made the approximation that $\phi_1 = \phi$ which leads to an extra term $e^2/2akT$ in his equation (36) as compared with the expression used by Sugden and Thrush (1951). The ionization potential corresponding to the process 1-1 is thus

$$\phi_z = \phi + \left(z - \frac{1}{2}\right)\frac{e^2}{c} \tag{1-5}$$

where z may be negative or zero as well as positive. In eq. 1-5, ze^2/c is the potential energy of an electron due to the charge ze on the particle and the term $e^2/2c$ is due to the image force. For a spherical particle of radius a, $c = a$. Thus, using eq. 1-5, eq. 1-3 becomes

$$K_z(T) = N_s \exp\left(\frac{\alpha}{2}\right) \exp(-z\alpha) \tag{1-6}$$

where

$$N_s(T) = 2\frac{(2\pi mkT)^{3/2}}{h^3} \exp\left(\frac{\phi}{kT}\right) \tag{1-7}$$

and

$$\alpha = e^2/akT$$

It may be seen that N_s is the saturation electron density in equilibrium with a pure conducting bulk solid at the temperature T (Sugden and Thrush 1951).

From eq. 1-2 it follows by iteration that

$$N_z = \frac{K_z}{N_e} N_{z-1} = \frac{K_z}{N_e}\frac{K_{z-1}}{N_e} N_{z-2} \cdots \frac{K_1}{N_e} N_0$$

$$= N_0 \prod_{j=1}^{j=z} \left(\frac{K_j}{N_e}\right) \tag{1-8}$$

where N_0 is uncharged particle density. From eqs. 1-6 and 1-8 we get

$$\prod_{j=1}^{j=z} \left(\frac{K_j}{N_e}\right) = \left(\frac{N_s}{N_e}\right)^z \exp\left(\frac{z\alpha}{2}\right) \exp[-\alpha(1 + 2 + \cdots z)]$$

$$= \left(\frac{N_s}{N_e}\right)^z \exp\left(-\frac{z^2\alpha}{2}\right) \tag{1-9}$$

Further, from eqs. 1-8 and 1-9 it follows that

$$\frac{N_z}{N_0} = \exp\left(\frac{x^2}{2\alpha}\right) \exp\left[-\frac{\alpha}{2}\left(z - \frac{x}{\alpha}\right)^2\right] \tag{1-10}$$

where

$$x = \ln\left(\frac{N_s}{N_e}\right)$$

Equation 1-10 expresses the charge distribution on solid particles. This is a Gaussian distribution of the charge but with a discontinuous nature due to the discrete nature of the charge, i.e., the integral multiple of the electronic charge. The center of the Gaussian distribution is given by

$$z_0 = \frac{x}{\alpha} = \frac{akT}{e^2} x, \tag{1-11}$$

the average charge per particle, $\bar{z}$, does not differ much from z_0, except for large values of α (very small particles or at low temperatures). From the distribution eq. 1-10,

$$\bar{z} = \frac{\sum_{z=-\infty}^{z=\infty} zN_z}{\sum_{z=-\infty}^{z=\infty} N_z} = \frac{x}{\alpha} - \frac{h(x, \alpha)}{\alpha} \tag{1-12}$$

where

$$h(x, \alpha) = 4\pi \frac{\sum_{n=1}^{\infty} n \exp(-2\pi^2 n^2/\alpha) \sin(2\pi n x/\alpha)}{1 + 2\sum_{n=1}^{\infty} \exp(-2\pi^2 n^2/\alpha) \cos(2\pi n x/\alpha)} \tag{1-13}$$

It has been pointed out by Smith that the factor $h(x, \alpha)$ may be neglected as compared to x/α if α is smaller than 6 (i.e., the particles are larger than 10 Å in radius at $T \sim 2500$).

If the only charged species present are solid particles and electrons, the requirement of charge balance leads directly to an equation for the electron density. Further, if the total concentration of the particles is N_p, one obtains

$$N_p = \sum_z N_z$$

Then the charge neutrality condition leads to

$$N_e = \bar{z} N_p$$

Combining this with eq. 1-12, and defining

$$W = \frac{N_p}{\alpha N_s}$$

we find

$$\frac{N_e}{N_s} = \exp(-x) = xW - Wh(x, \alpha) \tag{1-14}$$

The solutions of eq. 1-14, for various values of α, are shown in Figure 1, as a log-log plot of N_e/N_s vs. W. For small particles (large α) these curves have an extensive straight line segment, of slope $\frac{1}{2}$, characterizing the single ionization of species $[A]$ represented by

$$A \rightleftharpoons A^+ + e$$

However, at high values of the particle concentration, i.e., large W, the curves asymptotically approach unity (because of the formation of negatively charged particles); at low values of W they oscillate more and more closely about the curve for $\alpha = 0$, each oscillation corresponding to the predominance of particles of successively higher charge, A^{z+}. Hereafter we shall assume that

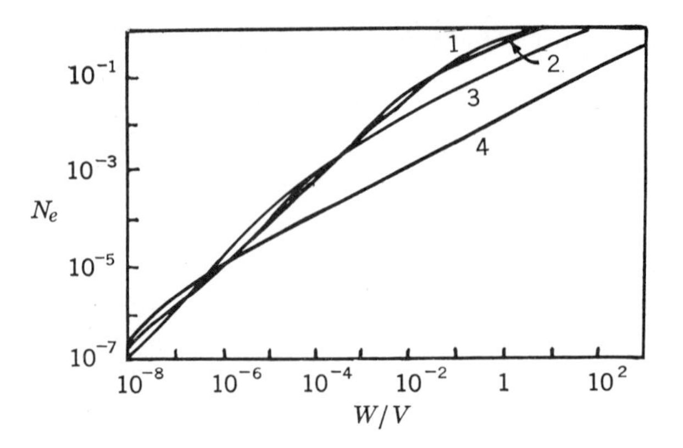

Fig. 1. Dependence of electron density on particle size ($U=0$): (1) $\alpha=0$; (2) $\alpha=2\pi$; (3) $\alpha=4\pi$; (4) $\alpha=24$. (After Smith 1959.)

the particles are large enough to neglect the correction $h(x, \alpha)$ which leaves us with the smooth curve represented by the equation

$$\frac{N_e}{N_s} + W \ln \left(\frac{N_e}{N_s}\right) = 0 \tag{1-15}$$

B. *Statistical Mechanics of Gaseous Dust Suspension*

Smith (1959) has also analyzed the case of gas-dust suspension when the effect of molecular ionization is also important. If we assume that only singly ionized states are important at flame temperatures, the reactions and their equilibrium constants can be expressed by

$$B_j \rightleftharpoons B_j^+ + e$$

$$B_j^- \rightleftharpoons B_j + e$$

$$K_j = \frac{N_e[B_j^+]}{[B_j]} \quad \text{and} \quad L_j = \frac{Ne[B_j]}{[B_j^-]}$$

The condition of charge balance is

$$\sum_j [B_j^+] = N_e + \sum_j [B_j^-]$$

If we introduce the definitions,

$$U = \frac{1}{N_s^2} \sum_j K_j[B_j] = \frac{N_e}{N_s^2} \sum_j [B_j^+]$$

$$V = N_e + \sum_j [B_j^-]/N_e = 1 + \sum_j \frac{[B_j]}{L_j}$$

the charge balance equation becomes

$$\left(\frac{N_e}{N_s}\right)^2 = \frac{U}{V} \tag{1-16}$$

When particles and molecules are present together, the charge balance condition requires

$$\sum_s N^s \bar{z}^s + \sum_j [B_j^+] = N_e + \sum_j [B_j^-] \tag{1-17}$$

where s represents sth species of particles,

$$N^s = \sum_z N_z^s$$

$$N_e = \sum_s \sum_z \bar{z}^s N_z^s$$

and

$$W = \sum_s W^s = \sum_s \frac{N^s}{\alpha^s N_s}$$

Using the previous definitions, we find that eq. 1-17 becomes

$$U\left(\frac{N_s}{N_e}\right) + W \ln\left(\frac{N_s}{N_e}\right) - V\left(\frac{N_e}{N_s}\right) - \sum_s W^s h(x, \alpha^s) = 0 \tag{1-18}$$

If we ignore the small correction term h, it becomes

$$U\left(\frac{N_s}{N_e}\right) + W \ln\left(\frac{N_s}{N_e}\right) - V\left(\frac{N_e}{N_s}\right) = 0 \tag{1-19}$$

Various solutions of eq. 1-19 are shown in Figure 2. In this equation, the effect of the large particles enters through W, that of the positive ions through U, and of negative ions through V.

Of the parameters entering into eq. 1-19 and influencing the electron density, W is the only one that depends on the size and the number of the

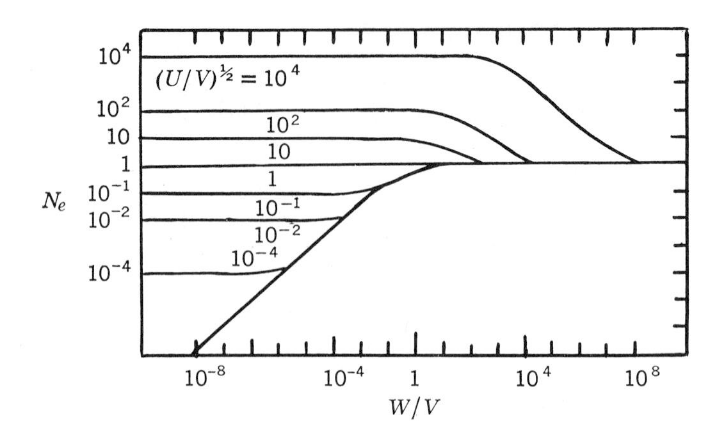

Fig. 2. Electron density in a mixture. (After Smith 1959.)

solid particles. In the region where N_e is sensitive to variations in W (a region roughly defined by the conditions $(UV)^{1/2} < W < 1$), a measurement of N_e and of the temperature, which is required to compute N_s and the ratio N_e/N_s, allows one to calculate from eq. 1-19 or 1-15. From W one can find the quantity

$$G = \sum_s N^s c^s = \frac{kTN_s(T)}{e^2} W \tag{1-20}$$

the total capacitance per unit volume due to the particles in the gas. This gives a measure of the size distribution of the particles because the capacitance of a particle is proportional to its radius and depends to some extent on its shape (e.g., for spherical particles, $c^s = a^s$). Combined with other information, such as a knowledge of the total amount of solid particles or the measurement of a different parameter dependent on the size distribution (as might be found from light scattering), an interpretation of G can be made in terms of an average particle size, deviation from sphericity, etc. Assuming spherical particles, Einbinder (1957) has interpreted Shuler and Weber's (1954) observations of electron density in rich oxyacetylene flames. It may be easily seen that, at low particle concentrations (small W), the electron density is determined by atomic ion formation and, as W increases, N_e passes through a transition region and approaches N_s.

Smith (1959) has also analyzed the influence of ionization on particle structure.

II. Thermal Ionization of Solid Particles: Electron Emission and Recombination

In the analyses of Einbinder (1957), Smith (1958, 1959), and Arshinov and Musin (1958a, b) the validity of Saha's expression for the equilibrium constant of molecular ionization has been assumed to extend to the case of solid particles. This assumption can at best be considered as clever intuition; it was considered worthwhile by Sodha (1963) to attempt an analysis of the rates of electron emission and recombination and thus the derivation of an expression for the equilibrium constant, applicable to the case of solid particles. Here we follow the treatment given by Sodha (1961, 1963).

A. *Thermionic Emission from Spherical Particles*

Sodha (1961, 1963) has assumed the validity of the Sommerfeld model of a metal, in which the energy states available to the free electrons are uniformly distributed in the momentum space and the probability of occupation of a state is given by the Fermi function. Thus the number of electrons per unit volume having velocity components between ω_x and $\omega_x + d\omega_x$, ω_y and $\omega_y + d\omega_y$, and ω_z and $\omega_z + d\omega_z$ is given by

$$d^3n_1 = \frac{2m^3}{h^3}\left\{1 + \exp\left[\frac{m}{2kT}(\omega_x{}^2 + \omega_y{}^2 + \omega_z{}^2) - \frac{E_F}{kT}\right]\right\}^{-1} d\omega_x\, d\omega_y\, d\omega_z \tag{2-1}$$

where the x-axis is normal to the surface. The velocity distribution of electrons (which are emitted), striking the surface per unit area per unit time, is given by

$$d^3n_2 = 0 \qquad \text{for } \omega_x < \omega_0$$

$$d^3n_2 = \omega_x d^3n_1 \qquad \text{for } \omega_x > \omega_0$$

where $\frac{1}{2}m\omega_0{}^2 = \psi$, the energy required by an electron to overcome the surface barrier. In considering emission from a spherical particle of radius a, with a charge $(z - 1)e$ before and a charge ze after the electron emission, we can conveniently distinguish between the following two cases.

Case 1: $z \leq 0$. In this case, in the classical approximation, the electrons having a normal energy $\frac{1}{2}m\omega_x{}^2 > \psi$ and striking the surface will escape. If u denotes the velocity of a particle just outside the surface

$$\tfrac{1}{2}mu_x{}^2 = \tfrac{1}{2}m\omega_x{}^2 - \psi, \qquad u_y = \omega_y, \text{ and } u_z = \omega_z$$

Hence the velocity distribution of emitted electrons per unit area per unit time just outside the surface is given by

$$d^3n_{\text{th}}(z-1) = \frac{2m^3}{h^3}\left\{1 + \exp\left[\frac{m}{2kT}(u_x^2 + u_y^2 + u_z^2) + \frac{\phi}{kT}\right]\right\}^{-1} u_x\, du_x\, du_y\, du_z \tag{2-2}$$

where $\phi = \psi - E_F$ is the work function of the metal. By making the usual assumption that the exponential term is much greater than unity, eq. 2-2 can be simplified to

$$d^3n_{\text{th}}(z-1) = \frac{2m^3}{h^3} u_x \exp\left(-\frac{\phi}{kT}\right) \times \exp\left[-\frac{m}{2kT}(u_x^2 + u_y^2 + u_z^2)\right] du_x\, dy_y\, du_z \tag{2-3}$$

The number of electrons $n_{\text{th}}(z-1)$ emitted per unit area per unit time by a particle with a charge $(z-1)e$ before electron emission and the mean energy of an emitted electron at the surface of the particle $\varepsilon_{\text{th}}(z-1)$ are given by

$$n_{\text{th}}(z-1) = \int_0^\infty \int_{-\infty}^\infty \int_{-\infty}^\infty d^3n_{\text{th}}(z-1) = \frac{4\pi m k^2 T^2}{h^3} \exp\left(-\frac{\phi}{kT}\right) \tag{2-4}$$

and

$$\varepsilon_{\text{th}}(z-1) = \int_0^\infty \int_{-\infty}^\infty \int_{-\infty}^\infty \tfrac{1}{2}mu^2\, d^3n_{\text{th}}(z-1)/n_{\text{th}}(z-1) = 2kT \tag{2-5}$$

If $\varepsilon'_{\text{th}}(z-1)$ denotes the mean energy at a large distance from the particle,

$$\varepsilon'_{\text{th}}(z-1) = \varepsilon_{\text{th}}(z-1) - \frac{ze^2}{a} = 2kT - \frac{ze^2}{a}$$

Equation 2-4 is Richardson's equation.

Case 2: $z \geq 0$. In this case, an electron will escape the potential (Page 1935) due to the charge on the particle or be emitted when $\frac{1}{2}mu^2 > ze^2/a$ (not $\frac{1}{2}mu_x^2 > ze^2/a$). Hence, to calculate $n_{\text{th}}(z-1)$, $d^3n_{\text{th}}(z-1)$ given by eq. 2-3 should be integrated in the limits $0 < u_x < \infty$, $-\infty < u_y < \infty$ and

$-\infty < u_z < \infty$ with the additional restriction that $\frac{1}{2}mu^2 > ze^2/a$ in the region of the integration. Thus, following Sodha (1961),

$$n_{\text{th}}(z-1) = \frac{4\pi mk^2T^2}{h^3}\left(1 + \frac{ze^2}{akT}\right)\exp\left[-\left(\frac{\phi + ze^2/a}{kT}\right)\right] \tag{2-6}$$

or

$$\frac{n_{\text{th}}}{n_0} = (1 + z\alpha)\exp(-z\alpha) \tag{2-7}$$

where n_0 is the rate of thermionic emission for neutral particles. Using the above expression, Sodha (1961) has discussed the kinetics of the charging of a single particle.

The mean energy of emitted electrons at the surface is given by

$$n_{\text{th}}(z-1)\varepsilon_{\text{th}}(z-1) = \int_0^\infty \int_{-\infty}^\infty \int_{-\infty}^\infty \tfrac{1}{2}mu^2\, d^3n_{\text{th}}(z-1)$$

where the condition $\frac{1}{2}mu^2 > ze^2/a$ is satisfied over the range of integration. Thus

$$\varepsilon_{\text{th}}(z-1) = \left(\frac{2 + z\alpha}{1 + z\alpha} + z\alpha\right)kT \tag{2-8}$$

and

$$\varepsilon'_{\text{th}}(z-1) = \varepsilon_{\text{th}}(z-1) - \frac{ze^2}{a} = \left(\frac{2 + z\alpha}{1 + z\alpha}\right)kT$$

B. *Collisions of Electrons with Charged Particles*

In Figure 3, if v_a is the velocity of the collision of an electron with a solid

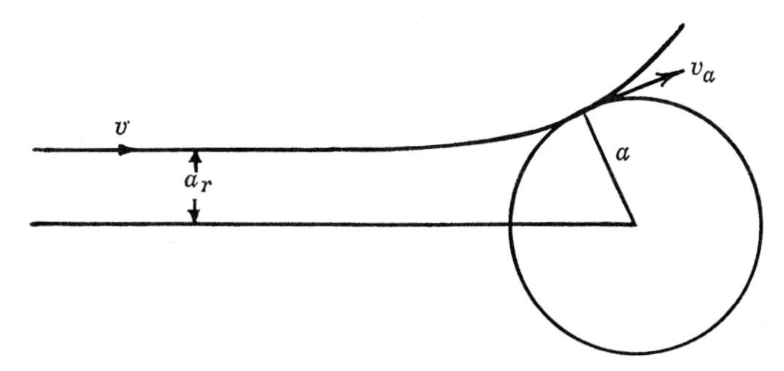

Fig. 3. Grazing collision between an electron and a charged particle.

particle of radius a and charge ze, the laws of conservation of momentum and energy require that

$$\frac{1}{2}mv^2 = \frac{1}{2}mv_a{}^2 - \frac{ze^2}{a}$$

and

$$mva_r = mv_a a$$

where v is the velocity of the electron at a great distance from the particle and $\pi a_r{}^2$ is the cross section of the recombination collision. From the above equations the cross section for the recombination collision is given by

$$\sigma(z) = \pi a_r{}^2 = \pi a^2 \left(1 + \frac{2ze^2}{amv^2}\right) \tag{2-9}$$

Hence the number of particles of charge ze recombining with electrons to produce particles of charge $(z-1)e$ per unit time per unit volume,

$$N_z' = N_z N_e \langle v\sigma(z)\rangle$$

and the average energy of the electrons recombining with the particles of charge ze is

$$\varepsilon'(z) = \langle \tfrac{1}{2}mv^2 v\sigma(z)\rangle / \langle v\sigma(z)\rangle$$

where $\langle\ \rangle$ denotes the average over the velocity distribution of electrons which may be assumed to be Maxwellian corresponding to an electron temperature T_e.

For the evaluation of $\varepsilon'(z)$ and $\langle v\sigma(z)\rangle$ we again distinguish between the two cases.

Case 1: $z \leq 0$. In this case, the limits of integration in the calculation of number of collisions and the total energy of colliding electrons should be $(-2ze^2/am)^{1/2} < v < \infty$ and for the remaining range $\sigma(z) = 0$. Thus we obtain

$$\langle v\sigma(z)\rangle = \left(\frac{8kT_e}{m\pi}\right)^{1/2} \pi a^2 \exp(z\alpha_e) \tag{2-10}$$

and

$$\varepsilon'(z) = 2kT_e - ze^2/a$$

where

$$\alpha_e = e^2/akT_e$$

The mean energy of the colliding electrons at the surface is given by

$$\varepsilon(z) = \varepsilon'(z) + \frac{ze^2}{a} = 2kT_e \tag{2-11}$$

Case 2: $z \geq 0$. In this case, the limits of integration are $0 < v < \infty$ and we obtain

$$\langle v\sigma(z)\rangle = \pi a^2 \left(\frac{8kT_e}{m\pi}\right)^{1/2} (1 + z\alpha_e) \tag{2-12}$$

$$\varepsilon'(z) = \left(\frac{2 + z\alpha_e}{1 + z\alpha_e}\right) kT_e$$

and

$$\varepsilon(z) = \left(\frac{2 + z\alpha_e}{1 + z\alpha_e} + z\alpha_e\right) kT_e \tag{2-13}$$

C. *Equilibrium Thermal Ionization of Solid Particles*

In thermal equilibrium $T = T_e$ and the number of particles of charge $(z - 1)e$ acquiring a charge ze per unit time per unit volume by electron emission must be equal to the number of particles of charge ze acquiring a charge $(z - 1)e$ per unit time per unit volume due to recombination with electrons. Hence

$$N_{z-1} 4\pi a^2 n_{\text{th}}(z - 1) = N_z N_e \langle v\sigma(z)\rangle \tag{2-14}$$

Substituting for $n_{\text{th}}(z - 1)$ and $\langle v\sigma(z)\rangle$ from eqs. 2-4 and 2-10 or 2-6 and 2-12 in the above equation, we obtain (in both cases $z \leq 0$ and $z \geq 0$):

$$K_z = \frac{N_z N_e}{N_{z-1}} = 2\left(\frac{2\pi mkT}{h^2}\right)^{3/2} \exp\left[-\left(\frac{\phi + ze^2/a}{kT}\right)\right] \tag{2-15}$$

where K_z is the equilibrium constant of the reaction

$$P_{z-1} \rightleftharpoons P_z + e$$

Equation 2-15 is identical with the expression for the equilibrium constant of the ionization of molecules first proposed by Saha, in which the statistical weights of the initial and final states of the particle are equal and the ionization energy is replaced by $\phi + ze^2/a$ which is actually the energy required by an electron to escape. This expression was used on intuitive considerations by Einbinder (1957) in his analysis of charge distribution of solid particles in thermal equilibrium. The present derivation provides a sound basis for the analysis of Einbinder (1957) and Arshinov and Musin (1958a, b) and the ability to derive it gives confidence to this treatment of the ionizing and deionizing processes. We may recapitulate that the work function ϕ_1 of the neutral particle is related to the bulk work function ϕ by eq. 1-4.

D. *Nonequilibrium Steady-State Thermal Ionization*

In general, the temperature of the electrons and the particles may be different. However, if the temperatures are changing slowly, steady-state ionization considerations may be applied. Equating the number of particles per unit volume per unit time of charge $(z-1)e$, which acquire a charge ze after electron emission, to the number of particles per unit volume per unit time of charge ze, which acquire a charge $(z-1)e$ after recombination with an electron, we obtain

$$K_z = \frac{N_z N_e}{N_{z-1}} = K \exp(-z\alpha_e) \qquad \text{when } z \leq 0 \tag{2-16}$$

and

$$K_z = K \exp(-z\alpha)\left(\frac{1+z\alpha}{1+z\alpha_e}\right) \qquad \text{when } z \geq 0 \tag{2-17}$$

where

$$K = 2\left(\frac{2\pi mk}{h^2}\right)^{3/2} T^2 T_e^{-1/2} \exp\left(-\frac{\phi}{kT}\right)$$

From eqs. 2-16 and 2-17 it can be shown that

$$N_z = N_0 \left(\frac{K}{N_e}\right)^z \exp\left\{-\frac{z(z+1)}{2}\alpha_e\right\} \qquad \text{when } z \leq 0 \tag{2-18}$$

$$N_z = N_0 \left(\frac{K}{N_e}\right)^z \prod_0^z \left(\frac{1+z\alpha}{1+z\alpha_e}\right) \exp(-z\alpha) \qquad \text{when } z \geq 0 \tag{2-19}$$

$$N_p = \sum_{-\infty}^{M} N_z \tag{2-20}$$

and

$$N_e = \sum_{-\infty}^{M} zN_z \tag{2-21}$$

where M is the number of free electrons in a solid particle. Substituting for N_z from eqs. 2-18 and 2-19 in the equation

$$\frac{N_e}{N_p} = \frac{\sum_{-\infty}^{M} zN_z}{\sum_{-\infty}^{M} N_z} \tag{2-22}$$

which is obtained from eq. 2-20 and 2-21, we have an algebraic equation that can be solved for N_e. Knowing N_e, we can obtain N_z from eqs. 2-18 and 2-19.

E. *Ionization Kinetics*

The kinetics of the thermal ionization process is described by

$$\frac{dN_z}{dt} = 4\pi a^2 N_{z-1} n_{\text{th}}(z-1) + N_{z+1} N_e \langle v\sigma(z+1)\rangle$$
$$-4\pi a^2 N_z n_{\text{th}}(z) - N_z N_e \langle v\sigma(z)\rangle \quad (2\text{-}23)$$

and eqs. 2-20 and 2-21. If we neglect radiative transfer of heat, a particle of charge ze can exchange energy with other particles in two ways:

1. Thermionic emission followed by recombination with an electron.
2. Recombination with an electron followed by thermionic emission.

Hence the conservation of energy may be expressed as

$$\frac{4}{3}\pi a^3 \rho s \frac{dT_z}{dt} = E_z = 4\pi a^2 n_{\text{th}}(z) \quad [\varepsilon(z+1) - \varepsilon_{\text{th}}(z)]$$
$$+ N_e \langle v\sigma(z)\rangle [\varepsilon_{\text{th}}(z-1) - \varepsilon(z)] \quad (2\text{-}24)$$

and

$$\frac{3}{2} k N_e \frac{dT_e}{dt} = \sum_{z=-\infty}^{z=M} N_z E_z \quad (2\text{-}25)$$

where ρ is the density and s the specific heat of the particles. From eqs. 2-5, 2-8, 2-11, and 2-13 it is seen that, when $T_z = T_e$, $dT_z/dt = 0$, as expected.

F. *Limitations of Present Analysis*

In the treatment of emission of electrons from a nonmetal, it is customary to introduce the concept of reflection coefficient r_0 of the boundary to an electron. If this is done expression for n_{th}, and $\langle v\sigma(z)\rangle$ should be multiplied by $(1 - r_0)$. Even when this is done, it is not expected that the results should hold in the case of materials for which Sommerfeld's model of a metal is not valid.

When $z \leq 0$, we have assumed with Spitzer (1941, 1948) that the probability that an electron with $\frac{1}{2}m\omega_x^2 < \psi$ will cross the surface is zero. This is true in the classical approximation. Actually, due to the negative Coulomb potential, it is expected that Schottky and field emission will be important under certain conditions. Therefore, in Section III, the electron emission-recombination from negatively charged particles will be considered, taking into account the effect of image force; thus considerations similar to those of the Schottky effect have been made. In Section V the field emission from negatively charged solid particles is considered where the tunneling of the electrons through the potential barrier is the main mechanism of the emission.

III. Electron Recombination and Thermionic Emission in Negatively Charged Solid Particles

A. *Potential Energy of an Electron*

In the earlier analysis of Sodha (1963) and Sodha et al. (1963), discussed in Section II, the potential energy V of the electron at a distance r from the center of a spherical solid particle of radius a and charge $(-ze)$ was assumed to be given by

$$V = -W_a + ze^2/a \qquad \text{for } r < a$$

and

$$V = \frac{ze^2}{a} \qquad \text{for } r > a \tag{3-1}$$

A qualitative illustration of this idealized variation is given in Figure 4*a*.

Since the electric field on the surface of the particle may be large, considerations similar to those made in discussions of the Schottky effect should also be made in the case of negatively charged spherical particles. By using the well-known results (e.g., Ramsay 1960) for the potential energy of a point charge near a spherical conductor, and following formulation of the problem similar to that of Seitz (1940), eq. 3-1 can be replaced by

$$V = \frac{-e^2 a}{2(r^2 - a^2) + (e^2 a/W_a)} + \frac{e^2 a}{2r^2} + \frac{ze^2}{r} \qquad \text{for } r > a \tag{3-2a}$$

and

$$V = -W_a + \frac{e^2}{2a} + \frac{ze^2}{a} \qquad \text{for } r < a \tag{3-2b}$$

The variation of V with r is illustrated qualitatively in Figure 4*b*.

The term $e^2 a/W_a$ in the denominator of the first term of eq. 3-2a is included just to obtain an expression valid for both the cases, namely, $r \gtrless a$. This inclusion makes the expression for $V(r)$ vary continuously with r.

Sodha and Sharma (1967) have calculated the coefficient of recombination of electrons with negatively charged particles and the rate of thermionic emission from such particles, using the appropriate relationship given by eq. 3-2 instead of the idealized behavior expressed by eq. 3-1. However, in this study the tunneling of electrons through the potential barrier is not considered; this effect is discussed in Section V.

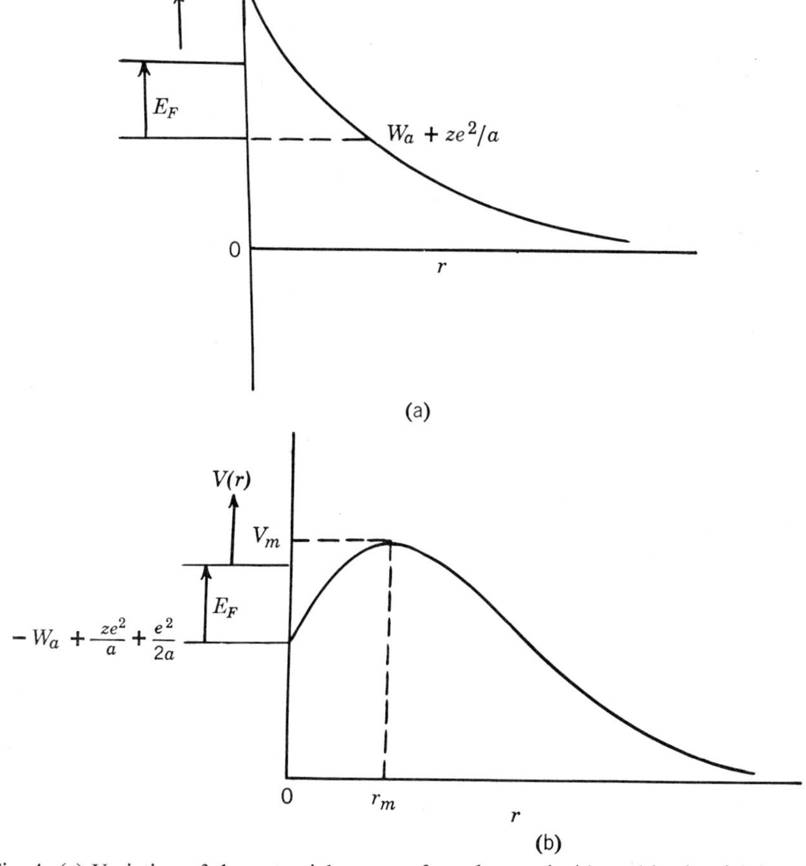

Fig. 4. (*a*) Variation of the potential energy of an electron inside and in the vicinity of a charged particle carrying a charge ($-Ze$), when the image force is neglected; (*b*) when the image force term is included.

B. *Thermionic Emission and Recombination*

Following the method given in Section II-B, one gets

$$R(-z) = \frac{\langle v\sigma(-z)\rangle}{\pi a^2(8kT_e/m\pi)^{1/2}} = \{1 + \alpha_e(U_m - U_a)\} \exp(-\alpha_e U_m) \quad (3\text{-}3)$$

where

$$U_a = \frac{V_a}{(e^2/a)} = -\frac{W_a}{(e^2/a)} + z + \frac{1}{2}$$

$$U_m = \frac{V_m}{(e^2/a)} = f(-z) = -\frac{1}{2(\xi_m{}^2 - 1)} + \frac{1}{2\xi_m{}^2} + \frac{z}{\xi_m}$$

the maximum of the potential energy, and $\xi_m = r_m/a$, the position of the maximum of the potential energy.

To obtain an appreciation of the variation of ξ_m, U_m, and R with the relevant parameters, see the tables given in Sodha and Sharma (1967).

Referring to Fig. 4*b*, one can see that the condition that an electron having a velocity v inside the particle can escape or be thermionically emitted is simply $\frac{1}{2}mv^2 > V_m - V_a$ because all particles that satisfy this condition will describe an open orbit around the particle (Page 1935). This condition is very different from that obtained by considering Figure 4*a* which represents the potential energy as a step function of r; in this case the condition for thermionic emission is $\frac{1}{2}mv_r{}^2 > W_a$, where v_r represents the radial component of the electron velocity. Thus, proceeding in a manner similar to that given in Section II-A, one obtains

$$\begin{aligned} \rho_{\text{th}}(-z-1) &= \frac{n_{\text{th}}(-z-1)}{4\pi mk^2T^2h^{-3}\exp(-\phi_0/kT)} \\ &= [1+\alpha(U_m - U_a)]\exp[\alpha(z+\tfrac{1}{2}-f(-z))] \end{aligned} \tag{3-4}$$

where $\phi_0 = W_a - E_F$ is the work function of the material. Thus the effective work function may be defined as

$$\phi(-z-1) = \phi_0 - \frac{e^2}{a}F(-z)$$

where

$$F(-z) = z + \tfrac{1}{2} - f(-z) \tag{3-5}$$

The dependence of $\rho_{\text{th}}(-z-1)$ on α and z has been calculated for $W_a/kT = 30$, 100, and 300; the results are presented in Table I.

Table II presents the dependence of the dimensionless decrease in work function $F(-z)$ on z.

When $z \gg 1$, it is shown that

$$\xi_m{}^2 \simeq 1 + (1+z)^{-1/2}$$

and

$$\frac{e^2}{a}F(-z) \simeq \frac{e^2}{a}z^{1/2} = e^{3/2}E_0^{1/2}$$

where $E_0 = ze/a^2$ is the magnitude of the electric field at the surface of the particle. This expression for $e^2/aF(-z)$ is identical with that obtained in the case of the plane surface (Seitz 1940).

TABLE I

Variation of the Dimensionless Rate of Thermionic Emission $\rho_{th}(-z-1)$ with α and z for three values of W_a/kT

W_a/kT	α	z = 1	3	10	40	100
30	10^{-3}	31.0	31.0	31.1	31.2	31.3
	10^{-2}	31.3	31.5	31.9	32.8	34.0
	10^{-1}	33.7	34.4	41.1	55.3	78.3
	1	72.9	14.5×10	5.3×10^2	1.0×10^4	3.4×10^5
100	10^{-3}	101.0	101.0	101.1	101.6	102.0
	10^{-2}	102.0	102.6	104.0	107.0	111.0
	10^{-1}	110.2	118.0	135.0	182.7	264.1
	1	242.3	492.8	18.5×10^2	38.3×10^3	14.9×10^5
300	10^{-3}	301.0	301.0	301.4	302.8	304.0
	10^{-2}	304.0	305.8	310.3	319.0	331.0
	10^{-1}	328.6	351.9	402.9	546.8	792.0
	1	726.1	148.0	563.6	11.9×10^4	47.5×10^5

TABLE II

Change in the Height of the Surface Barrier $F(-z)$ with z

z	1	3	10	40	100
$F(-z)$	0.88	1.61	2.94	6.01	9.70

Another interesting consequence of the inclusion of the image force term to the potential energy of an electron due to a spherical particle is the change in the density of states of electron energy and hence the Fermi energy. The analysis of this effect has been made by Sodha and Dubey (1970) and is discussed in the next section, where it is shown that the Fermi energy depends upon the size of the particle.

IV. Dependence of Fermi Energy on Size

Recently, Sodha and Dubey (1970) have calculated the change in Fermi energy (from that calculated for the bulk metal), when the contribution of image force to the potential energy of an electron outside the metal is also included. In this section the case of spherical metallic particles is considered. It is seen that the work function for spherical particles of radius 100 Å differs by an amount 0.03 eV from the corresponding value 4.50 eV for the bulk tungsten metal.

A. *Effective Volume Available to Different Electrons*

Figure 5 illustrates the model of potential energy of an electron for the spherical particle. The appropriate expression for the potential energy of an

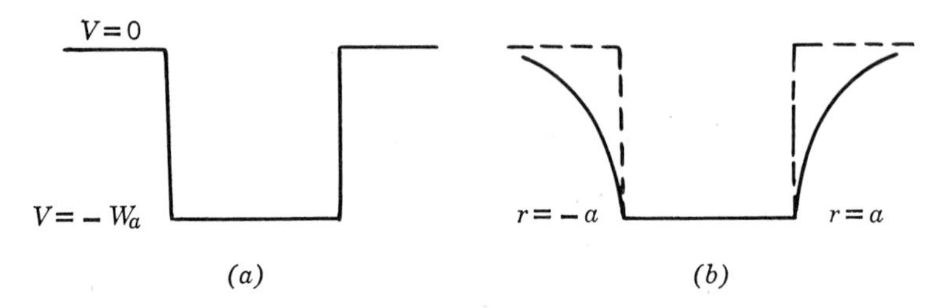

Fig. 5. Potential energy of electron (*a*) without image force; (*b*) with image force.

electron at a distance r from the center of the solid particle of radius a, taking into account the image force, is (Ramsay 1960):

$$V(r) = -W_a \qquad \text{for } \xi = \frac{r}{a} \leq 1$$

$$= \frac{e^2}{a}\, U(\xi) \qquad \text{for } \xi \geq 1 \tag{4-1}$$

where

$$U(\xi) = -\frac{1}{2\xi^2(\xi^2 - 1) + e^2/aW_a}$$

In contrast to the idealized barrier, one notes from Figure 5*b* that the volume available to electrons of different energies is different. Thus the effective volume corresponding to an electron energy E (measured with respect to $x = \infty$; hence it is negative) will be that of the sphere of radius r_{eff}, where

r_{eff} is the positive real root greater than a of the equation $E = V'(r_{eff})$ as a function of E. Hence

$$V'_{eff}(E) = \frac{V'}{2\sqrt{2}}\left\{1 + \left[1 - \frac{2e^2}{a}\left(\frac{1}{E} + \frac{1}{W_a}\right)\right]^{1/2}\right\}^{3/2} \qquad \text{for } -W_a < E < 0$$

$$= V' = \frac{4}{3}\pi a^3 \qquad \text{for } E = -W_a \qquad (4\text{-}2)$$

where $V'_{eff}(E)$ is the effective volume inside which the electrons of energy E can be assumed to move freely.

B. *Density of States and the Fermi Energy*

The density of states is given by (Seitz 1940):

$$g(p)\,dp = \frac{8\pi p^2\,dp}{h^3}\,V'_{eff}(p) \qquad (4\text{-}3)$$

where $V'_{eff}(p)$ is the effective volume available in the momentum range p and $p + dp$. Introducing a new variable $\varepsilon = E + W_a$, one obtains

$$g(\varepsilon)\,d\varepsilon = \frac{4\pi}{h^3}(2m)^{3/2}\varepsilon^{1/2}V'_{eff}(\varepsilon)\,d\varepsilon \qquad (4\text{-}4)$$

where $V'_{eff}(\varepsilon)$ is given by eq. 4-2. Evidently, the variable ε measures the electron energy with respect to the bottom of the well $(-W_a)$. The Fermi energy at absolute zero is to be determined by the condition

$$\int_0^{E_{FO}} g(\varepsilon)\,d\varepsilon = N \qquad (4\text{-}5)$$

where N is the total number of electrons in the metal. From eqs. 4-2, 4-4, and 4-5 the equation determining Fermi energy at absolute zero may be obtained as follows

$$N_e = \frac{4\pi}{h^3}(2m)^{3/2}\left[\frac{2}{3}E_{FO}^{3/2} + \frac{3}{2}\frac{e^2 W_a^{1/2}}{a}\right.$$

$$\left.\times\left\{\frac{1}{2}\ln\left(\frac{1+\sqrt{E_{FO}/W_a}}{1-\sqrt{E_{FO}/W_a}}\right) - \left(\frac{E_{FO}}{W_a}\right)^{1/2} - \frac{1}{3}\left(\frac{E_{FO}}{W_a}\right)^{3/2}\right\}\right] \qquad (4\text{-}6)$$

where $N_e = N/V'$ is the density of electrons (constant for a given metal).

The involved integration has been carried out in the approximation $2e^2/a(W_a - \varepsilon) \ll 1$ which is valid for a tungsten particle of radius $a \geqslant 100$ Å.

C. *Application to Thermionic Emission*

The modified values of Fermi energy obtained by the graphical solution of eq. 4-6 for spherical particles may be used in calculating the thermionic current. In the Richardson's equation, replacing the work function ϕ by its effective value calculated from the present model for spherical particles, the ratio of the thermionic current emitted by a spherical particle to that of bulk metal may be obtained:

$$\frac{I}{I_{\text{bulk}}} = \exp\left[-(\phi_{\text{eff}} - \phi)/kT\right] \tag{4-7}$$

Equation 4-6 gives the relation between N_e (fixed for a given metal) and the Fermi energy at absolute zero E_{F0} in terms of W_a (fixed for the metal) and radius (a) of the particle.

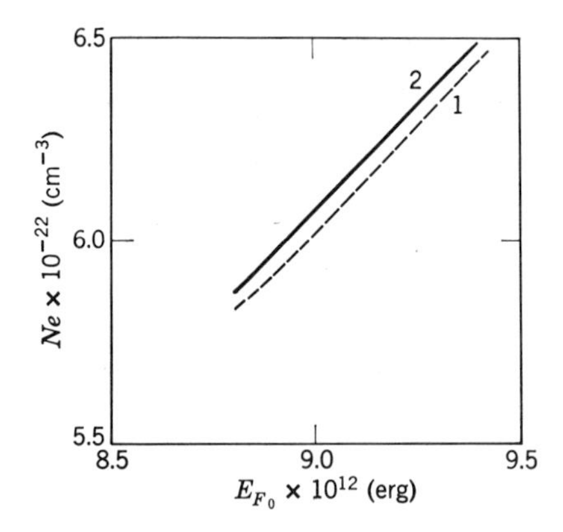

Fig. 6. Dependence of electron density on Fermi energy. (After Sodha and Dubey 1970.)

Figure 6 illustrates the variation of the Fermi energy with the electron density, N_e, for the following two cases:

1. Bulk metal.
2. Spherical metallic particle of radius 100 Å. For tungsten $N_e \simeq 6.3 \times 10^{22} \text{cm}^{-3}$, the values of Fermi energy are 5.80 and 5.77 eV, for the bulk and spherical particle, respectively; for the values $W_a = 10.3$ eV, the corresponding work functions are 4.50 and 4.53 eV.

An experimental check of the present model may be made by photoelectric measurements on spherical particles, in line with the Millikan condenser technique (Pope 1962a, b; Pope et al. 1965a, b) of measuring work function of particles.

An alternative rigorous approach to the present analysis would be the calculation of the density of states on the basis of the solution of Schrodinger's equation which takes into account the tunneling of the electronic wave function through the image force barrier into the region $V = 0$.

V. Field Emission from Negatively Charged Solid Particles

In Sections II and III the contribution to the number of emitted electrons that arises because of field emission from the charged dust particles is neglected. Sodha and Kaw (1968) and Dubey (1970) have investigated the field emission from a dust particle because of its negative charge and have studied the variation of this field-emission current with the charge, radius, and work function of the particle. However, Sodha and Kaw (1968) have neglected the effect of the image force on the field emission of the electrons. It was pointed out by Dubey (1970) that the neglect of the rounding effect of the image force on the potential energy barrier is justified only for electrons with energies much less than those corresponding to the peak of the barrier; but most of the field-emitted electrons have energies lying near the peak, so that taking into account the image force appreciably affects the calculated field emission.

A. *Transmission Coefficient of Electrons*

The time-independent Schrodinger's wave equation for an electron inside and outside the solid particle was solved (Dubey 1970) to calculate the radial transmission probability of electrons incident on the Coulomb energy barrier illustrated in Figure 4*b*. This equation is of the form

$$\nabla^2\psi(r, \theta, \phi) + \frac{2m}{\hbar^2}\left(\mathscr{E} + W_a - \frac{ze^2}{a} - \frac{e^2}{2a}\right)\psi(r, \theta, \phi) = 0 \qquad r \leq a \quad \text{(5-1a)}$$

and

$$\nabla^2\psi(r, \theta, \phi) + \frac{2m}{\hbar^2}\left(\mathscr{E} + \frac{e^2 a}{2(r^2 - a^2)} - \frac{e^2 a}{2r^2} - \frac{ze^2}{r}\right)\psi(r, \theta, \phi) = 0 \qquad r > a \quad \text{(5-1b)}$$

where $\mathscr{E}$ is the total kinetic energy of the electron measured with respect to the reference level at $r = \infty$ and the rest of the symbols have their usual meanings. Assuming a solution of the form

$$\psi(r, \theta, \phi) = rU(r)\, Y_{lm}(\theta, \phi) \qquad \text{(5-2)}$$

one obtains the following equation for $U(r)$:

$$\frac{d^2U}{dr^2} + \frac{2m}{\hbar^2}\left(E_r + W_a - \frac{e^2}{2a} - \frac{ze^2}{a}\right)U = 0 \qquad r \le a \tag{5-3a}$$

and

$$\frac{d^2U}{dr^2} + \frac{2m}{\hbar^2}\left(E_r + \frac{e^2 a}{2(r^2 - a^2)} - \frac{e^2 a}{2r^2} - \frac{ze^2}{r}\right)U = 0 \qquad r > a \tag{5-3b}$$

where E_r is the radial kinetic energy (referred to $r = \infty$). These equations are identical to the ones obtained for the one-dimensional problem of an electron with kinetic energy E_r incident on the potential energy barrier described by eq. 3-2. Making use of the slowly varying nature of the potential form, using WKB approximation, one obtains the following expression for the transmission probability (Kemble 1958):

$$T(E_r) = \exp\left[-\int_{r_1}^{r_2}\left\{\frac{8m}{\hbar^2}\left(-\frac{e^2 a}{2(r^2 - a^2)} + \frac{e^2 a}{2r^2} + \frac{ze^2}{r} - E_r\right)\right\}^{1/2} dr\right] \tag{5-4}$$

where $r_1 = a\xi_1$ and $r_2 = a\xi_2$ are the two positive real roots of the equation

$$-\frac{1}{2(\xi^2 - 1)} + \frac{1}{2\xi^2} + \frac{z}{\xi} = \frac{aE_r}{e^2} \tag{5-5}$$

where $\xi = r/a$ is chosen such that $\xi_1 < \xi_2$.

This formula is valid when E_r is appreciably less than the maximum of the barrier height, a range required for the field emission. This is a special case of a more general formula valid for all values of E_r; that has been discussed by earlier workers, e.g., Miller and Good (1953) and Kemble (1958). Equation 5-5 leads to the biquadratic equation

$$\xi^4 - \frac{z}{p}\xi^3 - \xi^2 + \frac{z}{p}\xi + \frac{1}{2p} = 0 \tag{5-6}$$

where

$$p = \frac{aE_r}{e^2},$$

Equation 5-6 is solved to obtain two real positive roots, each greater than one and such that $\xi_1 < \xi_2$. These two roots are used as limits in the integration of eq. 5-4.

B. *Complete Electron Emission*

Using the free electron model (Seitz 1940) for a metal and neglecting the thermionically emitted electrons (i.e., considering the emission at 0°K), the number of emitted electrons per unit area per unit time is given by

$$n_{em} = \frac{4\pi m}{h^3} \int_0^{E_F} (E_F - E)T(E)\, dE \tag{5-7}$$

where E is the radial kinetic energy of an electron measured with respect to the bottom level $-W_a + (ze^2/a) + (e^2/2a)$. The derivation of this equation involves the usual standard approximation, namely, the field emission electrons come from the neighborhood of the Fermi level.

Using the relation between E and E_r, given by

$$E_r = E - W_a + \frac{ze^2}{a} + \frac{e^2}{2a}$$

and combining eqs. 5-4 and 5-7, one arrives at

$$n_{em} = \frac{4\pi m\phi^2}{h^3} \int_0^{\eta_F} (\eta_F - \eta) \times \exp\left(-\int_{\xi_1}^{\xi_2} \left[p\left\{q\left(-\frac{1}{2(\xi^2-1)} + \frac{1}{2\xi^2} + \frac{z}{\xi}\right) - \eta + S\right\}\right]^{1/2} d\xi\right) d\eta \tag{5-8}$$

where $\eta = E/\phi$ is a new dimensionless variable, $\eta_F = E_F/\phi$, $p = 8ma^2\phi/\hbar^2$, $q = e^2/a\phi$ and $S = 1 + \eta_F - ze^2/a\phi - e^2/2a\phi$.

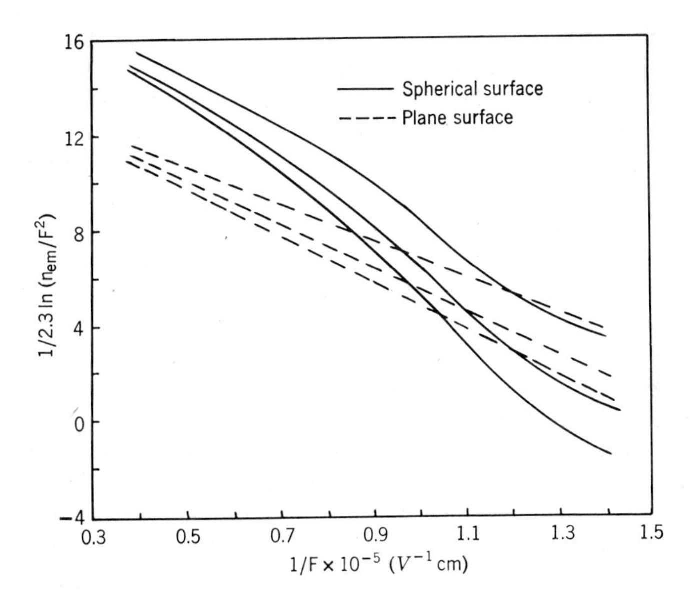

Fig. 7. Variation of $\log_{10}(nem/F^2)$ with $1/F$ for various values of ϕ. (After Dubey 1970.)

Equation 5-8 is solved numerically and the results are presented in Figure 7 in which the variation of $\log_{10}(n_{em}/F^2)$ has been shown with $1/F$; $F = ze/a^2$ being the field strength at the surface of the solid particle. The following values were chosen for the various parameters in order to cover the entire field range: $\phi = 4.1, 4.5, 4.7$ eV, z from 4 to 50, and a from 30 to 100 Å. One can see from Figure 7 that the dependence of n_{em} on z and a is reduced to the single dependence of F. For a fixed value of ϕ, with increasing z or decreasing a (i.e., increasing F) n_{em} increases in the manner shown in Figure 7; this variation is seen to have sizable deviations from the corresponding approximate straight line for plane surface. Also, the magnitude of field-emitted current density (i.e., en_{em}) is very different for the two types of surfaces.

VI. Effect of Solid Particles on Electron Density of Rocket Exhausts

A. *Equilibrium Ionization of Solid Particles in Rocket Exhausts*

Sodha, Palumbo, and Daley (1963) have used the analysis of Section II, to study the effect of solid particles on the electromagnetic properties of rocket exhausts. The need for alleviation of communication blackout resulting from rocket exhausts is becoming increasingly acute in view of the increased tempo of space programs. To maximize data acquisition, tracking, command system, and voice communications capability during the launch phase, an understanding of the electromagnetics of the rocket exhaust plasmas and the physics of possible techniques for alleviation of the problem is essential.

Sodha et al. (1963) have presented a discussion of electromagnetics of the exhaust plasmas and the physics of an alleviation technique, consisting of the addition of solid particles (of a material having a high work function) to the exhausts. The role of solid particles in modifying the electromagnetic properties of the exhausts has been analyzed in considerable detail, and representative numerical results based on an approximate model have been done. It is seen that the presence of solid particles of low work function material can account for the very high electron densities in rocket exhausts and that the electron density can be very considerably reduced by the addition of solid particles of a material, having a high work function. Similarly, Rosen (1962), Soo (1963a, b), Dimick and Soo (1964), and Anderson and Rosen (1965) have also analyzed this method for the suppression of free electrons in a gas-plasma (by the addition of dust of a refractory material). Their analyses are based on the effect of space charge on the thermal ionization of solid particles.

1. Composition of Rocket Exhaust Plasmas

If A_z^s denotes a particle of sth species (molecular, ionic, or solid particle) with a charge ze, then the composition of the system is described by the following equations:

$$\sum_{z=-\infty}^{z=m_s} [A_z^s] = A^s \qquad \text{(conservation of species)} \tag{6-1}$$

$$\sum_{s=1}^{s=l} \sum_{z=-\infty}^{z=m_s} z[A_z] = N_e \qquad \text{(conservation of charge)} \tag{6-2}$$

and

$$N_e \frac{[A_z^s]}{[A_{z-1}^s]} = K_z^s \qquad \text{(thermal equilibrium)} \tag{6-3}$$

where m_s denotes the number of electrons in A_0^s, l is the number of species and the lower limit $-\infty$ includes electron attachment. These equations are sufficient to determine the electron density and the composition of the system. Under conditions of thermodynamic equilibrium the expression for K_z^s pertaining to the molecular or atomic ionization was given by Saha (1920, 1921) as

$$K_z^s = \frac{\bar{\omega}_z^s}{\bar{\omega}_{z-1}^s} 2\left(\frac{2\pi mkT}{h^2}\right)^{3/2} \exp\left(=\frac{\varepsilon_z^s}{kT}\right)$$

where $\bar{\omega}_z^s$ is the statistical weight of A_z^s, and ε_z^s is the ionization potential of A_{z-1}^s.

2. Collisions of Electrons with Solid Particles—Molecules, Atoms, and Ions

a. *Solid Particles*

There are two types of electron-solid particle collision cross section. First, when a collision results in a direct hit (inelastic collision), the cross section $\sigma'_{A_z^s}$ is given by eq. 2-9. The second type of collision is elastic, in which the electron does not suffer a direct hit. The collision cross section is given by (following a method similar to that used in the calculation (Rose and Clark 1961) of ion-electron collision cross section),

$$\begin{aligned}\sigma''_{A_z^s}(v) &= 4\pi \int_{\chi_2}^{\chi_1} \frac{z^2 e^4}{m^2} \frac{1}{v^4 \sin(\chi/2)} \, d\left(\sin \frac{\chi}{2}\right) \\ &= 4\pi \frac{z^2 e^4}{m^2 v^4} \ln\left(\frac{\sin \chi_1/2}{\sin \chi_2/2}\right)\end{aligned}$$

where χ_2 refers to the scattering angle corresponding to an impact parameter equal to the Debye length D, and χ_1 refers to the scattering angle corresponding to the impact parameter a_r. The relationship between the scattering angle χ and the impact parameter b is given by

$$\cot\left(\frac{\chi}{2}\right) = \frac{mbv^2}{ze^2}$$

The total collision cross section is given by

$$\sigma_{A_z{}^s} = \sigma'_{A_z{}^s} + \sigma''_{A_z{}^s}$$

b. *Ions*

The cross section for collisions of an electron with ions is given by (Rose and Clark 1961, Gizburg and Gurevich 1960):

$$\sigma_{A_z{}^s,\,i}(v) = \frac{2\pi z^2 e^4}{m^2 v^4} \ln\left(1 + \frac{D^2 m^2 v^4}{e^4}\right)$$

c. *Atoms and Molecules*

No direct experimental data for the cross section of neutral species $A_0{}^s$ for elastic and inelastic collisions are available, but satisfactory determinations can be made by suitable interpretations of data for variation of electron mobility with electric field in pure gases. Inelastic collisions of electrons with neutral species should also be considered for a general case.

d. *Electron-Electron collisions*

Using the results of an analysis by Spitzer and Harm (1953), Sodha and Eastman (1958) have shown that the combined collision cross section of ion-electron and electron-electron scattering (when the electron velocity distribution is Maxwellian) is given by

$$\sigma_{A_z{}^s,\,e}(v) \simeq \frac{2\pi z^2 e^4}{4k^2 T^2} \ln\left(1 + \frac{9D^2 k^2 T^2}{e^4}\right) \frac{1}{\phi(u)}$$

where $u^2 = mv^2/2kT$ and $\phi(u)$ is a function of u tabulated by Spitzer and Harm (1953) for various values of $\sum_{\text{ions}} (1/N_e) z^2 [A_z{}^s]$. The net collision frequency of electrons is given by

$$\nu(v) = \sum_{s=1}^{s=l} \sum_{z=-\infty}^{z=m_s} \nu A_z{}^s(v) = \sum_{s=1}^{s=l} \sum_{z=-\infty}^{z=m_s} v\sigma_{A_z{}^s}(v)[A_z{}^s] \tag{6-4}$$

3. Approximate Model

In order to evaluate the effectiveness of solid particles in modifying the electromagnetics of hot gases, we can make the following assumptions.

1. The gaseous medium consists of only the gaseous atoms, singly charged gaseous ions, and free electrons.
2. There are two species of solid particles; all particles of a given species carry the same charge.
3. The ionization is an equilibrium phenomenon; in many studies, however, significant nonequilibrium effects have been inferred. This is discussed in Section VI-B.

Thus eqs. 6-1, 6-2, and 6-3, which determine the composition of the system, can be simplified to

$$N_0 + N_i = N \tag{6-5}$$

$$N_e = N_i + z_1 N_1 + z_2 N_2 \tag{6-6}$$

$$\frac{N_e N_i}{N_0} = K_0 = 2\left(\frac{2\pi mkT}{h^2}\right)^{3/2} \varpi_1/\varpi_0 \exp\left(-\frac{eV}{300kT}\right) \tag{6-7}$$

$$N_e = K_1 \exp(-z_1\alpha_1) \tag{6-8}$$

and

$$N_e = K_2 \exp(-z_2\alpha_2) \tag{6-9}$$

where N_0 is the number density of neutral gas atoms, N_i is the number density of ions, N_1 is the number density and $z_1 e$ is the charge on the solid particles of the first species, N_2 is the number density and $z_2 e$ is the charge on the solid particles of the second species,

$$\alpha_1 = \frac{e^2}{a_1 kT}, \quad \alpha_2 = e^2/a_2 kT$$

$$K_1 = 2\left(\frac{2\pi mkT}{h^2}\right)^{3/2} \exp\left(-\frac{e\phi_1}{300kT}\right)$$

$$K_2 = 2\left(\frac{2\pi mkT}{h^2}\right)^{3/2} \exp\left(-\frac{e\phi_2}{300kT}\right)$$

and ϕ_1 and ϕ_2 are the work functions of the two species of particles in volts.

Equations 6-8 and 6-9 are strictly valid when $|z| \gg 1$, so that the rates of recombination and electron emission for particles with charge ze are nearly equal to those of particles with charge $(z-1)e$. If this condition is not satisfied, alternate expressions can be obtained by using the expressions for rate of electron emission and recombination derived by Sodha (1963). See Section II.

From eqs. 6-5 to 6-9 we obtain

$$N_i = \frac{NK_0}{K_0 + N_e}$$

$$N_0 = N - N_i$$

$$z_1 = \frac{1}{\alpha_1} \ln \left(\frac{K_1}{N_e}\right)$$

$$z_2 = \frac{1}{\alpha_2} \ln \left(\frac{K_2}{N_e}\right)$$

and

$$N_e = \frac{NK_0}{K_0 + N_e} + \frac{N_1}{\alpha_1} \ln \left(\frac{K_1}{N_e}\right) + \frac{N_2}{\alpha_2} \ln \left(\frac{K_2}{N_e}\right)$$

which may be solved for N_e.

The effective collision frequency may be obtained by using the root mean square velocity $(3kT/m)^{1/2}$ for v in eq. 6-4. Thus

$$\nu = \nu_0 + \nu_i + \nu_1' + \nu_1'' + \nu_2' + \nu_2''$$

where

$$\nu_0 = N_0[v\sigma(v)]_{v=(3kT/m)^{1/2}} \tag{6-10}$$

$$\nu_i = N_i \frac{2\pi e^4}{m^{1/2}(3kT)^{3/2}} \ln \left(1 + \frac{9D^2k^2T^2}{e^4}\right)$$

$$\nu_{1,2}'' = N_{1,2} \frac{4\pi z_{1,2}^2 e^4}{9k^2T^2} \left(\frac{3kT}{m}\right)^{1/2} \ln \left(\frac{\sin \chi_1/2}{\sin \chi_2/2}\right) \tag{6-11}$$

$$\nu_{1,2}' = N_{1,2}\left(\frac{8kT}{m\pi}\right)^{1/2} \pi a_{1,2}^2 \exp (z_{1,2}\alpha_{1,2})$$

when $z_{1,2} < 0$.

$$\nu_{1,2}' = N_{1,2}\left(\frac{8kT}{m\pi}\right)^{1/2} \pi a_{1,2}^2(1 + z_{1,2}\alpha_{1,2}) \tag{6-12}$$

when $z_{1,2} > 0$.

$$\cot \left(\frac{\chi_2}{2}\right)_{1,2} = \frac{3DkT}{z_{1,2}e^2}$$

$$\cot \left(\frac{\chi_1}{2}\right)_{1,2} = \frac{3kT}{z_{1,2}e^2} a\left(1 + \frac{2z_{1,2}e^2}{3a_{1,2}kT}\right)^{1/2}$$

when $3a_{1,2}kT + 2z_{1,2}e^2 > 0$, and

$$\cot \left(\frac{\chi_1}{2}\right)_{1,2} = 0$$

when $3a_{1,2}kT + 2z_{1,2}e^2 < 0$. The expressions for $\nu'_{1,2}$ have been obtained by averaging (Sodha 1963) over the whole distribution, instead of setting $v = (3kT/m)^{1/2}$, since the averaged expressions are already available.

To get an appreciation of the effect of solid particles on the electromagnetics of gas (having an ionization potential of 10 V), we consider a gas at a temperature of 2318°K and a pressure of 3.16 atm., corresponding to a particle density of 10^{19}cm^{-3}. These numbers are typical of exhausts of rockets using high-performance propellants. Table III presents the variation

TABLE III

Electron Density, Electron Collision Frequency, Refractive Index and Absorption Coefficient (for a Wave of $\omega/2\pi = 2.84 \times 10^9 c/s$) in a Rocket Exhaust with Suspended Solid Particles

	Original				
S. No.	N_1 (cm^{-3})	N_e (cm^{-3})	ν (sec^{-1})	n	K (cm^{-1})
1	0	10^9	8.00×10^{11}	1.000	0.000
2	1.53×10^8	10^{10}	8.00×10^{11}	1.000	0.000
3	2.03×10^9	10^{11}	8.00×10^{11}	1.000	0.000
4	2.94×10^{10}	10^{12}	8.00×10^{11}	1.004	0.065
5	5.35×10^{11}	10^{13}	8.05×10^{11}	1.295	0.853
	After addition of N_2 particles per cm^3				
1	—	—	—	—	—
2	4.30×10^6	10^9	8.00×10^{11}	1.000	0.000
3	5.70×10^7	10^9	8.00×10^{11}	1.000	0.000
4	8.26×10^8	10^9	8.06×10^{11}	1.000	0.000
5	1.51×10^{11}	10^9	9.15×11^{11}	1.000	0.000

of electron density, electron collision frequency, refractive index, and absorption coefficient (for a wave of frequency $\omega/2\pi = 2.84 \times 10^9$ c/s) with the number of suspended solid particles (N_1 cm^{-3}) having a work function of 3 V and a radius of 480 Å; the effect of the addition of N_2 particles per cubic centimeter, having a radius of 480 Å and a work function of 15 V, to the mixture of the gas and solid particles of the first type (in different concentrations) has also been presented.

The ionization and deionization kinetics in the approximate model can be described by

$$\frac{dN_i}{dt} = k_1 N_0 - k_2 N_e N_i \tag{6-13}$$

$$\frac{dz_2}{dt} = 4\pi a_1{}^2 n_{\text{th}}(z_1) - N_e \langle v \sigma'_{A_{z2}}(v) \rangle \tag{6-14}$$

and

$$\frac{dz}{dt} = 4\pi a_2{}^2 n_{\text{th}}(z_2) - N_e \langle v \sigma'_{A_{z2^s}}(v) \rangle \tag{6-15}$$

together with eqs. 6-5 and 6-6, where k_1 is the ionization reaction rate, k_2 the recombination reaction rate, $n_{\text{th}}(z)$ the thermionic current per unit area from a particle with charge ze, and $\sigma'_{A_z{}^s}(v)$ as given by eq. 2-9. These equations have been used by Biasi et al. (1966). See Section VII.

It is seen that the presence of solid particles (of material having a low work function) in rocket exhausts can account for electron densities far in excess of that predicted by the application of Saha's equation to the gaseous medium. It is also remarkable that the addition of solid particles (of material having a high work function) causes a very large reduction of the electron density. Further, the effectiveness of solid particles in modifying the electron density increases with $N_p/\alpha(\propto(m/a^2))$ if m is the total mass of suspended solid matter per unit volume). Thus, for a given mass of suspended solid matter per unit volume, the effectiveness of particles in modifying the electron density increases with the decreasing size of the particles. In this study the additional electron density caused by the photoelectric emission from solid particles due to the solar radiation in the upper atmosphere has not been considered because Guha and Kaw (1968) have concluded that, for a system of argon laden with barium oxide dust, the photoelectric emission from the dust particles (for moderate intensities of the incident radiation) dominates the thermionic emission only up to about 900°K; for higher temperatures the reverse is true. Hence, in the temperature range considered in rocket exhausts, the photoelectric current is negligible compared to thermionic current.

Smith and Gatz (1963) have also drawn similar conclusions regarding the role of particles in changing the electron concentration in rocket exhausts.

B. *Nonequilibrium Ionization of Solid Particles in Rocket Exhausts*

In Section VI-A the effect of solid particles in modifying the electrons density in rocket exhausts has been analyzed by assuming equilibrium conditions of the gas-particle system. In the actual rocket exhaust, the assumption of thermal equilibrium is a very poor one for the supersonically expanding gases in and beyond the nozzle. In the vicinity of the throat, where temperature and pressure are high, a shifting equilibrium is more or less maintained. As the expansion continues, the chemical reactions can no longer keep up with it, and ultimately a frozen composition is to be expected. As a result, the ionization at the exit plane tends to be much higher than equilibrium calculations would predict (Tozzi 1965, Smith and Gatz 1963). Tozzi's 1965 results show that, in the exhaust zone, the ionization level is much higher than that obtained by equilibrium ionization process. Biasi et al. (1966) have studied the lowering of the ionization level of the same system due to the injection in the flow of high work function solid particles using the nonequilibrium effects connected with the expansion into account. This lowering depends on the position of the injection point and is considerably smaller than that found in the equilibrium case.

The system considered is made up of a monatomic inert gas, which contains a certain number of solid particles having a low work function, expanding supersonically in a Laval nozzle. Solid particles of a second kind with high work function are injected into the flow at various points. They have assumed that gas ionization is negligible everywhere; particle and electron temperatures are in equilibrium with the local gas temperature; and particles of each kind have all the same electric charge, equal to the average one. The assumption of the same charge on the particles is examined by Biasi and Tozzi (1967). See Section VI-C.

C. *Influence of Charge Distribution on Nonequilibrium Ionization of Solid Particles*

Biasi and Tozzi (1967) have examined the validity of the assumption made in Section VI-B (by Biasi et al. 1966) that all the particles had an electric charge equal to the local average one; this assumption, while correct when the departure from equilibrium is small or the average charge is much greater than one, is not a priori justified when the conditions are not satisfied. The study of the ionization state of the system (Biasi and Tozzi 1967) described in Section VI-B was made by taking into account the charge distribution among the particles and the distortions of the distribution itself due to nonequilibrium effects connected with the expansion.

Biasi and Tozzi (1967) have used eqs. 2-20, 2-21, and 2-23 which describe the kinetics of thermal ionization, assuming a negligible gas ionization. The solution was obtained by solving these equations in conjunction with that describing the expansion of the system in a hyperbolic axisymmetric nozzle. As a result, in the determination of the nonequilibrium ionization of gases containing solid particles, one can neglect the charge distribution among the particles, in first approximation, by introducing an average value; this procedure is more accurate the greater the ionization level of the system.

Experimental work has also been done by many workers to demonstrate the effect of the injection of particles on the electron concentration in a gas. Balwanz (1965) has showed that the addition of aluminum particles to a fuel raises the electron concentration by one to two orders of magnitude. Soo and Dimick (1963) have shown that the addition of particles (iron, alumina) quenches the glow discharge because of the removal of electrons by alumina particles. Later, Soo and Dimick (1965) made an experimental study with an arc flame of argon undergoing nonequilibrium recombination with or without the addition of various solid particles. This is a realistic model for the recombining set of a rocket after rapid expansion through a nozzle. The experiment shows that copper particles having a high work function increase the overall rate of recombination, while particles of alumina and magnesium oxide decrease it.

Johnson and Bullock (1964) gave further details on the Debye shielding distance and a critical temperature at which the sheath changes from that of ions to that of electrons, confirming the results of Soo and Dimick (1965).

VII. Electromagnetic Wave Propagation in Rocket Exhausts

Recently, Guha and Arora (1970) have modified the treatment of Sodha et al. (1963), using the kinetic approach of solving the Boltzmann's transfer equation to obtain the expressions for d-c and microwave conductivities in colloidal plasmas. This is in contrast to the analysis of Sodha et al. (1963) which was based on the phenomenological approach of using the effective collision frequency of an electron with the species of the medium.

A. *Electron Collision Frequency and Density*

If we assume a slightly ionized plasma, so that only electron collisions with the neutral gas molecules and charged solid particles are significant the total electron collision frequency is given by

$$\nu = \nu_0 + \nu' + \nu''$$

where ν_0, ν', and ν'' have been defined earlier. Using the relevant expressions for ν_0, ν', and ν'' (given in Sections II and VI-A) we get (for $z \geq 0$)

$$\nu = N_0 vQ_0 + N_p v \frac{\pi z^2 e^4}{m^2 v^4} \ln \left(\frac{\sin \chi_1/2}{\sin \chi_2/2}\right) + N_p v\pi a^2\left(1 + \frac{2ze^2}{amv^2}\right)$$

$$= \nu_0'(Ax^{1/2} + Bx^{-1/2} + Cx^{-3/2}) \tag{7-1}$$

where

$$Q_0 = \pi a_0^{\,2}$$

$$v_0' = \left(\frac{2kT}{m}\right)^{1/2} N_0 \pi a_0^{\,2}$$

$$A = 1 + \frac{N_1}{N_0}\left(\frac{a}{a_0}\right)^2$$

$$B = \frac{N_p}{N_0}\left(\frac{a}{a_0}\right)^2 z\alpha$$

and

$$C = \frac{N_p}{N_0}\left(\frac{a}{a_0}\right)^2 z^2\alpha^2 \ln\left(\frac{\sin \chi_1/2}{\sin \chi_2/2}\right)$$

In the above derivation ln (sin $x_1/2$/sin $x_2/2$) is a slowly varying function of v and thus can be assumed constant; hence it has been replaced by its root mean square value $(3kT/m)^{1/2}$.

The electron density in such a plasma can be given either by eq. 6-8 or 6-9.

B. *Conductivity and Current Density*

The conductivity of a plasma subjected to a microwave field of the type

$$E' = E \exp(i\omega t) \tag{7-2}$$

is given by (e.g., Sodha and Palumbo 1962):

$$\sigma = \frac{e^2 N_e}{3m}\left\langle \frac{1}{v^2}\frac{\partial}{\partial v}\left(\frac{v^3}{\nu + i\omega}\right)\right\rangle \tag{7-3}$$

where $\langle\ \rangle$ denotes the average over the velocity distribution.

Substituting the expression of ν (eq. 7-1) in eq. 7-3, one gets the current density as

$$J = \frac{e^2 N_e}{m\sqrt{\frac{5}{2}}}\frac{1}{v'_0}\left\{I_1 - i\frac{\omega}{v'_0} I_2\right\} E \exp(i\omega t) \tag{7-4}$$

where

$$I_1 = \int_0^\infty \frac{x^3(Ax^2 + Bx + C)e^{-x}\,dx}{(Ax^2 + Bx + C)^2 + (\omega/v_0')^2 x^3}$$

and

$$I_2 = \int_0^\infty \frac{x^{9/2}e^{-x}\,dx}{(Ax^2 + Bx + C)^2 + (\omega/v_0')^2 x^3}$$

If we put $\omega = 0$ in eq. 7-4, we obtain d-c conductivity,

$$\sigma_0 = \frac{e^2 N_e}{m\sqrt{\frac{5}{2}}\,\nu_0'} \int_0^\infty \frac{x^3 e^{-x}\,dx}{(Ax^2 + Bx + C)} \tag{7-5}$$

C. *Wave Propagation and Reflection Coefficient*

We assume that the electromagnetic wave is propagating in the z direction; then the wave equation can be written as

$$\frac{\partial^2 E'}{\partial z^2} = \frac{\varepsilon\mu}{c^2}\frac{\partial^2 E'}{\partial t^2} + \frac{4\pi\mu}{c^2}\frac{\partial J}{\partial t} \tag{7-6}$$

where ε and μ are the dielectric permittivity and magnetic permeability of the medium and c is the velocity of light in vacuum. Equation 7-6 can be reduced to the dimensionless form by substituting

$$\xi = (\varepsilon_0 \mu_0)^{1/2} \frac{\omega z}{c}$$

and

$$\mathscr{E} = \frac{E}{E_{00}}$$

where E_{00} is any normalizing field. Then it reduces to the form

$$\frac{\partial^2 \mathscr{E}}{\partial \xi^2} = \frac{1}{\omega^2}\frac{\partial^2 \mathscr{E}}{\partial t^2} + \frac{4\pi}{\varepsilon\omega^2}\frac{\partial}{\partial t}\left(\frac{J}{E_{00}}\right) \tag{7-7}$$

Substituting the expression for current density and electric field from eqs. 7-4 and 7-2, respectively, in eq. 7-7, one gets the wave equation,

$$\frac{\partial^2 \mathscr{E}}{\partial \xi^2} + \beta^2 \mathscr{E} = 0 \tag{7-8}$$

where

$$\beta = \left\{1 - i\,\frac{4\pi e^2 N_e}{\varepsilon\omega m \nu_0' \sqrt{\frac{5}{2}}}\left(I_1 - i\,\frac{\omega}{\nu_0'}\,I_2\right)\right\}^{1/2}$$

$$= (-n + iK)$$

n and K are the refractive index and absorption coefficient of the medium.

The general solution of eq. 7-8 is

$$\mathscr{E}_M = C_1 \exp(i\beta\xi) + C_2 \exp(-i\beta\xi) \tag{7-9}$$

Applying the appropriate boundary conditions,

$$\xi = 0,\ \mathscr{E} = \mathscr{E}_0,\ \text{and}\ \xi = \infty,\ \mathscr{E} = 0,$$

we find the solution of eq. 7-9:

$$\mathscr{E}_M = \mathscr{E}_0 \exp(i\beta\xi) \tag{7-10}$$

To obtain the expression of the field of the wave reflected from the plasma free-space interface, one has to apply the conditions that the electric and magnetic field vectors are continuous across the boundary.

The expression for the electric field in the free space is given by

$$\mathscr{E}_F = A_i \exp[i(\omega t - \xi)] + A_r \exp[i(\omega t + \xi)]$$

Using the above-mentioned boundary conditions the expression for the amplitude of the reflected wave is found to be

$$A_r = \frac{1+\beta}{1-\beta} A_i \exp[i(\omega t + \xi)]$$

Thus the reflection coefficient is given by

$$R = \left|\frac{A_r}{A_i}\right| = \left[\frac{(1-n)^2 + K^2}{(1+n)^2 + K^2}\right]^{1/2}$$

Calculations have been made for the conductivity, refractive index, and absorption and reflection coefficients in a special case in which argon is laden with barium oxide dust. The results obtained are presented in Tables IV to VII. It is seen that the d-c. and complex conductivities increase with the temperature of the gas, the size of the suspended solid particles, and the particle density. The reflection coefficient of the wave is found to increase with increasing values of particle density, temperature of the gas and with decreasing value of particle size.

TABLE IV

Variation of Conductivity, Propagation Parameters and Reflection Coefficient with Wave Frequency ω for $N_p = 10^{11}\text{cm}^{-3}$, $T = 1000°\text{K}$, $a = 10^{-6}\text{cm}$, and $\nu_0' = 3.2 \times 10^{10} s^{-1}$

ω rad. s^{-1}	$\sigma_r \times 10^{-8}$ $\Omega^{-1}\text{cm}^{-1}$	$\sigma_i \times 10^{-7}$ $\Omega^{-1}\text{cm}^{-1}$	n	K	R
0	5.5039				
3.2×10^8	5.5034	0.4798	4.6773	0.0865	0.4197
3.2×10^9	5.4542	4.733	1.4969	0.2712	0.0513
3.2×10^{10}	3.2973	25.211	0.7212	0.6235	0.1393
3.2×10^{11}	0.1074	7.169	0.7063	0.7059	0.1715

TABLE V

Variation of Conductivity, Propagation Parameters and Reflection Coefficient with Particle Density for $w/\nu_0' = 0.1$, $T = 1200°K$, $a = 10^{-6}$ cm, and $\nu_0' = 2.8 \times 10^{10} s^{-1}$

N_p cm^{-3}	$\sigma_0 \times 10^{-8}$	$\sigma_r \times 10^{-8}$	$\sigma_i \times 10^{-7}$	n	K	R
10^{10}	3.23	3.20	2.743	1.2436	0.3531	0.0357
10^{11}	22.96	22.03	18.14	3.1285	0.0310	0.2660
10^{12}	129.34	128.48	100.23	7.559	0.2284	0.5875
10^{13}	547.40	544.34	394.33	15.560	0.5308	0.7733

TABLE VI

Variation of Conductivity, Propagation Parameters and Reflection Coefficient with Particle Radius for $\omega/\nu_0' = 1$, $T = 1000°K$, $N_p = 10^{10} cm^{-3}$, and $\nu_0' = 3.2 \times 10^{10} s^{-1}$

$a \times 10^6$ cm	$\sigma_0 \times 10^{-8}$	$\sigma_r \times 10^{-8}$	$\sigma_i \times 10^{-8}$	n	K	R
1.0	1.1244	0.6719	0.5144	0.7094	0.6903	0.1651
2.5	2.1966	1.318	1.007	0.7199	0.6742	0.1588
5.0	3.5178	2.1241	1.617	0.7155	0.6539	0.1569

TABLE VII

Variation of Conductivity, Propagation Parameters and Reflection Coefficient with Temperature for $\omega/\nu_0' = 0.01$, $a = 2.5 \times 10^{-6}$ cm, and $N_p = 10^{11} cm^{-3}$

T °K	$\nu_0' \times 10^{10}$ s^{-1}	$\sigma_0 \times 10^{-8}$	$\sigma_r \times 10^{-8}$	$\sigma_i \times 10^{-6}$	n	K	R
1000	3.2	9.120	9.199	7.807	6.020	0.0572	0.5114
1250	2.8	44.915	44.912	35.022	14.125	0.0196	0.7530
1500	2.6	89.554	89.549	64.418	20.876	0.0512	0.8255

It may be mentioned here that, in the propagation of a wave through colloidal plasmas, one would expect scattering and dielectric losses due to the particles. In the particular case considered here, calculations have been made for the dielectric absorption and the scattering coefficient for solid particles (Van de Hulst 1957). It is found that the extinction coefficient due to this scattering is of the order of approximately 2.7×10^{-8} cm^{-1} (when $\omega = 3.2 \times 10^8$ radians sec^{-1}, $N_p = 10^{11}$ cm^{-3}, and $a = 10^{-6}$ cm), which is many orders of magnitude smaller than the absorption coefficients due to the electrons. The extinction coefficient due to dielectric absorption of the particles is found to be still much smaller. These results seem to be justified because the size of the particles is very small ($\sim 10^{-6}$ cm) and the wavelength of the electromagnetic wave is quite large (0.1–100 cm).

D-c. conductivity in dust suspensions has also been studied by Halasz et al. (1964) and Dennery (1964). The results of calculations of d-c. conductivity of argon laden with barium oxide dust of Sodha and Bendor (1964b) in the relevant range of parameters is satisfactorily represented by a power law,

$$\sigma_0 = C_0 \lambda^{\alpha'} p^{-z'} t^{y} \tag{7-11}$$

where p is the pressure of the gas, t the absolute temperature of the gas, and λ the ratio of the mass of the dust to the mass of the gas. Also, $C_0 = 1.39 \times 10^{-8}$, $\alpha' = 0.66722$, $z' = 0.13130$, and $y = 2.92191$.

Newby (1967) has presented a graph from which the electron solid particle collision frequency can be calculated for various values of the relevant parameters. This has been presented in Figure 8. This graph is useful in the calculation of d-c conductivity of colloidal plasmas.

Experimental studies have been conducted on a wide variety of rocket exhausts in an attempt to determine the sources of ionization; the electron capture mechanisms; and the spatial distribution of the electrons, as well as their collision frequencies and the parameters upon which they depend. A variety of components was added to the fuel to determine their influence upon the ionization of the exhausts (Balwanz 1965, Baldwin et al. 1964).

VIII. Application of Dust Suspension to Closed Cycle Magnetohydrodynamic Power Generation

The industrial application of closed cycle MHD generators, using a gas-cooled nuclear reactor, is mainly limited by the low electrical conductivity of the working gas (even after seeding with caesium) at temperatures (in the MHD duct) that are feasible with the present state of reactor technology.

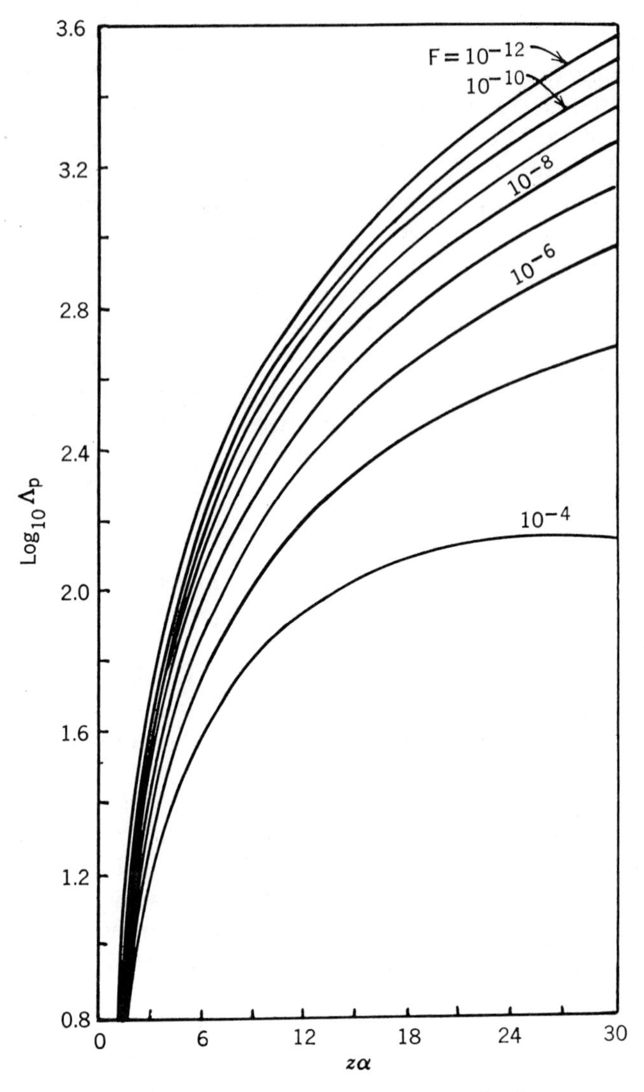

Fig. 8. Variation of Λ_p with $z\alpha$ for various values of F ($F = \frac{4}{3}\pi a^3 Np$, $\Lambda_p = \nu' + \nu''/Np\pi a^2 (3kT/m)^{1/2}$). (After Newby 1967.)

Hence a satisfactory means of enhancing the gas conductivity is very desirable, and many interesting concepts have been proposed to this end. One possible method of increasing the electron density, and hence the electrical conductivity of a gas, is by the addition of the fine dust of a material having a low-work function. Sodha and Bendor (1964a, b), Honma and Fushimi (1966), Hooper et al. (1966, 1967), Mori et al. (1966), Waldie and Fells (1967), and Honma et al. (1968) have investigated the application of this concept to MHD power generation compatible with gas-cooled reactor and material technology.

Experimental studies on dust suspensions have been reported by Zimin et al. (1966) and Waldie and Fells (1967). The latter authors have developed a technique for the preparation of submicron suspensions of barium oxide and other alkali earth oxides in inert gases. The highest conductivity measured was 0.1 mho/meter, at 1600°K for a barium oxide-argon suspension with a barium oxide/argon weight ratio of approximately 0.2 and an average particle size of 0.3 μm. The theoretical conductivity values for a weight ratio of 0.2 and with 0.3 μm particles are 0.50, 0.30, and 0.12 mho/meter for values of the work function taken as 1.7, 2.2, and 2.7 eV, respectively. Thus we find very good agreement with the theoretical value of conductivity which strengthens our conviction of the feasibility this concept.

A. *Flow of Gas-Dust Suspension*

The equation of one-dimensional gas dynamics was applied to two-phase flow by Sodha and Bendor (1964a, b), who used suitably modified density, specific heats, and other parameters. It can easily be shown that the use of the following quantities reduces the equations (and their solutions) to those for the "equivalent" gas:

$$\left.\begin{aligned} \rho &= (1+\lambda)\rho_g \\ R &= \frac{R_g}{1+\lambda} \\ C_p &= \frac{C_{pg}+\lambda C_{pp}}{1+\lambda} \\ \gamma &= \frac{\gamma_g(C_{pg}+\lambda C_{pp})}{C_{pg}+\gamma_g\lambda C_{pp}} \end{aligned}\right\} \tag{8-1}$$

where C_{pg} and C_{pp} are the specific heats at constant pressure of the gas and the material of the solid particles, respectively, and R_g is the unmodified gas constant. The well-known results for gas flow in ducts, nozzles, etc. can now be extended to the flow of gas-particle mixtures.

B. *Constant Mach Number MHD Generators*

The flow through constant Mach number MHD generator has been analyzed by Swift-Hook and Wright (1963). Denoting static and stagnation quantities by small and capital letters, respectively, and values at entry and exit to the magnetic field by suffixes 1 and 2, we find that their results are

$$\left.\begin{aligned} \frac{P}{P_1} &= \frac{p}{p_1} = \left(\frac{t}{t_1}\right)^{\beta} = \left(\frac{T}{T_1}\right)^{\beta} \\ \frac{u}{u_1} &= \left(\frac{T}{T_1}\right)^{1/2} \\ \frac{A}{A_1} &= \left(\frac{T}{T_1}\right)^{1/2-\beta} \end{aligned}\right\} \tag{8-2}$$

where

$$\beta = \gamma\{1 + \tfrac{1}{2}(1-K)(\gamma-1)(Ma)^2\}/(\gamma-1)K$$

Also, $K = V/dBu$ is the generator load factor, $(Ma) = u/(\gamma Rt)^{1/2}$ the Mach number, V the load voltage, d the distance between the electrodes, B the magnetic induction, and A the area of cross section of the duct. The terms K, V, and B are kept constant along the duct length. It was shown that

$$\frac{d}{dx}\left(\frac{T}{T_1}\right) = -K(1-K)\frac{\sigma}{\sigma_1}\frac{u}{u_1}\frac{\rho_1}{\rho}\,{x_0}^{-1}$$

where

$$x_0 = \frac{\rho_1 u_1 C_p T_1}{\sigma_1 {u_1}^2 B^2}$$

The variation of temperature with the duct station can be obtained from eqs. 7-11 and 8-2 and the above equations as

$$\frac{x}{x_0} = \frac{1-(T/T_1)^{\beta(z'+1)-y-1/2}}{K(1-K)[\beta(z'+1)-y-\frac{1}{2}]} \tag{8-3}$$

The conductivity at entry can be expressed by

$$\sigma_1 = C T_1^{\,y} P_1^{-z'}\left\{1 + \frac{(\gamma-1)}{2}(Ma)^2\right\}^{\gamma(z'-y)-y/(\gamma-1)} \tag{8-4}$$

where the constant C takes account of the fact that P_1 is expressed in meter-kilogram-second units (mks) and includes $\lambda^{\alpha'}$. Hence x_0 may be given by

$$x_0 = \frac{\gamma P_1^{z'+1} T_1^{-(y+1/2)}}{(Ma)B^2 C C_p^{1/2}(\gamma-1)^{3/2}}\left\{1 + \left(\frac{\gamma-1}{2}\right)(Ma)^2\right\}^{[2\gamma(y-z')+\gamma-2y-3/2(\gamma-1)]} \tag{8-5}$$

The duct length x_2 is found to be independent of the mass flow and is given in terms of stagnation pressure and temperature. The cross-sectional area at entry to the magnetic field is proportional to the mass flow $m = \rho_1 u_1 A_1$ and may be written as

$$A_1 = \frac{mT_1^{1/2}}{\gamma(Ma)P_1} \{(\gamma - 1)C_p\}^{1/2} \left\{1 + \left(\frac{\gamma - 1}{2}\right)(Ma)^2\right\}^{\gamma + 1/2(\gamma - 1)} \tag{8-6}$$

The cross-sectional area distribution along the duct follows from eqs. 8-2 and 8-3. Some estimate of the part of the heat lost to the duct walls that ultimately flows into the magnets has been made and is given by

$$\frac{q}{mc_p T_1} = \frac{(\gamma - 1)\alpha_0 BT_1 x_0}{\gamma V(1 - K)P_1} \left\{1 + \left(\frac{\gamma - 1}{2}\right)(Ma)^2\right\}^{1/(\gamma - 1)} \left\{\frac{1 - (T_2/T_1)^{3/2 - y + \beta z'}}{\frac{3}{2} - y + \beta z'}\right\} \tag{8-7}$$

where q is the heat given up to the magnets and α_0 the heat transfer factor assumed to be constant.

C. *Cycle Analysis and Overall Efficiency*

We shall consider the typical steam turbine MHD generator topping cycle illustrated in Figure 9. Suffixes here refer to conditions at the corresponding points in the gas cycle, as noted in the block diagram. For a mass

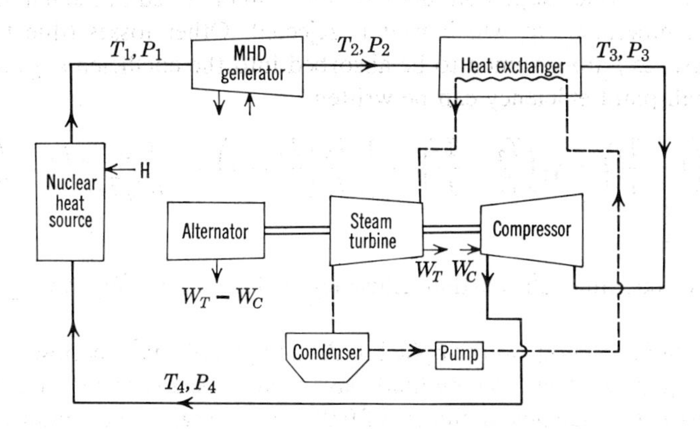

Fig. 9. Magnetohydrodynamic generator-steam turbine topping cycle. – – – steam water —— gas particle suspended. (After Sodha and Bendor 1964b.)

flux m of gas-particle mixture, the work and heat produced or absorbed by the components are

$$\left.\begin{aligned} W_M &= mC_p\eta_M(T_1 - T_2) \\ W_T &= mC_p\eta_T(T_2 - T_3) \\ W_c &= \frac{mC_p}{\eta_c}(T_4 - T_3) \\ H &= mC_p(T_1 - T_4) \end{aligned}\right\} \tag{8-8}$$

where η_c is the mechanical efficiency of the compressor and η_M the efficiency of the a-c. to d-c. conversion equipment associated with the MHD generator (it being assumed that an a-c output is required). The work produced by the turbine W_T is simply taken to be a fixed fraction η_T of the heat transferred from the gas in the heat exchanger.

The power required to remove the heat flux q from the magnet by a refrigeration cycle having carnot cycle coefficient of performance is obtained as

$$Q_M = mc_p T_1 \Phi \left\{ 1 - \left(\frac{T_2}{T_1} \right)^{3/2 - y + \beta z'} \right\} \tag{8-9}$$

where

$$\Phi = \frac{\varepsilon \alpha_0 T_R T_1^{1/2 - y} P_1^{z'} (\frac{3}{2} - y + \beta z')^{-1}}{T_M B(Ma) C C_p^{1/2} (\gamma - 1)^{1/2} V(1 - K)} \left(1 + \left(\frac{\gamma - 1}{2} \right) (Ma)^2 \right)^{\frac{2\gamma(y - z') + \gamma - 2y - 1}{2(\gamma - 1)}}$$

Also, $1/\varepsilon$ is the efficiency of refrigeration, and T_M and T_R, respectively, the temperature at which superconductivity can be maintained in the magnet coils and the temperature at which heat is rejected. Other losses (due to plant auxiliaries, etc.) are assumed to be absorbed into the coefficients η_M and η_T. The overall plant efficiency can be written

$$\eta = \eta_M \left(1 - \frac{T_2}{T_1} \right) + \eta_T \left(\frac{T_2}{T_1} - \frac{T_3}{T_1} \right) - \frac{1}{\eta_c} \frac{T_3}{T_1} \left(\frac{T_4}{T_3} - 1 \right) - \frac{Q_M}{mC_p T_1} \bigg/ \left(1 - \frac{T_3}{T_1} \frac{T_4}{T_3} \right) \tag{8-10}$$

The above equation shows that efficiency η depends mainly on T_2/T_1 and T_3/T_1.

The above methods were applied to the analysis of a typical power generating set (Figure 9). The working fluid was taken to be a suspension of barium oxide dust with a particle diameter of 0.05 μm in Argon, the conductivity of which is given by eq. 7-11. The following values of the various parameters were used:

1. Working fluid: $c_{pg} = 522$ J kg^{-1}/deg K^{-1}, $c_{pp} = 334$ J kg^{-1}/deg K^{-1}, and $\gamma_g = 1.667$.
2. MHD duct: $B = 10$ Wbm^{-2}, $V = 11{,}550$ v, $K = 0.8$, and $\varepsilon\alpha_0 T_R/T_M =$ 400 N m^{-1} sec^{-1} deg K^{-1}.
3. Component efficiencies: $\eta_M = 0.95$, $\eta_c = 0.98$, $\eta_p = 0.90$, $\eta_T = 0.45$, $K_A = 0.96$, and $K_H = 0.95$.

These efficiencies are typical in current practice. K_A and K_H are, respectively, the pressure drop factor across the heat exchanger and nuclear heat source.

To determine approximately the optimum temperature ratios, efficiencies were first computed for the system without including the power required to cool the magnets. The efficiency η was calculated for one (intermediate) value of $\lambda = 0.1$ only, for three Mach numbers (0.7, 1.0, and 1.3) and for a range of values of T_3/T_1 and T_2/T_1. The results for $(Ma) = 0.7$ are illustrated in Figure 10. For other Mach numbers it was found that η decreases slightly

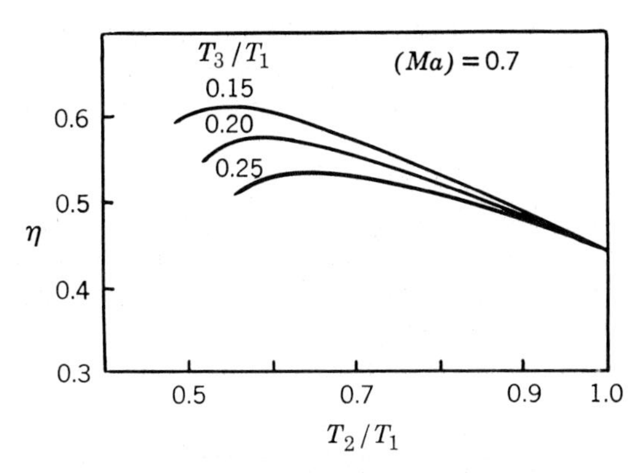

Fig. 10. Variation of efficiency of a steam turbine—MHD generator topping cycle with T_3/T_1 and T_2/T_1. (After Sodha and Bendor 1964b.)

with increasing (Ma) for fixed T_3/T_1 and T_2/T_1. The effect of the Mach number is due to the increase in β with increasing (Ma), leading to a larger pressure drop for a given generator output.

Two important points worthy of note appear in these results: (1) η increases with decreasing T_3/T_1, and (2) the value of T_2/T_1 for optimum efficiency increases with increasing T_3/T_1.

It is of some interest to consider the limiting case $\eta_T \to 0$. Overall cycle efficiencies for $\eta_T = 0$ and three Mach numbers were therefore also calculated, again neglecting the loss due to the magnets. The results for (Ma) = 0.7

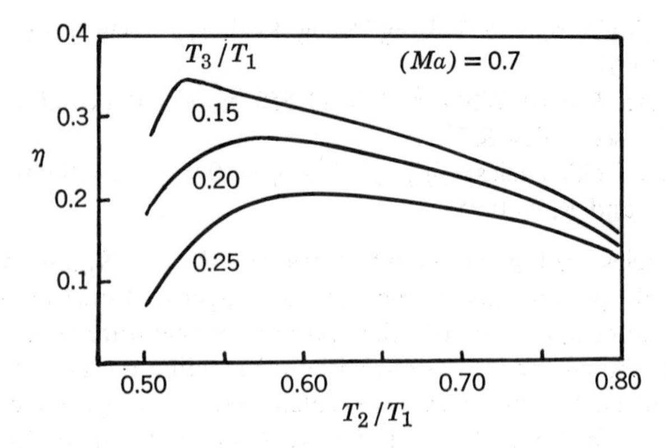

Fig. 11. Variation of efficiency of a closed cycle MHD power generation system ($\eta_T = 0$) with T_3/T_1 and T_2/T_1. (After Sodha and Bendor 1964b.)

which are to be compared with Figure 10 are illustrated in Figure 11. The optimum temperature ratios T_2/T_1 for given T_3/T_1 are nearly the same as those for $\eta_T = 0.45$. There was again a decrease in η with increasing (Ma) for fixed T_2/T_1 and T_3/T_1.

Such a scheme has applications in space power systems and ground mobile power stations (perhaps also stationary power systems in relatively inaccessible places), where lower weight rather than higher efficiency is the primary goal. The heat is passed from the working fluid to the coolant in the heat exchanger and then radiated away.

For the complete MHD-steam turbine cycle, it may be shown that

$$\frac{\partial}{\partial \eta_T}(\eta - \eta_T) = \left(\frac{T_2}{T_1} - \frac{T_3}{T_1}\right) \bigg/ \left(1 - \frac{T_3}{T_1}\frac{T_4}{T_3}\right) - 1 \simeq -0.25$$

for a representative case such as that given in Table VIII. Thus the extent to which the efficiency of a steam plant can be enhanced by the addition of the MHD topping cycle becomes more pronounced at lower steam cycle efficiencies. The selection of the somewhat high value $\eta_T = 0.45$ therefore leads to an underestimate of the advantage accruing from the inclusion of the MHD generator.

1. A Typical Cycle

To investigate the effect of λ and (Ma) in greater detail, some cycle calculations retaining the loss Q_M were carried out for a range of values of these parameters. The most prominent effect of λ is on the duct length x_2 which

TABLE VIII

Results for a selected cycle: $T_2/T_1 = 0.7$, (Ma) $= 1.0$, $\lambda = 0.3$.

T_1 (°K)	T_2 (°K)	T_3 (°K)	T_4 (°K)	P_1 (atm)	P_2 (atm)	P_3 (atm)	P_4 (atm)	σ_1 (mho m^{-1})	σ_2 (mho m^{-1})	x_2 (m)
1669	1167	367	635	7.00	1.73	1.65	7.28	6.85	2.95	6.35
	$\frac{W_M}{W}$	$\frac{W_T}{W}$	$\frac{W_C}{W}$	$\frac{Q_M}{W}$	$\frac{H}{W}$	$\left(\frac{A_1}{W}\right)10^{10}$ [a]		$\left(\frac{m}{W}\right)10^{7}$ [b]	$\left(\frac{\Omega}{W}\right)10^{8}$ [c]	η
$\eta_T = 0.45$	0.845	0.652	0.490	0.001	1.832	39.0		37.0	4.67	0.543
$\eta_T = 0$	2.428	—	1.408	0.029	5.253	109.5		106.2	13.15	0.19

[a] Units $m^2\omega^{-1}$.
[b] Units kg J^{-1}.
[c] Units $m^3\omega^{-1}$.

is independent of W_M for a fixed T_2/T_1. The variation of x_2 with A and (Ma) for $T_2/T_1 = 0.58$ is illustrated in Figure 12.

The particle concentration, and therefore the conductivity enters the calculation of η through the heat loss Q_M which was found to account for a 1% increase in plant efficiency as λ is increased from 0.03 to 0.3. It was estimated, e.g., that a 400-Mw MHD duct with an entry stagnation temperature of 2000°K and a duct length of 10 meters would lose about 20 Mw of heat if the walls were maintained at 1000°K. (The fractional loss of heat increases with decreasing generator power.) Thus the reduction of operating temperature made possible by the addition of thermionically emitting dust has an important effect on efficiency. The significance of short duct length, as far as efficiency is concerned, lies in the reduced heat loss for a given wall temperature.

The results of a complete cycle calculation are given in Table VIII for a reactor outlet temperature of 1669°K (2500°F). The value of $T_2/T_1 = 0.7$ was chosen so as to produce a steam turbine output somewhat higher than the compressor power. A reduction of T_2/T_1 to its optimum value of about 0.58 gives 1% to 2% greater efficiency, but then $W_T < W_c$, which is an undesirable feature in plant operation.

Dust of barium oxide and other materials with a particle size of 0.05 μm (or lower) is commercially available. Various methods for the uniform suspension of this fine dust in the gas have been discussed. Special mention should be made of the experimental work conducted in connection with studies of the dust suspension reactor (Schluderberg, Whitelaw, and Carlson 1962) and the dust fuel reactor (Krucoff 1959) which leads one to believe that closed cycle reactor system using gas-suspension fluids are practicable.

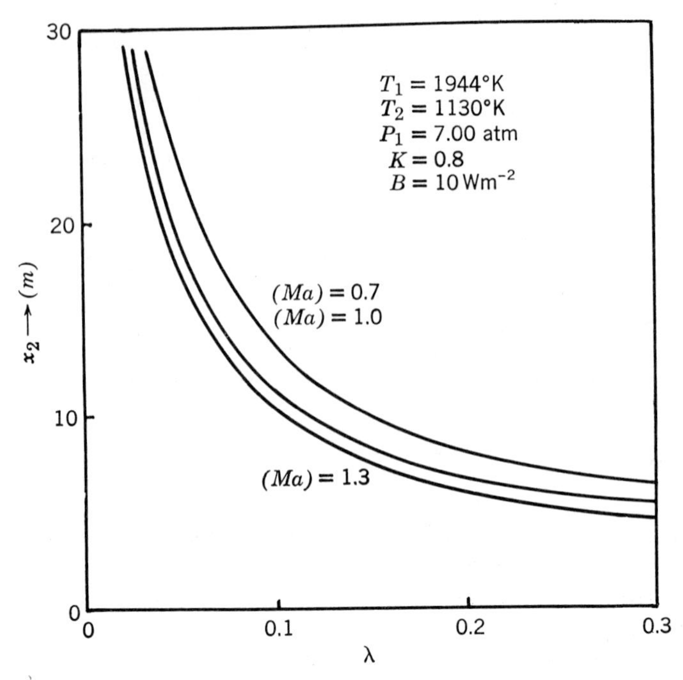

Fig. 12. Variation of duct length x_2 with λ and *Ma*. (After Sodha and Bendor 1964*b*.)

Honma and Fushimi (1966) have shown that the effect of "ion slip" in the current equation is considerably reduced by the suspension which allowed the use of high magnetic field without the reduction of current.

The use of solid emitting particles in open-cycle MHD generators, where coagulation after the duct would probably be advantageous, was considered by Williams et al. (1963). They concluded that an initial particle diameter of not more than 0.02 μm would be needed.

Another interesting possibility of the use of dust-laden gases has been proposed by Halasz and Szendy (1967). Since rare gases are practically transparent to radiant heat, whereas solid bodies absorb the greater part of the heat, it is possible, by blowing in the fine powder of a material having a low work function in a cold gas, to heat this gas to a high degree by the radiation of the high temperature ceramic material applied to the wall. This results in electron emission.

The procedure is based on a well-known fact, that for a short time (10^{-2}–10^{-3} sec) a great temperature difference (1000–3000°K) can exist between a gas and a fine solid particle ($2a = 10^{-7} - 10^{-8}$m) floating with zero relative velocity in it. This short time must be slightly longer than that during which the particle is traveling along the magnetized section of the generator (compressor) duct. The temperature of the solid particle is raised by fitting a ceramic or other suitable material, called a ceramic ring, into the duct wall, thermally insulating it from the remaining part of the wall structure and keeping it at a high temperature (2000–3300°K). Then its radiation will heat the particle, while the working fluid remains practically cold because it does not absorb radiation heat. The latter condition is almost completely fulfilled by the noble gas in a nuclear MHD power plant or the air in an MHD compressor. The whole inner surface of the generator (compressor) wall must be made of a material with high reflectivity, in order to prevent reradiation from the particles. Halasz and Szendy (1967) have analyzed quantitatively the feasibility of this system.

James and Vermeulen (1968) have reported setting up an experiment for the investigation of colloidal plasmas; however, no results have been reported yet.

IX. Nonequilibrium Ionization in Wet Alkali Metal Vapors

Smith (1965) has given a theoretical model for the transport of electrons in a wet vapor of alkali metal in which the condensate is considered to be of the form of homogeneously dispersed droplets of uniform but arbitrary size. The medium is considered to consist of electrons, neutral gas atoms, positively charged atomic ions, and liquid droplets that in the regime of interest here are negatively charged. This theory has been applied to MHD generator, for which the predominant effect of the droplets is found to be the absorption of free electrons from the system. The decrease of electron density is found to be most severe at a high percentage of moisture and for small droplets. This study gives another illustration of the usefulness of the gas-solid suspensions, in which the vapor droplets play the role of solid particles. We follow here the analysis of Smith (1965).

A. *Electron Collision Frequency and Electron Density*

In the electrical conduction in a wet alkali metal vapor, the following collision frequencies of the electrons are significant.

1. Electron neutral molecule collision (v_0)
2. Electron ion collision (v_i)
3. Electron charged droplet elastic collision (v'')
4. Electron droplet inelastic collision (v')

It is assumed that the droplet behaves as a metal particle, and considerations similar to those of solid particles are applied to the case of droplets.

The atom droplet collision frequency is considered to be entirely inelastic and of the hard sphere type, and is given by

$$\nu_{Ad} = N_d \pi a^2 \left(\frac{8kT_g}{m_A \pi}\right)^{1/2} \tag{9-1}$$

where m_A is the mass of an atom, N_d is the droplet density and T_g is the temperature of the gas. The expressions of the other collision frequencies are given in Section VI.

The electron density in the medium is a result of the combined effect of the ionization of gas atoms and the emission/recombination of electrons with the droplets. The ionization of alkali metal vapors is given by Saha's equation evaluated at the electron temperature as (Kerrebrock 1962)

$$\frac{N_e N_i}{N_A} = \left(\frac{2\pi mkT_e}{h^2}\right)^{3/2} \exp\left(-\frac{\phi_i}{kT_e}\right) \tag{9-2}$$

where ϕ_i is the ionization potential of alkali atom, N_i is the density of ions and N_A is the density of atoms. Considering a droplet to be a perfectly conducting sphere, a potential model for the removal of an electron from a negatively charged droplet has been given by Smith (1965) with the effective work function

$$\phi_s = \phi_0 - \frac{5}{8}\frac{e^2}{a} \tag{9-3}$$

where ϕ_0 is the work function of the bulk metal. The equation for the droplet charging is obtained by considering all droplets to possess the same average charge ze and then equating the random electron current to the droplet to the thermionic emission current from the droplet. Thus

$$N_e v' = \frac{J_d 4\pi a^2}{e} N_d \tag{9-4}$$

where J_d is the thermionic emission current, given by

$$J_d = \frac{4\pi emk^2 T_d^{\,2}}{h^3} \exp\left(-\frac{\phi_s}{kT_d}\right)$$

where T_d is the droplet internal temperature. Thus from eq. 9-4 the electron density can be obtained:

$$N_e = 2\left(\frac{2\pi mkT_d}{h^2}\right)^{3/2} \left(\frac{T_d}{T_e}\right)^{1/2} \exp\left(-\frac{\phi_s}{kT_d} + \frac{|z|e^2}{akT_e}\right) \tag{9-5}$$

and the equation for local charge neutrality is given by

$$N_e = N_i - |z|\, N_d \tag{9-6}$$

B. *Droplet Internal Temperature*

The equation for the droplet internal temperature is obtained by equating the rate with which the droplet is heated due to electron bombardment to the rate with which the droplet is cooled due to neutral atom bombardment (cooling because of ion bombardment is neglected with respect to that of the neutral atoms). Assuming a unit accommodation coefficient for both electrons and atoms, we obtain

$$N_e \nu'(T_e - T_d) = N_A \nu_{Ad}(T_d - T_g) \tag{9-7}$$

From the above equation, we get

$$T_d = \frac{\beta T_e + T_g}{1 + \beta} \tag{9-8}$$

where

$$\beta = \frac{N_e}{N_A}\left(\frac{m_A T_e}{m T_g}\right)^{1/2} \exp\left(-\frac{|z|\, e^2}{akT_e}\right) \tag{9-9}$$

C. *Electron Temperature*

An expression for the electron temperature as a function of the electromagnetic field in the gas is obtained by including the effect of energy loss resulting from the electron-droplet inelastic collisions and the electron neutral species elastic collisions. Hence the energy balance equation is given by

$$\mathbf{J} \cdot \mathbf{E}^* = \sum_s N_e \nu_{es} 2 \frac{m}{M_s} \frac{3}{2} k(T_e - T_g) + N_e \nu' 2k(T_e - T_d) \tag{9-10}$$

where

$$\mathbf{E}^* = \mathbf{E} + \mathbf{u} \times \mathbf{B}$$

and $\mathbf{J}$ is the electron current density, $\mathbf{E}$ the electric field, $\mathbf{u}$ the gas-flow velocity, and $\mathbf{B}$ the magnetic field, ν_{es} the electron collision frequency with the sth species.

For the segmented electrode MHD generator (Hurwitz et al. 1962), the electron temperature is given by

$$\frac{e}{m \sum_s \nu_{es}} u^2 B^2 = \sum_s 2\nu_{es} \frac{m}{M_s} \frac{3}{2} k(T_e - T_g) + N_e \nu' 2k(T_e - T_d) \tag{9-11}$$

As an example of the effect of droplets upon the conduction of electrons, a potassium-vapor system at 800°K, 0.1 atm (slightly supersaturated), and 75% moisture, was considered, for which

$$N_A \simeq \frac{p}{kT_g}$$

$$N_d \simeq \frac{N_A \times \%\,\text{moisture}}{n_d}$$

where n_d is the number of atoms per droplet. The following values of various constants have been taken: $\phi_0 = 2.15$ eV, the work function of the droplet was assumed equal to that of solid potassium, and $\phi_i = 4.34$ eV.

For the preceding conditions the droplet temperature was found to be equal to the gas temperature.

In Figure 13 the ratio of the electron density in the system to that with no droplets present is shown. An almost two order of magnitude decrease in the electron density is noted at the lower limit of electron temperatures considered. The electron density ratio then rises rapidly to reach approximately unity at

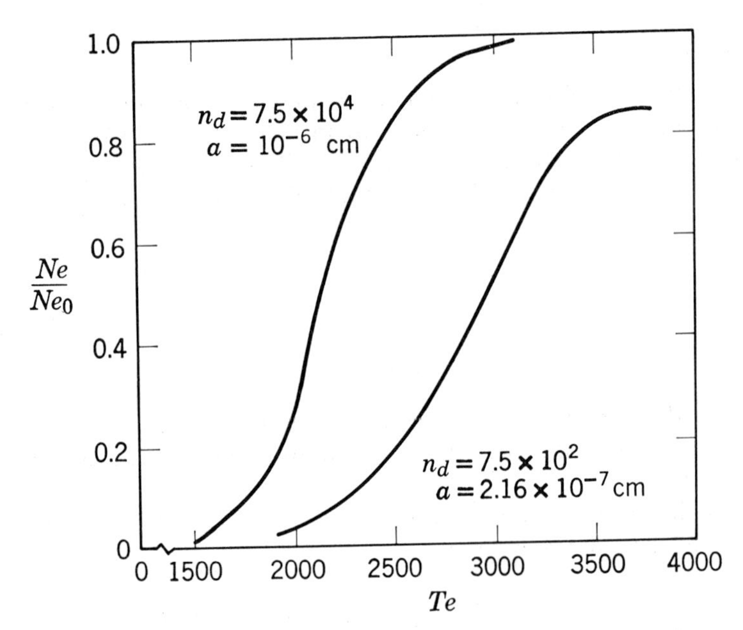

Fig. 13. Ratio of electron density with droplets to electron density without droplets vs. electron temperature. (After Smith 1965.)

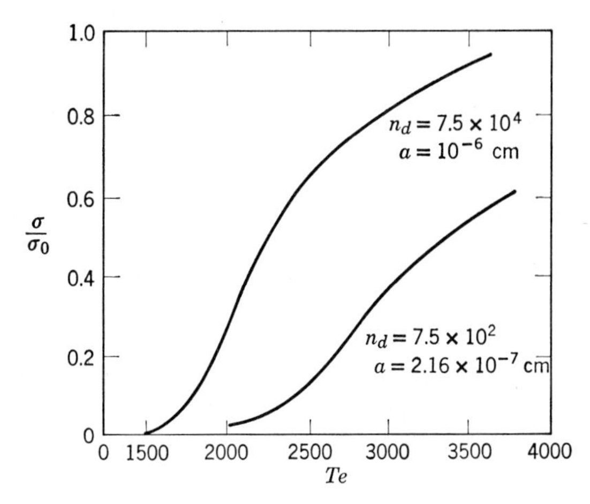

Fig. 14. Ratio of electrical conductivity with droplets to electrical conductivity without droplets vs. electron temperature. (After Smith 1965.)

the upper range of electron temperature, the decrease of the electron density being due to electron absorption by the droplets.

The ratio of the electrical conductivity in the system to that with no droplets is shown in Figure 14. The variation of the conductivity with the electron temperature is primarily a result of the electron density variation, inasmuch as the collision frequency is, within the range considered here dominated by electron neutral atom collisions. When the condensate is dispersed in small droplets, the conductivity is more severely depressed.

The effect upon the required magnetic field strength from the presence of the droplets is slight, and nearly independent of the electron temperature. The inelastic energy loss is insignificant, except at the lowest temperature at which its contribution is approximately 10%, compared to the elastic energy losses. The ratio of the magnetic field required to produce a given electron temperature with droplets to that without droplets was found to be approximately 1.08 for $a = 10^{-6}$ cm and 1.2 for $a = 2.16 \times 10^{-7}$ cm.

Instead of suppressing the electron density as in the work of Smith (1965), Kovacs et al. (1966) analyzed the use of condensed alkali metal vapors that form liquid droplets to enhance the electron density of the working gas in an MHD power generator. If the vapor, when changing phase, is heated to the point necessary for reaching the ionization potential or work function, it can satisfactorily ionize the working gas. The temperature drop required for spatial condensation of high-temperature metal vapor (1500–1700°K) is attainable by heat radiation or sudden pressure decrease. Kovacs et al. (1966) have proposed to investigate this effect experimentally.

X. Electrostatic Precipitation

In the electrostatic precipitation process, suspended particles in a gas are first electrically charged and then driven to the collecting electrodes under the force of the electric field. The process is distinguished from all mechanical or filtering processes, in that the separating forces are exerted directly on the particles themselves rather than on the gas as a whole. From a practical standpoint, this means that effective separation of particles can be achieved with low-power expenditure, with negligible draft loss, and with little or no effect on the gas.

Calculations of the electric separating forces require knowledge of the electric charges on the particles and of the electric field to which they are subjected. Many types of dispersoids are naturally charged as a result of their method of formation. However, these natural charges normally are inadequate for effective removal of the particles on a commercial basis; therefore, in practice, it is necessary to provide a positive means for charging the particles.

Of the possible charging methods, corona discharge is the most practical and is almost universally used in electrical precipitation practice. It has been found that negative corona usually takes the form of a unipolar gas discharge emanating from a wire or series of wires designated as emitting or discharge electrodes, and terminating on pipes or plates designated as collecting electrodes (Figure 15).

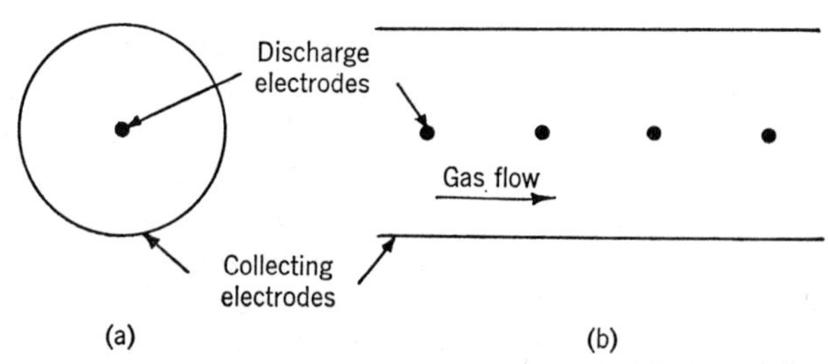

Fig. 15. Schematic illustration of electrode systems for electrical precipitators: (*a*) Pipe type; (*b*) plate or duct type.

A. *Corona Charging*

Extensive details of corona discharge and corona charging of particulate matter at low temperatures (therefore thermionic emission can be neglected) are presented in treatises by Cobine (1958) and White (1951, 1961). Theoretical studies indicate that two particle charging mechanisms are present in the corona discharge:

1. Bombardment of the suspended particles by ions under the force of the electric field in the region between the corona wires and collecting electrodes. This is also known as field charging process.
2. Attachment of ions to the particles by ion diffusion.

In general, process 1 is of primary importance in electrostatic precipitation, while process 2 is of limited interest and is important only for particles smaller than about 0.2-μ diameter.

1. Field Charging

Basic theoretical equations for the charging of small spherical particles in a unipolar corona discharge were developed in detail by Deutsch (1922), Arendt and Kallman (1925), Ladenburg (1930), and Pauthenier and Moreau-Hanot (1932). The problem is essentially one in electrostatics, and the solution is based on the distortion of the electric field in the neighborhood of the particle. Ions move along the lines of force; some will strike the particle and transfer their charge. This charge produces a repulsive force that reduces the rate of charging. Ultimately, sufficient charge is received to counteract completely the external field, so that the electric flux entering the particle is reduced to zero and charging ceases. Here we follow the analysis of Soo (1967). When a conducting sphere of radius a is placed in a uniform electric field E_0 with an infinitely uniform unipolar ion density n_0, the potential V distribution is given by Poisson's equation

$$\nabla^2 V = -\frac{qn_z}{\varepsilon_0} \tag{10-1}$$

where ε_0 is the permittivity, q is the charge per ion ($q = -e$ for an electron), and n_z the distributed ion density. The boundary conditions are $E = -\nabla V = E_0$ at infinity and $E = -ze/4\pi\varepsilon_0 a^2$ for z charges on the surface at a given time. It is readily shown that at any point on the sphere,

$$E_a = 3E_0 \cos\theta - \frac{ze}{4\pi\varepsilon_0 a^2} \tag{10-2}$$

where θ is the azimuthal angle in the spherical coordinate.

For particle radius a much larger than the mean free paths of the ions ($a \gtrsim 1\mu$), the random motion of the ions does not need to be considered, and the total electric flux ψ entering the sphere is given by

$$\begin{aligned} \psi &= \int_0^{\pi} 4\pi\, \varepsilon_0\, E_a\, 2\pi a^2 \sin\theta \, d\theta \\ &= 12\pi\varepsilon_0 a^2 E_0 \left(1 - \frac{ze}{12\pi\varepsilon_0 E_0 a^2}\right)^2 \end{aligned} \tag{10-3}$$

At saturation, $\psi = 0$, and

$$z_s = 12\pi\varepsilon_0 E_0 a^2/e \tag{10-4}$$

The rate of charging is given by the charging current i,

$$i = \frac{n_z qK\psi}{4\varepsilon_0} = \frac{d}{dt}(ze) \tag{10-5}$$

where K is the ion mobility. Integration gives

$$-\frac{n_z qKt}{4\varepsilon_0} = \frac{z/z_s}{1 - z/z_s}$$

or

$$z = \frac{z_s t}{t + 1/n_z qK} \tag{10-6}$$

This equation gives the charge z acquired after t seconds by a spherical particle in an electric field E_0. The saturation charge z_s is seen to be directly proportional to the electric field and to the surface area of the particle. $1/n_z qK$ is called the particle charging time constant. White (1961) showed that for $n_z = 5 \times 10^{14}$ m^{-3}, $K = -2.2$ (cm/sec)/(V/cm) and $q = -e$, the time to reach $z/z_s = \frac{1}{2}$ is 2 msec. This charging time is very small compared with the total treatment time. When applied to dielectric particles, eq. 10-4 is modified to

$$z_s = \frac{12\pi\varepsilon_0 E_0 a^2[1 + 2(\varepsilon_r - 1)/(\varepsilon_r + 2)]}{q} \tag{10-7}$$

where ε_r is the dielectric constant; $\varepsilon_r = \varepsilon/\varepsilon_0$, ε is the permittivity of the material.

The conclusion from the theoretical and experimental studies of particle charging is that all particles become highly charged in a few hundredths of a second or almost immediately on entering the precipitator field. Hence particle charging under normal conditions is fundamentally an efficient and rapid process.

Experimental studies of the field charging process have been made by numerous workers. These include the measurements of White (1951, 1961), Gillespin and Langstroth (1952), Penney and Lynch (1957), Hewitt (1957) and Drozin and LaMer (1959). These experiments show an order of magnitude agreement with theory.

2. Charging by Diffusion in an Electric Field

For a radius of the order of 0.5 μ or less, the random motion of ions must be accounted for and the contribution of the external electric field becomes less significant. Studies were made by Arendt and Kallmann (1925), White (1951), Gunn (1954), Natanson (1960), and Liu et al. (1967a), considering the particle charging process as a macroscopic diffusion process in which ions are assumed to diffuse continuously in a quasisteady state toward the particle under the action of a concentration gradient. This charging mechanism is independent of the externally applied field. The accumulation of electric charge on the particle tends to repel further ions, and progressively decreases the charging rate. The number of charge z acquired after a time t sec,

$$z = \frac{akT}{e^2} \ln\left(1 + \frac{\pi a \bar{v}_i n_z e^2 t}{kT}\right) \tag{10-8}$$

where $\bar{v}_i$ is the root mean square velocity of the ions. The relative importance of the ion bombardment and ion diffusion mechanisms of charging is shown in Table IX (Lowe and Lucas 1953) which shows the theoretical number of charges acquired by particles after various times of exposure of each mechanism. The conditions on which the calculations are based are $T = 300°\text{K}$, $n_z = 5 \times 10^7$ ions/cm³, $E_0 = 2$ kV/cm, $3\varepsilon_r/\varepsilon_r + 2 = p = 1.8$, and are representative of a wire in tube assembly inner and outer radii 0.15 in and 5 in, respectively, operating in air under atmospheric conditions at 40 kV with a discharge current of 40 μA/ft.

TABLE IX

Number of Charges Acquired by Particles

Particle radius	Ion bombardment process				Ion diffusion process			
	Period of exposure (sec)				Period of exposure (sec)			
μ	0.01	0.1	1	∞[a]	0.01	0.1	1	10
0.1	0.7	2	2.4	2.5	3	7	11	15
1.0	72	200	240	250	70	110	150	190
10	7,200	20,000	24,400	25,000	1,100	1,500	1,900	2,300

[a] Limiting charge.

Reference to Table IX shows that the ion diffusion process is predominant for the smallest particles, and ion bombardment for the larger. Fair agreement between observed and theoretical charging rates has been found by several workers (White 1951, Liu et al. 1967b).

The problem of interaction of a spherical particle with a slightly ionized gas was treated by Rosen (1962) and Dimick and Soo (1964). The latter extended the method of Murphy et al. (1959) for the case of zero external field to interaction with an ionized gas in the presence of ions of both signs.

The study of the particle charging by small ions is also of considerable importance in the ionic equilibrium of the atmosphere, since the mechanism of particle charging affects both the removal from the atmosphere of small ions produced by ionizing radiation and the resulting statistical distribution of electrical charge on aerosol particles (Gunn 1954, 1955). The mechanism of particle charging by thermal movement of ions with the particle has been invoked to describe the origin of charges on solid dusts and cloud droplets in the atmosphere (Gunn 1955).

Another aspect of charge collection of interest is the interaction of a spacecraft with the ionosphere. Davis and Harris (1961) computed the ion trajectories around a charged satellite in the ionosphere, neglecting the earth's magnetic field, with the solution of Poisson's equation for satellites in numbers (10, 25, etc.) of Debye length. The drag on a charged satellite was studied by Jastrow and Pearse (1957) and Wyatt (1960).

B. *Separation Velocities of Charged Particles and Efficiency*

The motion of a charged suspended particle in an electric field is determined basically by the Coulomb force that drives the particles and by the frictional retarding force of the gas. If we equate the driving and resistive forces on the particle, using Stokes' law and introducing the particle charge previously calculated, $z_s e = 12\pi\varepsilon_0 E_0 a^2$ (eq. 10-4), the particle velocity

$$v = \frac{2\varepsilon_0 E_0 a E_p}{\eta} \tag{10-9}$$

where E_p is the magnitude of the collecting field and η is the gas viscosity. This quantity v is frequently referred to as the precipitation rate. The precipitation efficiency is given by (White 1955):

$$E = 1 - \exp\left(-\frac{A}{V}v\right) \tag{10-10}$$

where A is the total collecting surface of the electrode and V is the rate of gas flow through the precipitator. Munden (1952) has shown that the efficiency of the electrostatic precipitators is far greater than that of commercial test filters.

XI. Photoelectric Ionization of Solid Particles

The physical processes occurring in a cloud of solid particles have been of great interest to astrophysicists because a substantial fraction of interstellar matter exists in this form. A number of scientists have been engaged in the study of interaction of these solid particles with radiation and elementary particles. An enlightening and brief discussion of the early work has been given by Spitzer (1941, 1948) who made an exhaustive investigation of the ionization of solid particles due to the photoelectric effect of the incident radiation. However, Spitzer made the simplifying assumption that all the particles carry the same charge and neglected the effect of the charge of the particles on electron emission, presumably because of his preoccupation with solid particles of negative charge, in which case there is no effect in the semi-classical approximation. Sodha (1963) has analyzed photoelectric ionization of solid particles, using the Sommerfeld model of a metal.

A. *Photoelectric Emission*

Sodha (1963) has used the following assumptions made by Fowler (1936):

1. The probability of absorption of a photon by an electron at the surface is independent of the initial state of the electron.
2. All the absorbed photon energy increases the normal kinetic energy of the electron; the velocity components parallel to the surface remain unchanged.
3. Electrons with normal energy greater than the surface barrier cross the surface.

Case 1: $z \leq 0$. In this case, the velocity distribution of number $n_{\text{ph}}(z-1)$ of photoelectrons emitted per unit area per unit time by a particle with charge $(z-1)e$ before and the charge ze after electron emission is given by (following eq. 2-1):

$$d^3 n_{\text{ph}}(z-1) = \beta \Lambda \frac{2m^3}{h^3} u_x \times \left\{1 + \exp\left[\frac{m}{2kT}(u_x{}^2 + u_y{}^2 + u_z{}^2) - \xi\right]\right\}^{-1} du_x\, du_y\, du_z \quad (11\text{-}1)$$

where β is characteristic of the surface and the radiation, Λ is the number of photons incident per unit area per unit time on the surface, $\xi = (h\nu - \varphi)/kT$, and ν is the frequency of the radiation. Integrating $d^3 n_{\text{ph}}(z-1)$ as given by eq. 11-1 in the limits $0 < u_x < \infty$, $-\infty < u_y < \infty$ and $-\infty < u_z < \infty$, we obtain

$$n_{\text{ph}}(z-1) = B\Lambda T^2 \Phi(\xi) \quad (11\text{-}2)$$

where

$$B = \frac{4\pi m k^2}{h^3} \beta$$

and $\Phi(\xi) = \int_0^{\exp(\xi)} \Omega^{-1} \ln (1 + \Omega)\, d\Omega$ is a well-known tabulated function. The mean energy of emitted electrons at the surface is given by

$$\varepsilon_{ph}(z-1) = \frac{kT}{\Phi(\xi)} \int_0^\infty \int_0^\infty (\xi_1 + \xi_2)\{1 + \exp(\xi_1 + \xi_2 - \xi)\}^{-1}\, d\xi_1\, d\xi_2 \quad (11\text{-}3)$$

and the mean energy at a large distance from the particle is

$$\varepsilon'_{ph}(z-1) = \varepsilon_{ph}(z-1) - \frac{ze^2}{a} \quad (11\text{-}4)$$

Case 2: $z \geq 0$. In this case, only electrons with $\frac{1}{2}\, mu^2 > ze^2/a$ can escape from the surface. Hence $n_{ph}(z-1)$ can be obtained by integrating $d^3 n_{ph}(z-1)$ as given by eq. 11-1 in the limits $0 < u_x < \infty$, $-\infty < u_y < \infty$ and $-\infty < u_z < \infty$, so that $\frac{1}{2}\, mu^2 > ze^2/a$ in the range of integration. Thus, after a little algebra, we obtain

$$n_{ph}(z-1) = B\Lambda T^2 \Psi(\xi, z\alpha) \quad (11\text{-}5)$$

where

$$\Psi(\xi, z\alpha) = \int_0^\infty \int_0^\infty \{1 + \exp(\xi_1 + \xi_2 - \xi)\}^{-1}\, d\xi_1\, d\xi_2 \qquad \xi_1 + \xi_2 > z\alpha$$

$$= z\alpha \ln\{1 + \exp(\xi - z\alpha)\} + \Phi(\xi - z\alpha)$$

Similarly,

$$\varepsilon_{ph}(z-1) = \frac{kT}{\Psi(\xi, z\alpha)} \int_0^\infty \int_0^\infty (\xi_1 + \xi_2) \times \{1 + \exp(\xi_1 + \xi_2 - \xi)\}^{-1}\, d\xi_1\, d\xi_2 \qquad \xi_1 + \xi_2 > z\alpha \quad (11\text{-}6)$$

and

$$\varepsilon'_{ph}(z-1) = \varepsilon_{ph}(z-1) - \frac{ze^2}{a} \quad (11\text{-}7)$$

For a wide spectral distribution of radiation, eqs. 11-2 and 11-5 can be generalized to

$$n_{ph}(z-1) = \int_{\nu_1}^{\nu_2} B\Lambda'(\nu) T^2 \Phi(\xi)\, d\nu \qquad \text{for } z \leq 0 \quad (11\text{-}2)'$$

and

$$n_{ph}(z-1) = \int_{\nu_1}^{\nu_2} B\Lambda'(\nu) T^2 \Psi(\xi, z\alpha)\, d\nu \qquad \text{for } z \geq 0 \quad (11\text{-}5)'$$

where $\Lambda'(\nu)\, d\nu$ is the number of photons incident per unit area per unit time on the surface, having frequencies between ν and $\nu + d\nu$. Equations 11-3 and 11-6 can be generalized likewise.

B. *Steady-State Photoelectric Ionization*

In the steady state the number of particles with charge $(z-1)e$, acquiring a charge ze by photoelectric emission, must be equal to the number of particles with charge ze recombining with electrons to acquire a charge $(z-1)e$ per unit volume per unit time. Hence (following eq. 2-14),

$$N_{z-1} 4\pi a^2 n_{\text{ph}}(z-1) = N_z N_e \langle v\sigma(z)\rangle$$

or

$$K_z = \frac{N_z N_e}{N_{z-1}} = \frac{4\pi a^2 n_{\text{ph}}(z-1)}{\langle v\sigma(z)\rangle}$$

where $\langle v\sigma(z)\rangle$ is given by eq. 2-10 or 2-12. Knowing K_z, we can proceed to solve eq. 2-22 for N_e with

$$N_z = N_0 \prod_1^z \left(\frac{K_z}{N_e}\right) \tag{11-8}$$

and also use eq. 2-20 to obtain N_0 and then eq. 11-8 to obtain N_z. The ionization kinetic equations (Section II) are valid for photoelectric case if we substitute n_{ph} for n_{th}.

C. *Photoelectric Enhancement of Electrical Conductivity in Dust-laden Gases*

Guha and Kaw (1968) have investigated analytically the possibility of using photoelectric emission from dust particles as a means of increasing the electron density (and hence the electrical conductivity) of a dust-laden gas. It has been shown that, for the particular system of argon laden with barium oxide dust, the photoelectric emission from the dust particles (for moderate intensities of the incident photoelectric radiation) dominates the thermionic emission to about 900°K; for higher temperatures the reverse is true.

1. Electron Density

Let us consider a gas in which the solid particles are in thermal equilibrium with the gas molecules. It is also assumed that the ionization potential of the gas molecules is so much higher than the work function of the solid particles that the entire electron density may be ascribed to the emission from the solid particles. Assuming for simplicity that all the particles carry the same positive charge ze, we find that n_{ph} is given by eq. 11-5

$$n_{\text{ph}} = \frac{4\pi m k^2 T^2}{h^3} \Lambda(\nu)\beta(\nu)[z\alpha \ln\{1 + \exp(\xi - z\alpha)\} + \Phi(\xi - z\alpha)] \tag{11-9}$$

where $\beta(\nu)$ may be defined as the ratio of the average number of photons or electrons leading to photoelectric absorption ($cm^{-2}\ sec^{-1}$) to the product of the number of photons incident on the surface ($cm^{-2}\ s^{-1}$) and the average number of electrons incident on the surface from inside ($cm^{-2}\ s^{-1}$). Following Spitzer (1948) and Sodha (1963), we may write

$$\beta(\nu) = \frac{Q(\nu)\chi(\nu)}{2m^3/h^3 \int_0^\infty \int_{-\infty}^\infty \int_{-\infty}^\infty \{1 + \exp[m/2kT(u_x^2 + u_y^2 + u_z^2) - E_F/kT]\}^{-1} u_x \, du_x \, du_y \, du_z}$$

where $Q(\nu)$ is the fraction of the incident energy being absorbed and $\chi(\nu)$ is the photoelectric efficiency. Equation 11-9 may be rewritten as

$$n_{\text{ph}} = \frac{Q(\nu)\chi(\nu)\Lambda(\nu)}{\Phi(E_F/kT)} \{z\alpha \ln[1 + \exp(\xi - z\alpha)] + \Phi(\xi - z\alpha)\} \tag{11-10}$$

In the steady state the electron density is given by (including the thermionic emission from the particles):

$$N_p(z)N_e\langle v\sigma(z)\rangle = N_p(z)4\pi a^2 n_{\text{th}} + N_p(z)\pi a^2 n_{\text{ph}} \tag{11-11}$$

where $N_p(z) \equiv N_p$, $\sigma(z)$ is the cross section for electron recombination with solid particles. In eq. 11-11, the cross section for photoelectric emission per particle has been taken as πa^2, rather than $4\pi a^2$, because the light-causing photoelectric emission is incident from one side only. Equations 11-10 and 11-11 lead to the steady-state electron density N_e

$$N_e = \frac{Q(\nu)\chi(\nu)\Lambda(\nu)}{(8kT/m\pi)^{1/2}\Phi(E_F/kT)(1 + z\alpha)} \{z\alpha \ln[1 + \exp(\xi - z\alpha)] + \Phi(\xi - z\alpha)\} + 2\left(\frac{2\pi mkT}{h^2}\right)^{3/2} \exp\left(-\frac{\phi}{kT}\right) \exp(-z\alpha) \tag{11-12}$$

where $z = N_e/N_p$.

The use of the same $\Lambda(\nu)$ for each of the N_p solid particles (i.e., the same intensity of incident radiation) in eq. 11-12 is not justified, in general, because one expects scattering of incident radiation by the peripheral dust particles which leads to obscuration of the central region. The following calculation shows that for many cases of interest the effect of this scattering is not very important. We shall consider small nonabsorbing dust particles of radii less than about $\frac{1}{10}$ of the wavelength of incident radiation. The scattered intensity in this case is given by the Rayleigh scattering formula. In general, the intensity of the radiation, after it travels a distance x through a gas containing a suspension of N_p dust particles per cm^3 (each of radius a), is given by

$$I = I_0 \exp(-N_p \pi a^2 Q_{sc} x)$$

where Q_{sc} is the efficiency factor for scattering defined by Van de Hulst (1957). The light intensity is therefore reduced to $1/e$ of its original value after traveling a distance

$$x_0 = (N_p \pi a^2 Q_{sc})^{-1}$$

For Rayleigh scattering by spherical particles (Van de Hulst 1957),

$$Q_{sc} = \frac{128\pi^4 a^4}{3\lambda^4} \left(\frac{n^2 - 1}{n^2 + 2}\right)^2$$

where n is the refractive index of the material of the particles, so that

$$x_0 = \frac{3\lambda^4}{128\pi^5 N_p a^6} \left(\frac{n^2 + 2}{n^2 - 1}\right)^2$$

For BaO particles, $n \simeq 1.99$. Choosing $\lambda = 5000$ Å, $a = 250$ Å, and $N_p = 10^{11}$ cm^{-3} (actual range considered 5×10^{10} to 5×10^{11} cm^{-3}), one obtains $x_0 \simeq 80$ cm. Thus, for this special case, the light intensity is reduced to $1/e$ of its value in a distance of 80 cm. This shows that, in general, this effect is not very important for many cases of practical interest.

2. Electrical Conductivity

The electrical conductivity of a dust-laden gas is given by

$$\sigma = \frac{e^2 N_e}{m \nu_c}$$

where ν_c is the total electron collision frequency, which for most cases of interest involves mainly electron collisions with solid particles and the neutral gas molecules. Then ν_c is given by

$$\nu_c = \nu_0 + \nu_p$$

where ν_0 is electron molecule collision frequency given by eq. 6-10 and ν_p is the electron particle collision frequency given by eqs. 6-11 and 6-12.

Numerical calculations have been performed for the conductivity of argon arising as a result of photoelectric emission from barium oxide dust. The results are tabulated in Tables X to XII. In these tables the number

TABLE X

Variation of Photoelectric Conductivity with Radiation Intensity[a]

Pressure (atm)	Temperature (°K)	Conductivity ($\Omega^{-1}m^{-1}$)		
		300 Wcm^{-2}	400 Wcm^{-2}	500 Wcm^{-2}
1	300	1.65	1.85	2.02
1	600	1.89	2.33	2.51
1	900	2.19	2.48	2.76
1	1200	2.22	2.68	2.87
2	300	1.21	1.39	1.54
2	600	1.41	1.69	1.84
2	900	1.51	1.77	2.02
2	1200	1.60	1.89	2.12
3	300	0.97	1.14	1.28
3	600	1.11	1.32	1.48
3	900	1.21	1.43	1.63
3	1200	1.26	1.52	1.72

[a] $\lambda = 3000$ Å, $\lambda_p = 0.1$.

TABLE XI

Variation of Photoelectric Conductivity with Radiation Wavelength[a]

Pressure (atm)	Temperature (°K)	Conductivity ($\Omega^{-1}m^{-1}$)		
		3000 Å	4000 Å	5000 Å
1	300	2.02	1.51	0.89
1	600	2.51	1.70	1.02
1	900	2.76	1.88	1.10
1	1200	2.87	1.99	1.16
2	300	1.54	1.02	0.58
2	600	1.84	1.18	0.66
2	900	2.02	1.29	0.69
2	1200	2.12	1.38	0.73
3	300	1.28	0.78	0.44
3	600	1.48	0.93	0.49
3	900	1.63	1.00	0.52
3	1200	1.72	1.04	0.54

[a] $I = 500\ Wcm^{-2}$, $\lambda_p = 0.1$.

TABLE XII
Variation of Photoelectric Conductivity with Dust Particle Concentration[a]

Pressure (atm)	Temperature (°K)	Conductivity ($\Omega^{-1}m^{-1}$) $\lambda_p = 0.1$	$\lambda_p = 0.075$	$\lambda_p = 0.05$
1	300	2.02	1.78	1.48
1	600	2.51	2.21	1.78
1	900	1.76	2.43	2.00
1	1200	2.87	2.58	2.13
2	300	1.54	1.36	1.15
2	600	1.84	1.53	1.37
2	900	2.02	1.80	1.52
2	1200	2.12	1.94	1.57
3	300	1.28	1.13	0.96
3	600	1.48	1.34	1.05
3	900	1.63	1.46	1.24
3	1200	1.72	1.55	1.33

[a] $\lambda = 3000$ Å, $I = 500$ Wcm^{-2}.

density N_p of solid particles enters through the parameter λ_p, which is defined as the ratio of the mass of the dust to the mass of the gas and is given by

$$\lambda_p = \frac{\frac{4}{3}\pi a^3 \rho_p N_p}{N_0 m_g}$$

Some of the parameters that are constant throughout the calculations are: $a = 250$ Å, $\phi = 1.7$ eV, $E_F = 1$ eV, $\rho_p = 5.72$ gm cm^{-3}, $m_g = 39.94$ amu, $\pi a_0{}^2 = 3.0 \times 10^{-17}$ cm^2, $\chi = 10^{-2}$ (Spitzer 1948), and $Q = 1$ (Spitzer 1948).

It is noted from Tables X to XII that the photoconductivity increases with the increase of intensity of the photoelectric radiation, gas temperature, and the concentration of the solid particles and with the decrease of gas pressure and wavelength of the photoelectric radiation. The study of the effect of the material of the dust particles showed, as expected, that the electron density increases as the work function of the particles decreases. All these conclusions are readily understandable on physical grounds.

The contributions to the electron density from photoelectric and thermionic emission from the solid particles for the temperature range 900–1200°K and for the same wavelength and intensities as used above are compared in Table XIII. It is noted that up to about 900°K the photoelectric emission gives the dominant contribution and that at higher temperatures the reverse is true.

TABLE XIII

Comparison of the Thermionic and Photoelectric Contribution to Electron Density

$\lambda_p = 0.1$, $\lambda = 3000$ Å, $I = 300$ Wcm^{-2}

Pressure (atm)	Temperature (°K)	Electron density (cm^{-3}) $(N_e)_{ph} \times 10^{-11}$	$(N_e)_{th} \times 10^{-11}$
1	1200	3.94	6.18
2	1200	5.15	10.42
3	1200	5.93	13.91
1	900	4.35	0.342
2	900	5.64	0.368
3	900	6.62	0.380

$\lambda_p = 0.1$, $\lambda = 5000$ Å, $I = 500$ Wcm^{-2}

Pressure (atm)	Temperature (°K)	Electron density (cm^{-3}) $(N_e)_{ph} \times 10^{-11}$
1	1200	1.83
2	1200	2.27
3	1200	2.50
1	900	2.00
2	900	2.49
3	900	2.79

From the foregoing conclusions it appears that the photoelectric effect as a means of enhancing the electrical conductivity of a dust-laden gas is not efficient in the range of temperature (>900°K) useful for closed cycle MHD power generation. It is, however, an interesting effect and provides an improvement over some of the proposed MHD systems using the photoionization of seed vapors (Balfour and Harris 1964). Further it may also prove useful for some propulsion and open cycle MHD power generation systems.

This analysis further demonstrates the possibility of increasing the electrical conductivity of gases by low work function dust suspensions. The novel feature of the present investigation is that a substantial increase in electrical conductivity can be obtained, even at room temperatures, thus this method circumvents the difficulties encountered in high-temperature experiments.

XII. Solid-State Colloidal Plasmas

A. *Colloids in Alkali Halide Crystals*

If an additively colored alkali halide crystal containing F centers is heated to 150°–450°C, the peak of the F band diminishes and the optical absorption bands peaking on the long wavelength side of the F band (between 700 and 800 mμ for KCl, i.e., in the infrared region) are formed (Scott and Smith 1951, Scott et al. 1953, Schulman and Compton 1962). Scott et al. (1953), Doyle (1958, 1960), and Penley and Witte (1962) have identified the centers responsible for the infrared bands with the small colloidal particles of the alkali metal in the crystal lattice.

Scott et al. (1953) and Jain and Sootha (1965, 1968) have found that there is a heterogeneous equilibrium in the temperature range 300 to 450°C between the F centers and the centers responsible for the 730 mμ band in a KCl crystal, suggesting that this 730 mμ band is due to a different phase of the metal.

Further, Jain and Sootha (1968) have made electron spin resonance (ESR) measurements on highly pure KCl crystals containing different concentrations of excess potassium metal after quenching the crystal in the dark from 300°C. They have observed a narrow ESR line superimposed on the broad F center ESR line and have attributed the narrow ESR line to the conduction electron spin resonance in the colloidal particles of potassium metal. The observed Lorentzian shape of ESR line is also found consistent with the metallic nature of the centers (Dyson 1955, Feher and Kip 1955, Catterall 1965, Pressley and Berk 1965, Walsh et al. 1966). They have also estimated the size of the particles in KCl crystals with an upper limit of 700 atoms per colloid.

The half-width of the optical absorption band, as well as the CESR absorption line, remains constant under heat treatment in the dark; the colloid size remains constant in highly pure KCl crystal (Jain and Sootha 1968).

Savostianova (1930) and Karlson and Beckman (1967) have analyzed the absorption and scattering of light by spherical alkali metallic particles in an alkali halide crystal, using Mie's theory. Their results are in good qualitative agreement with the experimental observations.

Further confirmation of the existence of these colloidal metal particles has been obtained from the measurements of electron diffraction (McLennan 1951), photoconductivity (Gyulai 1926a, b), X-ray scattering (Smallman and Willis 1957), nuclear magnetic resonance (Ring et al. 1958) and ESR (Doyle et al. 1959).

Jain and Sootha (1965, 1968) have observed that the electrical conductivity of an additively colored KCl crystal is many times larger than that of a normal crystal. Kothari and Jain (1964), and later on Jain and Sootha (1965, 1968), attributed this observed excess conductivity ($\sigma_c - \sigma_n$) in the temperature range 300° to 450°C in a KCl crystal to thermionic emission of electrons from the potassium metallic particles into the conduction band of the crystal. Guha's (1968a) analysis of this phenomenon makes use of the work of Smith (1958), and Sodha et al. (1965) on the conductivity of dust-laden gases.

B. *Electronic Conductivity of Additively Colored Alkali Halide Crystals*

The electronic conductivity σ_e of the crystal is given by

$$\sigma_e = eN_e\mu \tag{12-1}$$

where μ is the mobility of an electron in the crystal.

1. Electron Density

The electron density in the conduction band of the crystal is given by eq. 1-15, where α is defined as $e^2/\varepsilon akT$, ε the dielectric constant of the crystal, and N_s is the saturation electron density near the particles, given by

$$N_s = 2\left(\frac{2\pi mkT}{h^2}\right)^{3/2} \exp\left(-\frac{\phi - \chi}{kT}\right) \tag{12-2}$$

Here ϕ is the work function for the bulk solid and χ is the electron affinity of the crystal. The work function ϕ in Smith's treatment has been replaced by $(\phi - \chi)$ (as suggested by the work of Kothari and Jain (1964)) to yield eq. 12-2.

When $N_p/\alpha N_s \gg 1$, which is true for conditions of reported experiments, eq. 1-15 reduces to

$$N_e = N_s \tag{12-3}$$

Kothari and Jain (1964), however, have assumed that

$$N_e = SN_s \tag{12-4}$$

where S is the surface area of all the colloids in unit volume. Jain (private communication) has pointed out that eq. 12-4 is dimensionally incorrect and should be modified.

Following Kothari and Jain (1964), we find that the number of metallic colloids at a temperature T is given by

$$N_p = \frac{N_k - N_F(T)}{n} \tag{12-5}$$

where N_k is the concentration of total excess metal atoms per unit volume, $N_F(T)$ is the number of F center per unit volume, given by Scott and Smith (1951) and Scott et al. (1953), as

$$N_F(T) = C\exp\left(-\frac{\Delta H_c}{kT}\right) \tag{12-6}$$

where C is a constant for a given crystal; ΔH_c is the energy required to form one F center from the colloidal particles and is characteristic of the crystal; $n = \frac{4}{3}\pi a^3 \rho L/A$ is the number of atoms in a colloid; and ρ, L, and A are the density of particles, Avogadro's number, and the atomic weight of the metal, respectively.

By using the above equations, one can calculate the electron density in the conduction band of an alkali halide crystal for a given value of N_p and T.

2. Mobility of Electrons

Petritz and Scanlon (1955) have reviewed the various theories of mobility of electrons in polar crystals and concluded that the mobility μ is given by

$$\mu = \frac{8a_0 e}{3(2\pi mk\Theta)^{1/2}} \frac{\varepsilon\varepsilon_0}{(\varepsilon - \varepsilon_0)} \left(\frac{m}{m^*}\right)^{3/2} \frac{\chi_0(z')(e^{z'} - 1)}{z'^{1/2}} \tag{12-7}$$

where $a_0 = \hbar^2/me^2$, Θ the Debye temperature of the crystal, ε and ε_0 the static and high frequency dielectric constants of the crystal, and m^* the effective mass of the electron in the crystal. Also, $z' = \Theta/T$,

$$\chi_0(z') = 1 \qquad \text{when } z' \ll 1$$

and

$$\chi_0(z') = \tfrac{3}{8}(\pi z')^{1/2} \qquad \text{when } z' \gg 1$$

Using the results of Sodha et al. (1965), it may be seen that for all cases of interest the scattering of electrons by the charged metallic colloids is negligible.

3. Electronic Conductivity

Equations 12-1 to 12-3 and 12-7 have been used to study the variation of σ_e with temperature T in the range 300 to 450°C for large values of N_k when the electronic conductivity becomes independent of both N_k and the radius of colloids a (10–200 Å) because $N_p/\alpha N_s \gg 1$ (except very near the critical temperature T_c when $N_k = N_F(T_c)$). The range of N_k for which experimental data are available (Sootha 1967, Jain and Sootha 1968), namely, 4.0×10^{17} to 1.5×10^{18} cm^{-3}, satisfies this criterion. The calculated variation of σ_e with T and the experimental values of Sootha (1967) are presented in Table XIV.

TABLE XIV

Temperature (°C)		Experimental σ_e (Ω^{-1}cm^{-1})		Calculated σ_e (Ω^{-1}cm^{-1})
N_k (cm^{-3})	4.0×10^{17}	8.5×10^{17}	1.5×10^{18}	
T_c (°C)	383	473	560	
300	0.446×10^{-10}	1.858×10^{-10}	1.858×10^{-10}	9.713×10^{-12}
350	2.266×10^{-9}	3.520×10^{-9}	3.521×10^{-9}	1.134×10^{-10}
400	2.769×10^{-8}	6.540×10^{-8}	6.560×10^{-8}	9.793×10^{-10}
450	5.426×10^{-8}	3.336×10^{-7}	4.052×10^{-7}	6.274×10^{-9}

The following data have been used in the calculations: $\varepsilon = 4.68$, $\varepsilon_0 = 2.13$, $\phi = 2.2$ eV, $\chi = 0.7$ eV (Timusk 1961, this value gives the best agreement with the experiment) $\Theta = 300$°K and $m^*/m = 0.6$ (Redfield 1954), $\Delta H_c = 0.35$ eV (Scott and Smith 1951). Although Θ/T in the range of temperature of interest to us is not much less than unity, we have nevertheless used eq. 12-7 because the error involved in this account may not be larger than the other uncertainties in the theory.

It is clear from Table XIV that the calculated values of electronic conductivity are about 1/70th to 1/20th of the experimentally measured values. Further, the theory predicts that, at temperatures lower than the critical temperature, the electronic conductivity is independent of N_k. This is valid as an approximation, as can be seen from Table XIV. The disagreement between the theory and the experiment becomes more pronounced when the electron mobility is calculated from the value for mean free path, given by Seitz (1940). The theory is not expected to apply to the case when $N_k = 4.0 \times 10^{17}$ cm^{-3} and $T > T_c$ (for the temperatures 400°C and 450°C). From eqs. 12-5 and 12-6 it may be seen that $N_p = 0$ for $T > T_c$ and hence $N_e = 0$. Thus it is necessary to look for mechanisms to explain the conductivity other than the one proposed by Kothari and Jain (1964) and analyzed in this section.

4. Electron Affinity

It may be seen from eqs. 12-1, 12-2, and 12-7 that, if

$$\ln [\sigma_e \Theta^{1/2}/(e^{z'} - 1)\chi_0(z')T^2]$$

is plotted against $1/T$, the slope of the curve gives the value of 2.303 $(\phi - \chi)/k$ from which χ can be determined. Using the experimental values of Sootha (1967), we represent the plots for $N_k = 8.5 \times 10^{17}$ cm^{-3} and 1.5×10^{18} cm^{-3} in Figure 16. A measurement of the slopes gives $\chi = 0.38$ eV. This may be compared with the following values obtained by other workers: 0.07 eV, Mott

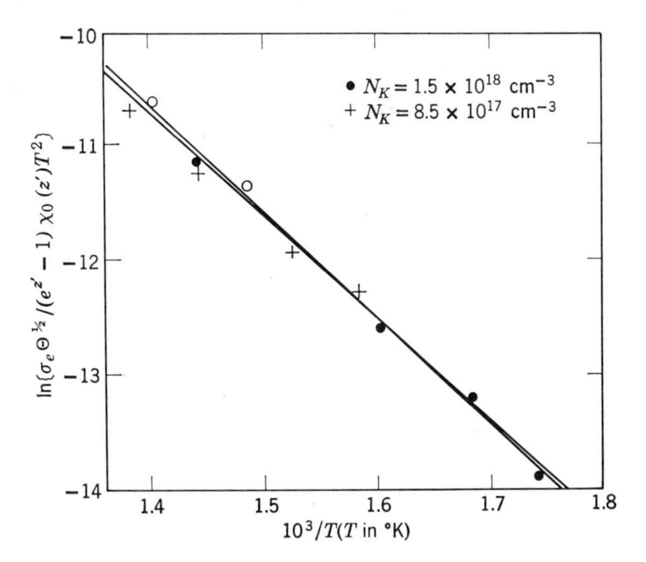

Fig. 16. Plot of $\log_{10}(\sigma_e \Theta^{1/2}/(e^{z'} - 1)\chi_0(z,)T^2$ a as function of the reciprocal of the absolute temperature T for two values of N_k in a KCl crystal. (After Guha 1968a.)

and Gurney (1948); 0.7 eV, Timusk (1961); 0.4 eV, Gilleo (1953); 0.46 eV, Kothari and Jain (1964); and 0.32 eV, Jain and Sootha (1965). The values of χ obtained by Kothari and Jain (1964) and Jain and Sootha (1965) are based on eq. 12-4, the validity of which has been discussed before.

In the calculation of electronic conductivity we have used the value $\chi = 0.7$ eV because, from all the values proposed by various workers, this gives the best agreement between measured and calculated values of σ_e. Lower values of χ will lead to poorer agreement. However, the value of χ obtained from the nature of variation of σ_e with T is 0.38 eV.

C. *Photoconductivity of Additively Colored Alkali Halide Crystals*

The measurements of optical absorption and photoconductivity spectrum with a natural blue-colored NaCl crystal were made by Gyulai (1926a, b). This study showed that the positions of the peaks of the two spectra do not coincide with each other (photoconductivity peak occurs at 370 mμ and the principal absorption bank peak at 550 mμ); this is in contrast to the behavior of a crystal having a yellow F band coloration. Gyulai also found that no change occurs in the absorption spectrum by illuminating the crystal with light that provoked photoconduction (like bleaching or the growth of other bands). This behavior is also different from that of the F-centered crystal, where illumination with F band light at room temperature causes bleaching as well as photoconductivity. He could also show that the blue coloration in a natural NaCl crystal duplicates the blue obtained in an additively colored NaCl crystal. Hence Gyulai has suggested that the photoconductivity is due to the photoelectric emission of electrons from the colloidal metal particles; the blue coloration is an evidence of the presence of such particles.

The curve of photoconducting current plotted against wavelength (Gyulai 1926a, b) is very similar to that for the photoelectric current from a clean sodium surface, but the cutoff wavelength is displaced 0.5 eV in the direction of lower energy. Mott (1938) has used this result to show that the bottom of the conduction band in NaCl lies at a depth of 0.5 eV, i.e., the electron affinity of the crystal is 0.5 eV.

Guha (1968b) has analyzed this phenomenon of photoconductivity in such crystals, when irradiated by light. This is a result of photoelectric emission of electrons from the colloids to the conduction band of the crystal.

1. Electron Density

Following Sodha (1963) (eq. 11-2), the number of photoelectrons emitted per unit area per second when light of frequency ν and of intensity I_0 falls on a plane metallic surface, is given by

$$n_{\text{ph}} = \beta(\nu)\,\frac{4\pi m k^2 T^2}{h^3}\,\frac{I_0}{h\nu}\,\Phi(\xi) \tag{12-8}$$

For the case $\xi \gg 1$, $\Phi(\xi)$ can be expanded (Fowler 1936):

$$\Phi(\xi) = \frac{\pi^2}{6} + \frac{\xi^2}{2} - \frac{e^{-\xi}}{1} + \frac{e^{-2\xi}}{2^2} \cdots$$

When the crystal is at room temperature, ξ becomes very much larger than one and hence $\Phi(\xi)$ reduces to

$$\Phi(\xi) \simeq \frac{\xi^2}{2} \tag{12-9}$$

Thus from eqs. 12-8 and 12-9, $\beta(\nu)$ is given by

$$\beta(\nu) = \frac{Y}{2\pi m/h^2\,(\nu - \nu_0)^2/\nu} \tag{12-10}$$

where $Y = n_{\text{ph}}/I_0$ is called the photoelectric yield of electrons.

Consider an additively colored alkali halide crystal containing N_p spherical metallic colloids per unit volume, each having a radius a. When irradiated with light of frequency ν, the number of photoelectrons emitted per unit time from solid metallic particles having charge $(+ze)$ is given by Sodha (1963), eq. 11-5,

$$n'_{\text{ph}} = \beta(\nu)\,\frac{4\pi m k^2 T^2 \pi a^2}{h^3}\,\frac{I_0}{h\nu}\,\Psi(\xi', z\alpha) \qquad \text{when } z \geq 0 \tag{12-11}$$

and

$$\xi' = \frac{h\nu - (\phi - \chi)}{kT}$$

Equation 12-11 implies that the temperature of the crystal and the colloids is the same.

At room temperature ξ' also becomes very much larger than one; hence $\Psi(\xi', z\alpha)$ can be reduced to the form

$$\Psi(\xi', z\alpha) \simeq \frac{\xi'^2 - z^2\alpha^2}{2} \tag{12-12}$$

The steady-state electron density N_e can be obtained by assuming that all the colloidal particles carry the same positive charge ze and by equating the rate of photoelectric emission of electrons from the colloidal particles to the rate of recombination of the electrons with them. Thus

$$N_p(z)n'_{\text{ph}} = N_p(z)N_e\langle v\sigma(z)\rangle \tag{12-13}$$

where $\langle v\sigma(z)\rangle$ is given by eq. 12-12. From eqs. 12-11 to 12-13 we obtain N_e as

$$N_e = \frac{\beta(\nu)\,\dfrac{4\pi m k^2 T^2}{h^3}\,\dfrac{I_0}{h\nu}\,(\xi'^2 - z^2\alpha^2)/2}{(8kT/m\pi)^{1/2}(1 + z\alpha)} \tag{12-14}$$

and $z = N_e/N_p$.

For the case of interest $z^2\alpha^2 \ll \xi'^2$, then

$$N_e = \frac{1}{2}\left\{-\frac{N_p}{\alpha} + \left[\left(\frac{N_p}{\alpha}\right)^2 + \frac{4N_pK}{\alpha}\right]^{1/2}\right\} \tag{12-15}$$

where

$$K = \beta(\nu)\,\frac{4\pi m k^2 T^2}{h^3}\,\frac{I_0}{h\nu}\,\frac{\xi'^2}{2}\left(\frac{m\pi}{8kT}\right)^{1/2}$$

The number of metallic coloids per unit volume is given by eq. 12-5. By using eqs. 12-14 and 12-5, one can calculate the electron density in the conduction band of an additively colored alkali halide crystal for a given value of N_k, T, ν, a, and I_0. The mobility of the electrons is given by eq. 12-7.

2. Photoconductivity

Using eq. 12-10, we can find the variation of $\beta(\nu)$ with the wavelength for sodium and potassium from the experimental work of Maurer (1940) and Dickey (1951), respectively; the results are presented in Tables XV and XVI. These values of $\beta(\nu)$ are used in the calculation of N_e.

Equations 12-1, 12-5, 12-7, and 12-14 are used for studying the variation of electronic conductivity with the wavelength of light λ, its intensity I_0, the radius of the colloid a and with initial concentration of F centers N_k; the

TABLE XV

Variation of $\beta(\nu)$ with Wavelength for Sodium

Wavelength λ (Å)	$\beta(\nu) \times 10^{32}$
4800	4.624
4400	4.940
4000	4.930
3600	4.337
3200	1.740
2800	0.518

TABLE XVI

Variation of $\beta(\nu)$ with Wavelength for Potassium

Wavelength λ (Å)	$\beta(\nu) \times 10^{32}$
5160	0.327
4950	0.510
4580	0.680
4270	1.282
4120	1.490
3530	0.413
3340	0.204
3260	0.137
3090	0.0932

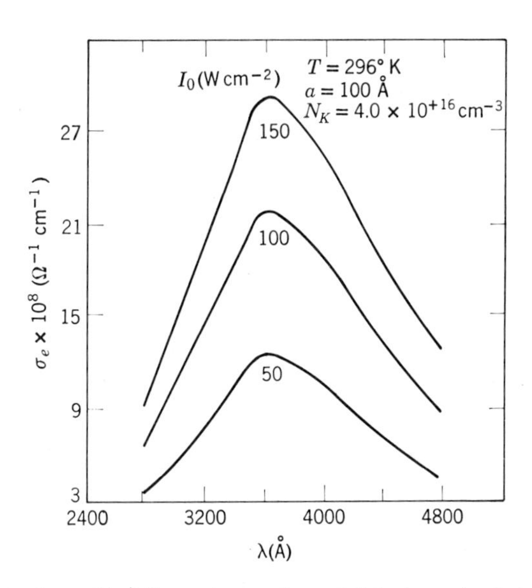

Fig. 17. Variation of σ_e with λ for various values of light intensity I_0 in a NaCl crystal. (After Guha 1968b.)

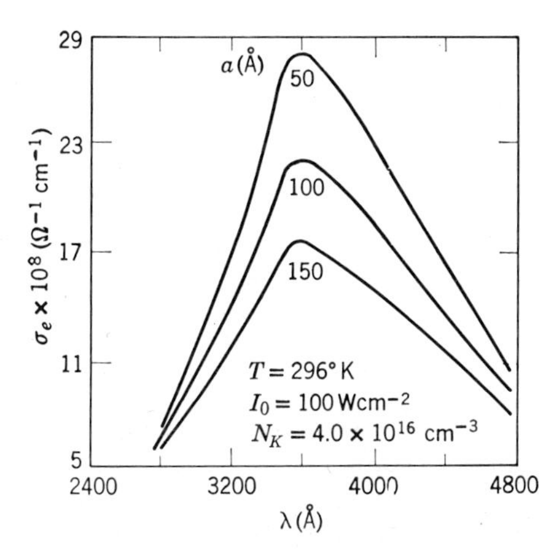

Fig. 18. Variation of σ_e with λ for various values of radius of colloids a in a NaCl crystal. (After Guha 1968b.)

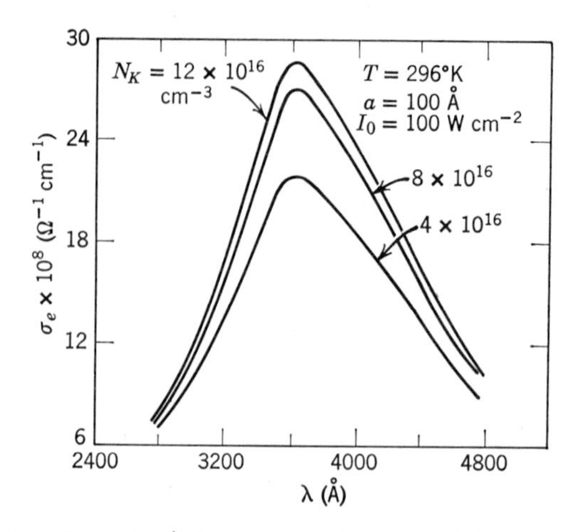

Fig. 19. Variation of σ_e with λ for various values of N_k in a NaCl crystal. (After Guha 1968b.)

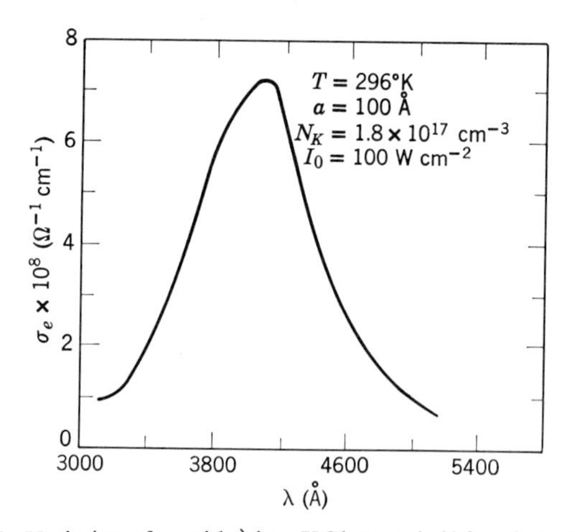

Fig. 20. Variation of σ_e with λ in a KCl crystal. (After Guha 1968b.)

results are presented in the form of graphs (see Figures 17–20). Some of the parameters used in the calculation, which are constant for a NaCl crystal, take the following values: $\varepsilon = 5.62$, $\varepsilon_0 = 2.25$, $\phi = 2.28$ eV, $\chi = 0.5$ eV (Mott 1938), $\Theta = 370$°K (Redfield 1954), $m^*/m = 0.55$ (Redfield 1954). For a KCl crystal the parameters take the same values as in Section XII-B. The values of $N_k(T)$ and $N_F(T)$ for NaCl and KCl crystals are taken from the experimental work of Scott et al. (1953).

It may be seen from the graphs that the electronic conductivity increases with the increase of intensity of incident light, with the concentration of F centers, and with the decrease of radius of the colloids. The nature of variation in a KCl crystal is similar to that in a NaCl crystal.

XIII. Experimental Measurements

A large number of workers have made experimental measurements of the size of the particles and the charge on them, the work function of the particles, and the conductivity of the dust suspensions. A brief description of some of the experimental methods and their usefulness to the present review are given in this section.

A. *Conductivity Measurements and Preparation of Particles*

Waldie and Fells (1967) have developed a technique of preparing submicron size particles of BaO or alkali earth oxides in argon at atmospheric pressure and have measured the electrical conductivity of dust suspensions.

1. Preparation of Submicron Particle Suspensions

A diagram of the apparatus is shown in Figure 21. A central vessel is made of recrystallized alumina, as are the connecting tubes that are jointed to the vessel with alumina cement. Before use, the vessel is heated to 1400°C for 24 hr while purging with argon dried to less than 0.1 ppm water. Barium metal chips, prepared under oil to prevent oxidation, washed with carbon tetrachloride and dried with argon, are inserted into the vessel under argon. The temperature of the vessel is regulated to keep the molten barium at a certain temperature, generally in the range of 1000 to 1400°C, chosen to give the desired vapor concentration in the argon stream.

The argon plus barium vapor in the exit tube meet oxygen added at a controlled rate through a concentric inner tube, and barium oxide particles leaves via *T*-piece outside the furnace enclosure. By regulating the vessel temperature and argon and oxygen flow rates, the particle loading of the suspension can be varied. Particle loadings were measured by collecting particles on a cellulose membrane filter and weighing this at intervals.

Samples of particles prepared in this way were collected on electron microscope grids coated with Formvar. The particles were found to be spherical in shape and there were a few examples of agglomeration.

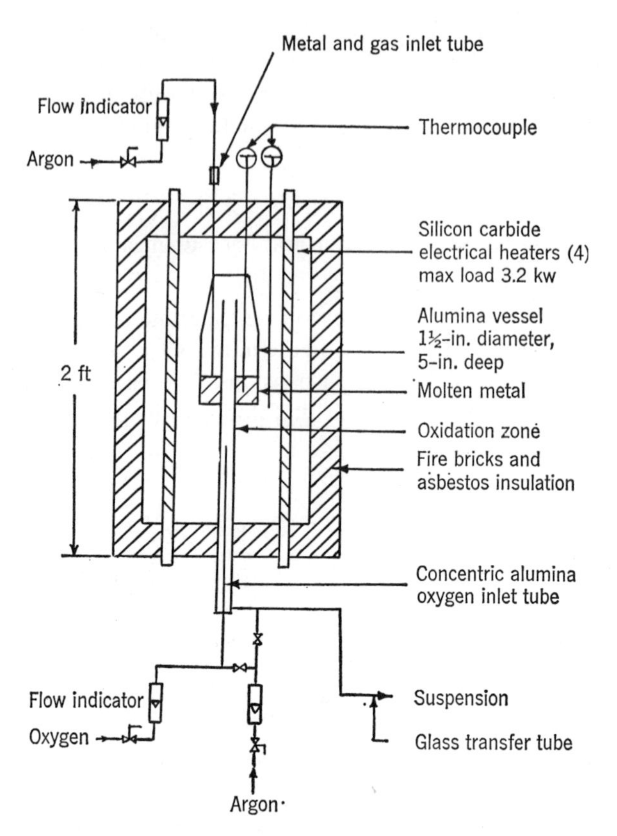

Fig. 21. Apparatus for preparation of suspensions of barium oxide in argon. (After Waldie and Fells 1967.)

2. Measurement of Conductivity

Figure 22 shows the apparatus used to measure the conductivity of dust suspensions. The suspension is passed through a heated alumina tube containing a pair of 1 in. × 1 in. zirconium electrodes. The suspension flow rate was such that the suspension attained the tube wall temperature before passing between the electrodes. An a-c conductivity bridge connected to the electrodes indicated the conductance of the suspension.

The suspension was found to be electrically conducting; the conductivity reading could be made to fluctuate by altering the argon and oxygen flow rates. The results of the measurements have been discussed in Section VIII.

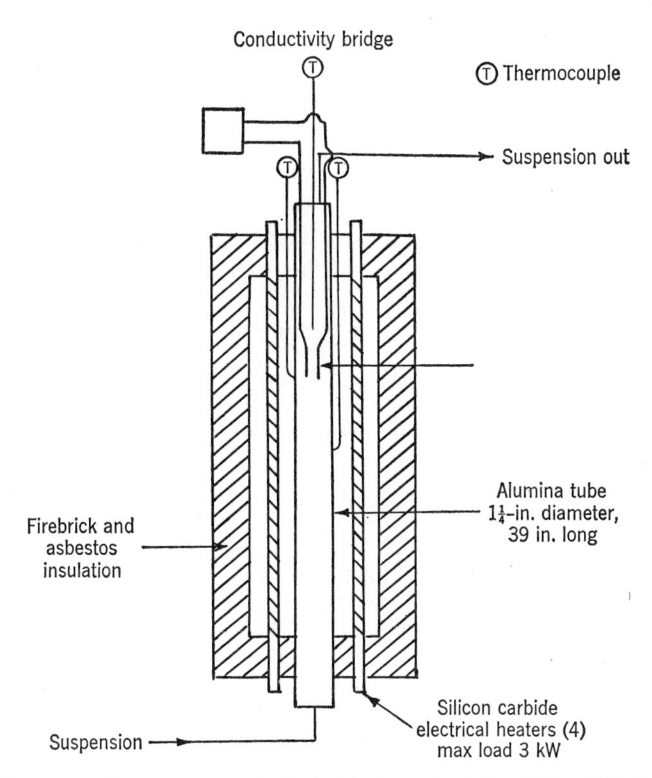

Fig. 22. Apparatus for measurement of electrical conductivity. (After Waldie and Fells 1967.)

B. *Work Function of Particles*

Pope (1962a, b) has described a method of determining the work functions of small particles based on the Millikan's method of determination of the charge of the electron. The powder is introduced into an argon-filled Millikan apparatus which is fitted with quartz windows. The capacitor plates are charged to high potential, the upper plate being negative. The particles falling between these plates are exposed to light. Two different wavelengths, both shorter than the threshold wavelength, are used. At each wavelength the particle stops falling and starts rising, indicating that electrons have been ejected from the particle, which leave it positively charged. Therefore, at each wavelength, the particle is balanced against the force of gravity by regulating the external field across the capacitor plates. Since ϕ is a constant for the material at a given temperature, the two wavelengths and the two balancing potentials are all that are required to determine ϕ.

The charged particle is held motionless in space when

$$Eq = \frac{Vq}{d} = mg = \beta \tag{13-1}$$

where E is the applied electric field, V the potential difference between the plates of distance d apart, m the mass of the particle, and q the particle charge.

When the electron emission is limited by the retarding potential, the equilibrium can be expressed by the Einstein photoelectric equation

$$h\nu = \phi + \psi \tag{13-2}$$

where ψ is the retarding potential energy due to the charge q on the particle. Hence

$$\psi = \frac{q}{C} \tag{13-3}$$

where C is the capacitance of the particle. From eqs. 13-1 and 13-2 we get

$$\frac{hc}{\lambda} = \phi + \frac{\beta d}{VC} \tag{13-4}$$

For two wavelengths shorter than the threshold, we measure the two corresponding balancing potentials, and derive

$$\phi = \frac{hc}{1}\left[(\lambda_2 V_1 - \lambda_1 V_2)/\lambda_1\lambda_2(V_1 - V_2)\right] \tag{13-5}$$

From this value of ϕ we can calculate a value for ψ at any particular wavelength, assuming an equipotential surface.

C. *Radius of the Particle*

If the particle is spherical, then its radius may be obtained from eqs. 13-1, 13-4, and 13-5 as

$$a^2 = \frac{3hc}{4\pi\rho g d}\left(\frac{V_1 V_2}{\lambda_1 \lambda_2}\right)\left(\lambda_2 - \frac{\lambda_1}{V_2} - V_1\right) \tag{13-6}$$

by using the value for the capacitance of a sphere $C = a$. Knowing a, we can also get C, q, ψ, and ψ/a, the electric field intensity at the surface of the particle. The wavelengths that can be chosen for this experiment depend on the dielectric strength of the medium in which the particles are suspended, and thus also on the size and shape of the particle. If the balancing of the particle could be done in a vacuum, this limitation on the shorter wavelength could vanish.

The above method has several advantages. For the determination of ϕ, no knowledge is required of the shape of the particle or the viscosity of the medium. Pope (1962b) has measured ϕ for graphite particle; the value determined is in agreement with that of the bulk graphite.

D. *Size Determination of the Particles*

A comprehensive account of the various methods of measurement of particle size has been given by Irani and Callis (1963), Fuchs (1964), and Soo (1967).

Some of the methods of the determination of size and charge of the particles are discussed below.

1. One of the methods is to use a microscope or an electron microscope to measure the diameter of particles precipitated on a slide by (1) gravitational sedimentation, (2) directed flow, (3) thermal precipitation, or (4) electrostatic precipitation. These methods have been used by a large number of workers.

2. Sedimentation Based on Stokes' Law: a. For large-size particles, one may use the gravitational settling method, if the settling distance as a function of time is determined. The measurement of total intensity of transmitted or scattered light with time (Kerker et al. 1955) or electric current from the precipitated charged particles, also with time, will give the particle size distribution (Drozin and LaMer 1959).

b. Centrifugal sedimentation: In this method variation of the settling distance with time for a given set of parameters (angular velocity, dimensions of apparatus, etc.) is determined. Then, for size distribution, the weight of the samples taken at different times or at different radii of rotation (Brown 1944, Robinson and Martin 1948) is taken. One may also do microscopic counting of particles on a slide or quantitative chemical analysis of particles placed on certain portions of a slide (Sawyer and Walton 1950), or determine time dependence of the electric current from charged particles precipitated on the conducting surface (Drozin and LaMer 1959).

c. Electrostatic precipitation: For size determination, the distance traveled by a charged particle as a function of time is measured. For size distribution, the electric current from the plate, where particles of a given size are precipitated (Daniel and Bracket 1951) or the electric current in a precipitator as a function of time (Drozin and LaMer 1959) has to be determined.

3. a. Counting the electric pulses of given amplitude produced by charged particles hitting a wire connected to the grid of an amplifier will determine the particle size (Geist et al. 1951).

b. One counts the electric pulses of a given amplitude in a photomultiplier by light scattered by individual particles as a function of the particle size.

4. The measurement of angle-dependent scattering intensity as a function of particle size using Mie's theory and diffraction theory (Kerker and LaMer 1950) gives the size of the particle; for size distribution, one measures the total intensity of scattered light as a function of the angle of observation.

The adoption of one specific method in preference to another must be based on such things as precision and accuracy required, the time per analysis, and the size range of the particles. For example, electron microscope and electron photomicrograph technique give accurate measurement of the particle size, but X-ray diffraction and light scattering reveal the average size of the particles.

E. *Charge Determination*

Similarly, various methods have been used to determine the charge on the particle. Some of the methods used follow.

1. Gillespie and Langstroth (1952) have developed an instrument to permit the measurement of the electric charges on a relatively large number of particles during a comparatively short period. It is done by depositing a sample on microscope slides, in such a way that the charge on any particle at deposition can be determined in a subsequent examination from the size of the particle and its position on the slide. They have shown that the act of sampling with this instrument does not cause spurious charging of particles.

2. Another apparatus, based on the Hopper and Laby's apparatus (for determining the electronic charge), has been designed by Kunkel and Hansen (1950). The main feature of this method is to record photographically successive positions of the particles as they settle under the influence of gravity in a horizontal electric field. From the observed rate of fall, it is simple to calculate the Stokes' law diameter; then, from the simultaneous horizontal deflection, one can calculate the approximate charge of the particles. The most serious shortcoming of the arrangement is the tendency to develop convection currents inside the analyzer.

3. Hinkle et al. (1954) have developed a method and an apparatus that permit rapid evaluation of the amounts of negatively and positively charged particles in a system by measurement of the total deflection of the dust system in an electric field. The principal advantages of this method are the rapidity and ease of obtaining data and the elimination of the need for numerous samples to give statistically accurate results.

4. Hendrick's method (1962) has been improved by Min et al. (1963). It consists of a shielded hollow metal cylinder through which the particles are allowed to pass, one at a time. The voltage pulse resulting from the induced charges is measured by means of an oscilloscope. This method, in conjunction

with a suitable pulse counting system, is capable of measuring the charge spectrum of particles in a localized region of a dust-gas flow system.

Some interesting experiments have been performed by Balwanz (1965) and Soo and Dimick (1963, 1965). Balwanz (1965) studied ionization in rocket exhausts by absorption of a focused beam of electromagnetic wave. Soo and Dimick (1963, 1965) have studied the interaction of solid particles with an ionized gas in an arc flame of argon. The detailed experimental techniques are given in their papers.

Acknowledgment

S. Guha gratefully acknowledges financial support from the Council of Scientific and Industrial Research, India.

References

Anderson, W. L., and Rosen, G., 1965, *J. Appl. Phys.*, **36**, 3760.
Arendt, P., and Kallman, H., 1925, *Z. Physik*, **35**, 421.
Arshinov, A. A., and Musin, A. K., 1958a, *Sov. Phys.—Dokl.*, **3**, 99.
———1958b, *Sov. Phys.—Dokl.*, **3**, 588.
———1962, *Radio Eng. and Electronics Phys.*, **7**, 842.
Baldwin, K., and Balwanz, W. W., 1964, *IEEE Trans. Nucl. Sci.*, **11**, 51.
Balfour, D., and Harris, J. H., 1964, *Symposium on MHD Electrical Power Generation*, Vol. 1, ENEA, Paris, 1964, p. 159.
Balwanz, W. W., 1965, *Proc. 10th Symposium (Int.) on Combustion*, (Combustion Institute, Pittsburgh) p. 685.
Biasi, L., Giannolio, L., and Tozzi, A., 1966, *Brit. J. Appl. Phys.*, **17**, 525.
Biasi, L., and Tozzi, A., 1967, *Brit. J. Appl. Phys.*, **18**, 219.
Brown, C., 1944, *J. Phys. Chem.*, **48**, 246.
Burbank, P. B., 1965, NASA Tech. Note D-2747.
Catterall, R., 1965, *J. Chem. Phys.*, **43**, 2262.
Cobine, J. S., 1958, *Gaseous Conductors*, Dover, New York.
Daniel, J. H., and Bracket, I. S., 1951, *J. Appl. Phys.*, **22**, 542.
Davidson, J. R., and Sandorff, P. E., 1963, NASA Tech. Note D-1493.
Davis, A. H., and Harris, I., 1961, NASA Tech. Note D-704.
Dennery, F., 1964, *Symposium on MHD Electrical Power Generation*, Vol. 1, ENEA, Paris, 1964, p. 195.
Deutsch, W., 1922, *Am. Phys. Series*, **4**, 68, 355.
Dickey, J., 1951, *Phys. Rev.*, **81**, 612.
Dimick, R. C., and Soo, S. L., 1964, *Phys. Fluids*, **7**, 1638.
Dingman, E. H., 1965, *IEEE Trans. Nucl. Sci.*, **12**, 544.
Doyle, W. T., 1958, *Phys. Rev.*, **111**, 1067.
Doyle, W. T., Ingram, D. J. E., and Smith, M. J. A., 1959, *Phys. Rev. Letters*, **2**, 494.
Doyle, W. T., 1960, *Proc. Phys. Soc.* (London), **75**, 649.

Drozin, V. G., and LaMer, V. K., 1959, *J. Colloid Interface. Sci.*, **14**, 74.
Dubey, P. K., 1970, *J. Phys. D.*, **3**, 145.
Dyson, F. J., 1955, *Phys. Rev.*, **98**, 349.
Einbinder, H., 1957, *J. Chem. Phys.*, **26**, 948.
Feher, G., and Kip, A. F., 1955, *Phys. Rev.*, **98**, 337.
Fowler, R. H., 1936, *Statistical Mechanics* Cambridge University Press, London p. 358.
Friedman, R., Fagg, L. W., Miller, T. K., Charles, W. D., and Hughes, M. C., 1963, *Progress in Astronautics and Aeronautics*, Vol. 12, K. E. Shuler and J. B. Fenn, Eds., Academic Press, New York, p. 379.
Friichtenicht, J. F., 1962, *Rev. Sci. Instrum.*, **33**, 209.
Fristrom, R. M., Oyhus, F. A., and Albrecht, G. H., 1962, *Am. Rocket Soc. J.*, **32**, 1729.
Fuchs, N. A., 1964, *The Mechanics of Aerosols*, Pergamon Press, New York.
Geist, T. U., York, T. L., and Bard Brown, G. G., 1951, *Ind. Eng. Chem.*, **43**, 1371.
Gilleo, M. A., 1953, *Phys. Rev.*, **91**, 534.
Gillespie, T., and Langstroth, G. O., 1952, *Can. J. Chem.*, **30**, 1056.
Ginzburg, V. L., and Gurevich, A. V., 1960, Sov. Phys—Usp., **3**, 115, 175.
Guha, S., 1968a, *Brit. J. Appl. Phys.*, **1**, 1571.
———1968b, *Brit. J. Appl. Phys.*, **1**, 1615.
Guha, S., and Kaw, P. K., 1968, *Brit. J. Appl. Phys.*, 1, 193.
Guha, S., 1970, communicated.
Guha, S., and Arora, A. K., 1970, *Appl. Sci. Res.*, **22**, 777.
Gunn, R., 1954, *J. Meteorol.*, **11**, 339.
———1955, *J. Colloid Interfac. Sci.*, **10**, 107.
Gyulai, Z., 1926a, *Z. Physik*, **35**, 411.
———1926b, *Z. Physik*, **37**, 889.
Halasz, D., Szendy, C., and Kovacs, C. P., 1964, *Symposium on MHD Electrical Power Generation*, Vol. 1, ENEA, Paris, 1964, p. 339.
Halasz, D., and Szendy, C., 1967, *Acta Tech. Hungr.*, **58**, 169.
Hendrick, C. D., Jr., 1962, *J. Colloid Interfac. Sci.*, **17**, 249.
Hewitt, C. W., 1957, *Trans. AIEE*, (II), **76**, 300.
Hinkle, B. L., Clyde, Orr, Jr., and Dalla Valle, J. M., 1954, *J. Colloid Interfac. Sci.*, **9**, 70.
Hofmann, U., and Wilm, D. 1936, *Z. Elektrochem*, **42**, 504.
Honma, T., and Fushimi, K., 1966, *Jap. J. Appl. Phys.*, **5**, 238.
Honma, T., Nomura, O., and Kanai, A., 1968, *Bull-Electrotech. Lab. (Japan)*, **32**, 83.
Hooper, A. T., Newby, D., and Russell, A. H., 1966, *Electricity from MHD*, Vol. 1, IAEA Vienna, p. 631.
Hooper, A. T., and Newby, D., 1967, UKAEA Rept. AEE, W-M 755.
Hurwitz, H., Sutton, G. W., and Tamor, S., 1962, *Am. Rocket J.*, **32**, 1237.
Irani, R. R., and Callis, C. F., 1963, *Particle Size: Measurement, Interpretation and Application*, Wiley, New York.
Jain, S. C., and Sootha, G. D., 1965, *J. Phys. Chem. Solids*, **26**, 267.
Jain, S. C., and Sootha, G. D., 1968, *Phys. Rev.*, **171**, 1075.
James, C. R., and Vermeulen, F., 1968, *Can. J. Phys.*, **46**, 855.
Jastrow, R., and Pearse, C. A., 1957, *J. Geophys. Res.*, **62** (3), 413.
Johnson, C. C., and Bullock, C. K., 1964, *J. Appl. Phys.*, **35**, 2804.
Karlson, A. V., and Beckman, O., 1967, *Solid State Communication*, **5**, 795.
Kemble, E. C., 1958, *The Fundamental Principles of Quantum Mechanics*, Dover, New York, p. 109.
Kerker, M., and LaMer, V. K., 1950, *J. Amer. Chem. Soc.*, **72**, 3516.

Kerker, M., Lucile Cox, A., and Shoenberg, M. D., 1955, *J. Colloid Interfac. Sci.*, **10**, 413.
Kerrebrock, J., 1962, *Second Symposium on Engineering Aspects of MHD*, Columbia University Press, New York, p. 327.
Kothari, D. S., and Jain, S. C., 1964, *Phys. Letters*, **13**, 203.
Kovacs C., P., Halasz, and D., Szendy, C., 1966, *Acta Tech. Hungr.*, **56**, 383.
Krucoff, D., 1959, *Nucleonics*, **17**, 100.
Kunkel, W. B., and Hansen, J. W., 1950, *Rev. Sci. Instrum.*, **21**, 308.
Ladenburg, R., 1930, *Ann. Phys.* (Academic Press, New York) **4**, 863.
Lawton, J., 1964, *Brit. J. Appl. Phys.*, **15**, 935.
Liu, B. Y. H., Whitby, K. T., and Yu, H. S., 1967a, *J. Colloid Interface. Sci.* **23**, 367.
———1967b, *J. Appl. Phys.*, **38**, 1592.
Lowe, H. J., and Lucas, D. H., 1953, *Brit. J. Appl. Phys.*, **4**, 340.
Maurer, R. J., 1940, *Phys. Rev.*, **57**, 653.
McLennan, D. E., 1951, *Can. J. Phys.*, **29**, 122.
Miller, S. C., Jr., and Good, R. H., Jr., 1953, *Phys. Rev.*, **91**, 174.
Min, K., Chao, B. T., and Wyman, M. E., 1963, *Rev. Sci. Instrum.*, **34**, 529.
Mori, F., Fushimi, K., and Honma, T., 1966, *Electricity from MHD* Vol. 1 IAEA, Vienna, p. 643.
Mott, N. F., 1938, *Trans. Faraday Soc.*, **34**, 500.
Mott, N. F., and Gurney, R. W., 1948, *Electronic Processes in Ionic Crystals*, Oxford University Press, London, pp. 146, 168.
Munden, D. L., 1952, *J. Appl. Chem.*, **2**, 65.
Murphy, A. T., Adler, F. T., and Penney, G. W., 1959, *AIEE Trans.*, **78**, 318.
Natanson, G. L., 1960, *Sov. Phys.—Tech. Phys.*, **5**, 538.
Newby, D., 1967, *Brit. J. Appl. Phys.*, **18**, 383.
Norris, W. T., 1966, *Elect. Times*, Aug. 11, 208.
Page, L., 1935, *Introduction to Theoretical Physics*, Van Nostrand Reinhold, New York, p. 88.
Parker, W. G., and Wolfhard, H. G., 1950, *J. Chem. Soc.*, 2038.
Pauthenier, M. M., and Moreau-Hanot, M., 1932, *J. Phys. Radium*, Series 7, **3**, 590.
Penley, J. C., and Witte, R. S., 1962, *J. Appl. Phys.*, **33**, 2875.
Penney, G. W., and Lynch, R. D., 1957, *Trans. AIEE*, **76**, (I), 294.
Pepperhoff, W., 1951, *Optik*, **8**, 354.
Petritz, R. L., and Scanlon, W. W., 1955, *Phys., Rev.*, **97**, 1620.
Pope, M., 1962a, *J. Chem. Phys.*, **36**, 2810.
Pope, M., 1962b, *J. Chem. Phys.*, **37**, 1001.
Pope, M., Kallman, H., and Giachino, J., 1965a, *J. Chem. Phys.*, **42**, 2540.
Pope, M., Burgos, J., and Giachino, J., 1965b, *J. Chem. Phys.*, **43**, 3367.
Pressley, R. J., and Berk, H. L., 1965, *Phys. Rev.*, **140**, A, 1207.
Ramsay, A. S., 1960, *Electricity and Magnetism*, Cambridge University Press, London, p. 120.
Redfield, A., 1954, *Phys. Rev.*, **94**, 537.
Ring, P. T., O'Keefe, J. G., and Bray, P. J., 1958, *Phys. Rev. Letters*, **1**, 453.
Robinson, M. E., and Martin, S. W., 1948, *J. Phys. Chem.*, **52**, 854.
Rose, D. J., and Clark, M., 1961, *Plasmas and Controlled Fusion*, Wiley, New York, pp. 62, 162.
Rosen, G., 1962, *Phys. Fluids*, **5**, 737.
Rossler, F., 1953, *Optik*, **10**, 531.
Saha, M. N., 1920, *Phil. Mag.*, **40**, 472.

———1921, *Z. Phys.*, **6**, 40.
Savostianova, M., 1930, *Z. Physik*, **64**, 262.
Sawyer, K. F., and Walton, W. M., 1950, *Rev. Sci. Instrum*, **27**, 272.
Schluderberg, D. C. Whitelaw, R. L., and Carlson, R. W., 1962, *Nucleonics*, **19**, 67.
Schulman, J. H., and Compton, W. D., 1962, *Color Centers in Solids*, Pergamon Press, New York.
Scott, A. B., and Smith, W. A., 1951, *Phys. Rev.*, **83**, 982.
Scott, A. B., Smith, W. A., and Thompson, M. A., 1953, *J. Phys. Chem.*, **57**, 757.
Seitz, F., 1940, *Modern Theory of Solids*, McGraw-Hill, New York, p. 161.
Shuler, K. E., and Weber, J., 1954, *J. Chem. Phys.*, **22**, 491.
Smallman, R. E., and Willis, B. T. M., 1957, *Phil. Mag.*, **8**, 1018.
Smith, F. T., 1958, *J. Chem. Phys.*, **28**, 746.
———1959, *Proc. Third Conf. on Carbon*, University of Buffalo, Buffalo, New York, 1957 (Pergamon Press, New York), p. 419.
Smith, F. T., and Gatz, C. R., 1963, *Progress in Astronautics and Aeronautics*, Vol. 12 Academic Press, New York, p. 301.
Smith, J. M., 1965, *AIAA J.*, **3**, 648.
Sodha, M. S., and Eastman, P. C., 1958, *Z. Phys.*, **150**, 242.
Sodha, M. S., 1961, *J. Appl. Phys.*, **32**, 2059.
Sodha, M. S., and Palumbo, C. J., 1962, *Proc. Phys. Soc.*, **80**, 1155.
Sodha, M. S., 1963, *Brit. J. Appl. Phys.*, **14**, 172.
Sodha, M. S., Palumbo, C. J., and Daley, J. T., 1963, *Brit. J. Appl. Phys.*, **14**, 916.
Sodha, M. S., and Bendor, E. 1964a, *Symposium on MHD Electrical Power Generation*, Vol. 2, ENEA, Paris, 1964, p. 289.
Sodha, M. S., and Bendor, E., 1964b, *Brit. J. Appl. Phys.*, **15**, 1031.
Sodha, M. S., Kaw, P. K., and Srivastava, H. K., 1965, *Brit, J. Appl. Phys.*, **16**, 721.
Sodha, M. S., and Sharma, S., 1967, *Brit. J. Appl. Phys.*, **18**, 1127.
Sodha, M. S., and Kaw, P. K., 1968, *Brit. J. Appl. Phys.*, **1**, 1303.
Sodha, M. S., and Dubey, P. K., 1970, *J. Phys. D.*, **3**, 139.
Soo, S. L., 1963a, *Phys. Fluids*, **6**, 145.
———1963b, *J. Appl. Phys.*, **34**, 1689.
Soo, S. L., and Dimick, R. C., 1963, *Proc. Multi-phase Flow Symposium*, ASME, p. 43.
Soo, S. L., and Dimick, R. C., 1965, *Proc. Tenth Symposium on Combustion*, (Combustion Institute, Pittsburgh, Penna.), p. 699.
Soo, S. L., 1967, *Fluid Dynamics of Multiphase Systems*, Blaisdell, Waltham, Mass.
Sootha, G. D., 1967, Ph.D. Thesis, University of Delhi.
Spitzer, L., 1941, *Astrophys. J.*, **93**, 369.
———1948, *Astrophys. J.*, **107**, 6.
Spitzer, L., and Harm, R., 1953, *Phys. Rev.*, **89**, 977.
Sugden, T. M., and Thrush, B. A., 1951, *Nature*, **168**, 703.
Swift-Hook, D. T., and Wright, J. K., 1963, *J. Fluid Mech.*, **15**, 97.
Timusk, T., 1961, *J. Phys. Chem. Solids*, **18**, 265.
Tozzi, A., 1965, *Chim. Ind.*, (*Milan*) **47**, 596.
Van de Hulst, H. C., 1957, *Light Scattering by Small Particles*, Wiley, New York.
Vedder, J. F., 1963, *Rev. Sci. Instrum.* **34**, 1175.
Waldie, B., and Fells, I., 1967, *Phil. Trans. Roy. Soc. London* **A261**, 490.
——— 1968, *Int. Symposium on MHD Electrical Power Generation*, Warsaw, 1968, IAEA Vienna, Paper SM 107/127.
Walsh, W. M., Jr., Rupp., L. W., Jr., and Schmidt, P. H., 1966, *Phys. Rev.*, **142**, 414.

White, H. J., 1951, *Trans. AIEE*, **76**, 1186.
———1955, *Ind. Eng. Chem.* **47**, 932.
———1961, *Industrial Electrostatic Precipitation*, Addison-Wesley, Reading, Mass.
Williams, H., Lewis, J. D., and Hobson, R. M., 1963, *Advances in MHD*, I. A., McGrath, R. G. Siddall, and M. W. Thring, Eds., Pergamon Press, London.
Wolfhard, H. G., and Parker, W. G., 1949, *Proc. Phys. Soc.*, **B62**, 523.
Wyatt, P. J., 1960, *J. Geophys. Res.*, **66**, 1673.
Zimin, E. P., Mikhnevich, Z. G., and Popov, V. A., 1966, *Electricity from MHD*, Vol. 3 IAEA Vienna, p. 97.

The Motion of a Charged Particle in a Strong Magnetic Field

IRA B. BERNSTEIN

Yale University, New Haven, Connecticut

Abstract

The reduced description in terms of drifts and adiabatic invariants of the motion of a charged particle in a strong magnetic field is derived. The demonstration employs systematically two time scales and an iteration scheme for each quasiperiodicity. This leads to a particularly expeditious derivation, as well as the details of the rapid oscillations at each stage. Moreover the motivation of each part is clear, as is the relation to simple problems in dynamics. The small parameters, the existence of which underlines the method, are displayed explicitly.

Introduction

A central problem in plasma physics is the derivation of a tractable description of the motion of a charged particle in a strong magnetic field. This task was initiated by Alfvén (1) on a physical basis, and carried to a high degree of mathematical sophistication by Kruskal (2). The formal considerations of the latter author provide a constructive technique for the development of a so-called "reduced description" of the motion in powers of an appropriate small parameter. The method applies to all dynamical system that exhibit one or more, almost periodic motions. A partial summary of both points of view has been given by Northrup (3), who combines several points of view. The basic notion underlying all treatments is that of the existence for each almost periodic motion of two time scales, one of which describes the rapid periodic aspect, and the other any slow perturbation of this. When this notion is applied systematically to the problem of the motion of a charged particle in a strong magnetic field, coupled with an appropriate iteration scheme, it is possible to derive, in a very efficient manner, all of the well-known results and, in addition, indicate explicitly what are the small parameters and the details of the reduced description.

Section I is devoted to developing the guiding-center description and the associated approximate constant of the motion or adiabatic invariant, the magnetic moment, to the lowest significant order. Section II is concerned with the derivation of the reduced description when the motion of the guiding center along the lines of force is periodic and the particle does not move much perpendicular to the line in one period. The second adiabatic invariant, the so-called longitudinal invariant, is found to the lowest significant order in this second small parameter, as well as a description of the rapid oscillation. In Section III, a final reduction is affected in the description when the energy of the guiding-center particle changes but little in the time required to circulate once on a magnetic surface. Here, in addition to a third adiabatic invariant the magnetic flux, the details of the motion in the constant flux surface are found.

The derivations presented here have the virtue of considerable analytical simplicity and conceptual unity. It is also clear from them explicitly what the small parameter in question is, and also how to proceed to the next order. Moreover, at all stages the treatment is simply related to a familiar problem in classical mechanics.

I. The Gyrating Particle and the Magnetic Moment Adiabatic Invariant

The equation of motion of a particle of charge q and mass m acted on by an electric field $\mathbf{E}(\mathbf{r},t)$, a magnetic field $\mathbf{B}(\mathbf{r},t)$ and a gravitational potential $G(\mathbf{r},t)$ is

$$\ddot{\mathbf{r}} = \mathbf{a} - \mathbf{\Omega} \times \dot{\mathbf{r}} \tag{1-1}$$

where

$$\mathbf{a}(\mathbf{r},t) = q\mathbf{E}(\mathbf{r},t)/m - \nabla G(\mathbf{r},t) \tag{1-2}$$

$$\mathbf{\Omega}(\mathbf{r},t) = q\mathbf{B}(\mathbf{r},t)/mc \tag{1-3}$$

When $\mathbf{a}$ and $\boldsymbol{\Omega}$ are constant in space and time, the solution of eq. 1-1 can be written (4):

$$\mathbf{r} = \mathbf{R} + \boldsymbol{\rho} \tag{1-4}$$

where

$$\mathbf{R}(t) = \mathbf{R}(0) + \mathbf{bb} \cdot [\dot{\mathbf{R}}(0)t + \tfrac{1}{2}\,\mathbf{a}t^2] + \mathbf{b} \times \mathbf{a}\,t/\Omega \tag{1-5}$$

$$\boldsymbol{\rho}(t) = [\mathbf{e}_2 \cos(\Omega t + \phi) + \mathbf{e}_1 \sin(\Omega t + \phi)]w/\Omega \tag{1-6}$$

In the above expressions the perpendicular speed w and the phase ϕ are constants, and we have introduced

$$\Omega = qB/mc \tag{1-7}$$

$$\mathbf{b} = \mathbf{B}/B \tag{1-8}$$

and the orthonormal right-handed set of Cartesian unit vectors, $\mathbf{e}_1$, $\mathbf{e}_2$, and $\mathbf{e}_3 = \mathbf{b}$. Clearly,

$$\dot{\mathbf{R}}(t) = \mathbf{bb} \cdot [\dot{\mathbf{R}}(0) + \mathbf{a}t] + \mathbf{b} \times \mathbf{a}/\Omega \tag{1-9}$$

$$\begin{aligned} \dot{\boldsymbol{\rho}} &= w[-\mathbf{e}_2 \sin(\Omega t + \phi) + \mathbf{e}_1 \cos(\Omega t + \phi)] \\ &= -\boldsymbol{\Omega} \times \boldsymbol{\rho} \end{aligned} \tag{1-10}$$

The solution is readily verified by substitution in eq. 1-1. The vector $\mathbf{R}(t)$ describes the trajectory of the so-called guiding center; the term in brackets in eq. 1-5 arises from the accelerated motion in the direction of the magnetic field; the term $\mathbf{b} \times \mathbf{a}/\Omega$ is designated the *drift velocity* perpendicular to the magnetic field.

Consider the case in which $\mathbf{a}$ and $\boldsymbol{\Omega}$ depend on space and time but do not change much in a distance w/Ω or a time $1/\Omega$, where w is the magnitude of the component of the velocity of the particle orthogonal to the magnetic field measured relative to the drift velocity, and Ω is the value of the gyration frequency that prevails at the point in question on the trajectory of the particle. It then seems plausible that the solution of the equation of motion will be very much like that given in eqs. 1-4 to 1-6. If this is so, one is led to seek a solution effectively in powers of the small parameter

$$\varepsilon = \left|\Omega^{-1}\left(\frac{\partial}{\partial t} + \dot{\mathbf{R}} \cdot \nabla\right) \ln aB\right| + |(w/\Omega)\nabla \ln aB| \tag{1-11}$$

This is the program adopted by Kruskal (2) and leads to an asymptotic representation.

An alternative method that is more expeditious for obtaining lowest significant order results consists in the introduction of an auxiliary variable

θ contrived to describe the rapid gyration indicated in eq. 1-6 for the constant field case and a suitable iteration scheme. We shall require periodicity in θ and choose the period to be unity, so that θ has the character of an angle variable in Hamilton Jacobi theory. For the case of constant fields one sees from eq. 1-6 that $\theta = \Omega t/2\pi$ is an appropriate choice. For the general case we write

$$\mathbf{r} = \mathbf{R}(t) + \boldsymbol{\rho}(\theta,t) \tag{1-12}$$

whence if we denote partial derivatives by subscripts and set $\dot{\theta} = v(t)$,

$$\dot{\mathbf{r}} = \dot{\mathbf{R}}(t) + v(t)\,\boldsymbol{\rho}_\theta(\theta,t) + \boldsymbol{\rho}_t(\theta,t) \tag{1-13}$$

Presumably,

$$v^2\boldsymbol{\rho}_\theta{}^2 \gg \boldsymbol{\rho}_t{}^2$$

and we expect that in the order of magnitude $v \sim \Omega/2\pi$.

In dealing with the equation of motion, since we anticipate that $\boldsymbol{\rho}^2 = w^2/\Omega^2$, we are led to expand $\boldsymbol{\Omega}(\mathbf{r},t)$ and $\mathbf{a}(\mathbf{r},t)$ in powers of $\boldsymbol{\rho}$ because this is effectively an expansion in powers of the small parameter ε of eq. 1-11. Thus one writes

$$\mathbf{a}(\mathbf{R}+\boldsymbol{\rho},t) = \mathbf{a}(\mathbf{R},t) + \boldsymbol{\rho}\cdot\nabla\mathbf{a}(\mathbf{R},t) + \tfrac{1}{2}\boldsymbol{\rho\rho}\colon\nabla\nabla\mathbf{a}(\mathbf{R},t) + \cdots \tag{1-14}$$

and a parallel expansion for Ω. When these expansions and the time derivative of eq. 1-13 are employed in the equation of motion 1-1, we obtain the result,

$$\begin{aligned}\ddot{\mathbf{R}} + v^2\boldsymbol{\rho}_{\theta\theta} + 2v\boldsymbol{\rho}_{\theta t} + \dot{v}\boldsymbol{\rho}_\theta + \boldsymbol{\rho}_{tt} = {} & \mathbf{a} + \boldsymbol{\rho}\cdot\nabla\mathbf{a} + \tfrac{1}{2}\boldsymbol{\rho\rho}\colon\nabla\nabla\mathbf{a} + \cdots \\ & -\Omega\times\dot{\mathbf{R}} - \boldsymbol{\rho}\cdot(\nabla\boldsymbol{\Omega})\times\dot{\mathbf{R}} - \tfrac{1}{2}\boldsymbol{\rho\rho}\colon(\nabla\nabla\boldsymbol{\Omega})\times\dot{\mathbf{R}} \\ & -\cdots \\ & -v\Omega\times\boldsymbol{\rho}_\theta - v\boldsymbol{\rho}\cdot(\nabla\boldsymbol{\Omega})\times\boldsymbol{\rho}_\theta - \tfrac{1}{2}v\,\boldsymbol{\rho\rho}\colon(\nabla\nabla\boldsymbol{\Omega})\times\boldsymbol{\rho}_\theta \\ & -\cdots \\ & -\Omega\times\boldsymbol{\rho}_t - \boldsymbol{\rho}\cdot(\nabla\boldsymbol{\Omega})\times\boldsymbol{\rho}_t - \tfrac{1}{2}\boldsymbol{\rho\rho}\colon(\nabla\nabla\boldsymbol{\Omega})\times\boldsymbol{\rho}_t \\ & -\cdots\end{aligned} \tag{1-15}$$

Note that in eq. 1-15 θ occurs only in $\boldsymbol{\rho}$ and its derivative, $\mathbf{a} = \mathbf{a}(\mathbf{R},t)$, $\boldsymbol{\Omega} = \boldsymbol{\Omega}(\mathbf{R},t)$, and ∇ denotes the gradient with respect to $\mathbf{R}$. We shall require that $\boldsymbol{\rho}$ be periodic in θ with period unity and the average of $\boldsymbol{\rho}$ over one period in θ vanish. That is, following Kruskal, one can write the Fourier series:

$$\boldsymbol{\rho}(\theta,t) = \sum_{n=1}^{\infty}[\boldsymbol{\rho}^{(n)}(t)e^{i2\pi n\theta} + \boldsymbol{\rho}^{(n)*}(t)e^{-i2\pi n\theta}] \tag{1-16}$$

Thus, if we integrate eq. 1-15 over one period of θ, we obtain

$$\ddot{\mathbf{R}} = \mathbf{a} + v\int_0^1 d\theta\,\boldsymbol{\rho}_\theta\times[\boldsymbol{\rho}\cdot(\nabla\boldsymbol{\Omega})] - \boldsymbol{\Omega}\times\dot{\mathbf{R}} + \cdots \tag{1-17}$$

As will be shown later, the dependence of $\boldsymbol{\rho}$ on θ to the lowest significant order is given by eq. 1-6 with $2\pi\theta$ replacing Ωt. Thus, if one carries out integrations over one period in θ, the integral of any quantity cubic in $\boldsymbol{\rho}$ or its derivatives will vanish to the lowest significant order. In particular, it follows from this observation that the terms indicated by dots above are smaller by a factor of the order ε^2 than the largest term retained explicitly.

When eq. 1-17 is subtracted from eq. 1-15, on regrouping terms and recognizing that $\dot{\mathbf{R}}$ is not a function of $\mathbf{R}$ but depends only on t, we find that

$$(\nu^2\boldsymbol{\rho}_\theta + \nu\boldsymbol{\Omega}\times\boldsymbol{\rho})_\theta = -2\nu\boldsymbol{\rho}_{\theta t} - \dot{\nu}\boldsymbol{\rho}_\theta + \boldsymbol{\rho}\cdot\nabla(\mathbf{a} + \dot{\mathbf{R}}\times\boldsymbol{\Omega}) + \boldsymbol{\rho}_t\times\boldsymbol{\Omega}$$
$$+ \nu\boldsymbol{\rho}_\theta\times(\boldsymbol{\rho}\cdot\nabla\boldsymbol{\Omega}) - \int_0^1 d\theta\, \nu\boldsymbol{\rho}_\theta\times(\boldsymbol{\rho}\cdot\nabla\boldsymbol{\Omega}) + \cdots \quad (1\text{-}18)$$

The terms indicated by dots on the right-hand side of eq. 1-18 are smaller by a factor of order ε than these indicated explicitly on the right. These latter in turn are smaller by a factor of the order ε than those written on the left-hand side of eq. 1-18. Thus, to the lowest significant order, we require that the left-hand side above vanish, whence on integration in θ

$$\nu^2\boldsymbol{\rho}_\theta + \nu\boldsymbol{\Omega}\times\boldsymbol{\rho} = \xi(t) \quad (1\text{-}19)$$

where $\xi(t)$ is the *constant of integration.* If one integrates eq. 1-19 over one period in θ, it follows that the left-hand side vanishes because of eq. 1-16. Thus $\xi = 0$, and if we resolve eq. 1-19 in the Cartesian coordinate system associated with the unit vectors introduced prior to eq. 1-19, we obtain the result:

$$\rho_{1\theta} - \frac{\Omega}{\nu}\rho^2 = 0 \quad (1\text{-}20)$$

$$\rho_{2\theta} + \frac{\Omega}{\nu}\rho_1 = 0 \quad (1\text{-}21)$$

$$\rho_{3\theta} = 0 \quad (1\text{-}22)$$

It follows from eq. 1-22 and the requirement that $\boldsymbol{\rho}$ have no part constant in θ that $\rho_3 = 0$. Moreover, if one adds i times eq. 1-21 to eq. 1-20,

$$(\rho_1 + i\rho_2)_\theta = -i\frac{\Omega}{\nu}(\rho_1 + i\rho_2) \quad (1\text{-}23)$$

whence

$$\rho_1 + i\rho_2 = i\rho(t)e\left[-i\left(\frac{\Omega}{\nu}\theta + \phi(t)\right)\right]^{1/2} \quad (1\text{-}24)$$

where we have introduced the real constants of integration, $\rho(t)$ and $\phi(t)$.

In order that $\boldsymbol{\rho}$, as determined from eq. 1-24, be periodic in θ with period unity, we must require that

$$\nu(t) = \Omega(\mathbf{R}(t),t)/2\pi \quad (1\text{-}25)$$

Thus, on rewriting these results in vector form, to the lowest significant order in ε, we obtain

$$\boldsymbol{\rho}(\theta,t) = \rho[\mathbf{e}_2 \cos(2\pi\theta + \phi) + \mathbf{e}_1 \sin(2\pi\theta + \phi)] \tag{1-26}$$

where ρ, ϕ, $\mathbf{e}_1$, and $\mathbf{e}_2$ all depend on time. Clearly eq. 1-26 reduces to eq. 1-6 for the case of constant fields. The time dependence of θ is now determined from

$$\theta(t) = \int dt\, \Omega(\mathbf{R}(t),t)/2\pi \tag{1-27}$$

Note, however, that to this order in ε, ρ, and ϕ are not yet determined as functions of time.

If one wishes to calculate to the next order in ε, it is adequate to drop the terms indicated by dots in eq. 1-18. Rather than solve the resulting equation completely, we shall be content to derive an approximate constant of motion, correct to that order in ε corresponding to dropping the dots in eq. 1-18. The derivation proceeds by forming the scalar product of eq. 1-18 with ρ_θ, after deleting the terms indicated by dots:

$$\left(\frac{1}{2}\, v^2 \boldsymbol{\rho}_\theta{}^2\right)_\theta + (v\rho_\theta{}^2)_t = \boldsymbol{\rho}_\theta \boldsymbol{\rho} \colon \nabla(\mathbf{a} + \dot{\mathbf{R}} \times \Omega) + \boldsymbol{\rho}_\theta \times \boldsymbol{\rho}_t \cdot \boldsymbol{\Omega} - \boldsymbol{\rho}_\theta \cdot \int_0^1 d\theta\, v\boldsymbol{\rho}_\theta \times (\boldsymbol{\rho} \cdot \nabla \boldsymbol{\Omega}) \tag{1-28}$$

If we integrate eq. 1-28 over one period in θ, we obtain

$$\left(\int_0^1 d\theta v \rho_\theta{}^2\right)_t = \int_0^1 d\theta \boldsymbol{\rho}_\theta \boldsymbol{\rho} \colon \nabla(\mathbf{a} + \dot{\mathbf{R}} \times \Omega) + \int_0^1 d\theta \boldsymbol{\rho}_\theta \times \boldsymbol{\rho}_t \cdot \boldsymbol{\Omega} \tag{1-29}$$

The integration over θ has removed the nominally large terms in eq. 1-28, and it is adequate to use the lowest significant order approximation eq. 1-26 in eq. 1-29. Thus

$$\begin{aligned}\int_0^1 d\theta \rho_\theta{}^2 &= 4\pi^2\rho^2 \int_0^1 d\theta[\cos^2(2\pi\theta + \phi) + \sin^2(2\pi\theta + \phi)] \\ &= 4\pi^2\rho^2\end{aligned} \tag{1-30}$$

and

$$\begin{aligned}\int_0^1 d\theta \boldsymbol{\rho}_\theta \boldsymbol{\rho} &= 2\pi\rho^2 \int_0^1 d\theta[-\mathbf{e}_2 \sin(2\pi\theta + \phi) + \mathbf{e}_1 \cos(2\pi\theta + \phi)] \\ &\qquad \cdot [\mathbf{e}_2 \cos(2\pi\theta + \phi) + \mathbf{e} \sin(2\pi\theta + \phi)] \\ &= 2\pi\rho^2 \int_0^1 d\theta\{(\mathbf{e}_1\mathbf{e}_1 - \mathbf{e}_2\mathbf{e}_2) \sin(2\pi\theta + \phi) \cos(2\pi\theta + \phi) \\ &\qquad + \mathbf{e}_1\mathbf{e}_2 \cos^2(2\pi\theta + \phi) - \mathbf{e}_2\mathbf{e}_1 \sin^2(2\pi\theta + \phi)\} \\ &= \pi\rho^2(\mathbf{e}_1\mathbf{e}_2 - \mathbf{e}_2\mathbf{e}_1)\end{aligned} \tag{1-31}$$

In order conveniently to reduce the remaining integral, we note that on combining eqs. 1-19 and 1-25 we can write

$$\boldsymbol{\rho}_\theta = -2\pi\, \mathbf{b} \times \boldsymbol{\rho} \tag{1-32}$$

But substituting **b** in eq. 1-32, and using the result eq. of 1-26 that $\mathbf{b} \cdot \boldsymbol{\rho} = 0$, we find that

$$\mathbf{b} \times \boldsymbol{\rho}_\theta = 2\pi\boldsymbol{\rho} \tag{1-33}$$

Thus

$$\begin{aligned}\int_0^1 d\theta \boldsymbol{\rho}_\theta \times \boldsymbol{\rho}_t \cdot \boldsymbol{\Omega} &= \Omega \int_0^1 d\theta \mathbf{b} \times \boldsymbol{\rho}_\theta \cdot \boldsymbol{\rho}_t \\ &= 2\pi\Omega \int_0^1 d\theta \boldsymbol{\rho} \cdot \boldsymbol{\rho}_t \\ &= 2\pi\Omega \frac{\partial}{\partial t} \int_0^1 d\theta \frac{1}{2} \boldsymbol{\rho}^2 \\ &= \pi\Omega \frac{\partial}{\partial t} \rho^2\end{aligned} \tag{1-34}$$

since $\boldsymbol{\rho}^2 = \rho^2$ is independent of θ. These results permit one to write eq. 1-29 as

$$\begin{aligned}(2\pi\Omega\rho^2)_t &= \pi\rho^2(\mathbf{e}_1\mathbf{e}_2 \cdot \nabla - \mathbf{e}_2\,\mathbf{e}_1 \cdot \nabla) \cdot (\mathbf{a} + \dot{\mathbf{R}} \times \Omega) + \pi\Omega(\rho^2)_t \\ &= -\pi\rho^2[(\mathbf{e}_1 \times \mathbf{e}_2) \times \nabla] \cdot (\mathbf{a} + \dot{\mathbf{R}} \times \boldsymbol{\Omega}) + (\pi\Omega\rho^2)_t - \pi\rho^2\Omega_t\end{aligned} \tag{1-35}$$

On using eqs. 1-2 and 1-3 on the right-hand side above, after transposing the term that is a multiple of that on the left-hand side, we obtain the result, since $\mathbf{e}_1 \times \mathbf{e}_2 = \mathbf{b}$,

$$\begin{aligned}(\pi\Omega\rho^2)_t &= -\pi\rho^2\{(\mathbf{b} \times \nabla) \cdot (\mathbf{a} + \dot{\mathbf{R}} \times \boldsymbol{\Omega}) + \Omega_t\} \\ &= -\pi\rho^2\{\mathbf{b} \cdot \nabla \times (\mathbf{a} + \dot{\mathbf{R}} \times \boldsymbol{\Omega}) + \Omega_t\} \\ &= -\pi\rho^2 \frac{q}{m}\left\{\mathbf{b} \cdot \nabla \times \left[\mathbf{E} + \frac{1}{c}\dot{\mathbf{R}} \times \mathbf{B}\right] + \frac{1}{c} B_t\right\}\end{aligned} \tag{1-36}$$

In eq. 1-36, B_t is to be interpreted as a time derivative holding θ fixed, i.e., a convective derivative following the guiding-center motion characterized by $\dot{\mathbf{R}}$, namely,

$$B_t = \frac{\partial B}{\partial t} + \dot{\mathbf{R}} \cdot \nabla B \tag{1-37}$$

But from the Maxwell equation,

$$c\nabla \times \mathbf{E} = -\frac{\partial \mathbf{B}}{\partial t} \tag{1-38}$$

it follows, since $\nabla \cdot \mathbf{B} = 0$ and $\nabla\dot{\mathbf{R}} = 0$, that

$$\begin{aligned}\mathbf{B}_t &= \frac{\partial \mathbf{B}}{\partial t} + \dot{\mathbf{R}} \cdot \nabla \mathbf{B} \\ &= -c\nabla \times \left[\mathbf{E} + \frac{1}{c}\,\dot{\mathbf{R}} \times \mathbf{B}\right] + \mathbf{B} \cdot \nabla\dot{\mathbf{R}} + \dot{\mathbf{R}}\nabla \cdot \mathbf{B} - \mathbf{B}\nabla \cdot \dot{\mathbf{R}} \\ &= -c\nabla \times \left[\mathbf{E} + \frac{1}{c}\,\dot{\mathbf{R}} \times \mathbf{B}\right] \qquad (1\text{-}39)\end{aligned}$$

Thus

$$\begin{aligned}\mathbf{b} \cdot \nabla \times \left[E + \frac{1}{c}\,\dot{\mathbf{R}} \times \mathbf{B}\right] + \frac{1}{c}\,B_t &= -\frac{1}{c}\,\mathbf{b} \cdot \mathbf{B}_t + \frac{1}{c}\,B_t \\ &= -\frac{1}{cB}\,\mathbf{B} \cdot \mathbf{B}_t + \frac{1}{c}\left(\frac{B^2}{2}\right)_t \\ &= 0 \qquad (1\text{-}40)\end{aligned}$$

Note that $\mathbf{E} + 1/c\,(\dot{\mathbf{R}} \times \mathbf{B})$ is just the electric field seen by an observer moving with the guiding center.

We can now conclude from eq. 1-36 that

$$(\pi\Omega\rho^2)_t = 0 \qquad (1\text{-}41)$$

whence

$$\mu = \frac{q}{c}\,\frac{\Omega}{2\pi}\,\pi\rho^2 = \frac{q}{c}\,\nu\pi\rho^2 \qquad (1\text{-}42)$$

the so-called magnetic moment, is an approximate constant of the motion. Such an approximate constant is conventionally termed an *adiabatic invariant*.

Let us now return to eq. 1-17. Note that the term therein involving $\nabla\Omega$, on using eq. 1-31, can be approximated by

$$\begin{aligned}v\int_0^1 d\theta\, \boldsymbol{\rho}_\theta \times [\boldsymbol{\rho} \cdot (\nabla\boldsymbol{\Omega})] &= \nu\pi\rho^2(\mathbf{e}_1\mathbf{e}_2 \cdot \nabla - \mathbf{e}_2\mathbf{e}_1 \cdot \nabla) \times (q\mathbf{B}/mc) \\ &= -\frac{\mu}{m}\,[(\mathbf{e}_1 \times \mathbf{e}_2) \times \nabla] \times \mathbf{B} \\ &= -\frac{\mu}{m}\,(\mathbf{b} \times \nabla) \times \mathbf{B} \\ &= -\frac{\mu}{m}\,[(\nabla\mathbf{B}) \cdot \mathbf{b} - \mathbf{b}\nabla \cdot \mathbf{B}] \\ &= -\frac{\mu}{m}\,\nabla B \qquad (1\text{-}43)\end{aligned}$$

Thus on dropping the terms indicated by dots, after dotting and substituting **b**, eq. 1-17 yields

$$\mathbf{b} \cdot \ddot{\mathbf{R}} = \mathbf{b} \cdot \mathbf{a} - \frac{\mu}{m} \mathbf{b} \cdot \nabla B \tag{1-44}$$

and

$$\begin{aligned} \dot{\mathbf{R}}_\perp &\equiv \mathbf{b} \times (\dot{\mathbf{R}} \times \mathbf{b}) \\ &= \frac{1}{\Omega} \mathbf{a} \times \mathbf{b} + \frac{\mu}{m\Omega} \mathbf{b} \times \nabla B + \frac{1}{\Omega} \mathbf{b} \times \ddot{\mathbf{R}} \\ &= c\frac{\mathbf{E} \times \mathbf{B}}{B^2} + \frac{1}{\Omega} \mathbf{b} \times \nabla G + \frac{\mu}{m\Omega} \mathbf{b} \times \nabla B + \frac{1}{\Omega} \mathbf{b} \times \ddot{\mathbf{R}} \end{aligned} \tag{1-45}$$

If we define

$$u = \mathbf{b} \cdot \dot{\mathbf{R}} \tag{1-46}$$

we can write eq. 1-44 in the form

$$m\dot{u} = \mathbf{b} \cdot [q\mathbf{E} - m\nabla G - \mu \nabla B] + m\dot{\mathbf{b}} \cdot \dot{\mathbf{R}} \tag{1-47}$$

since $\mathbf{b} \cdot \dot{\mathbf{b}} = 0$. Clearly, the acceleration along the magnetic field should not be so large as to change the magnetic field in a time comparable with Ω^{-1}; otherwise, the theory here developed is invalid.

Equation 1-45 can be solved by iteration, assuming that the acceleration **a** dominates, namely, to the lowest order,

$$\dot{\mathbf{R}}_\perp = \frac{1}{\Omega} \mathbf{a} \times \mathbf{b} \tag{1-48}$$

and to the next order

$$\dot{\mathbf{R}}_\perp = \frac{1}{\Omega} \mathbf{a} \times \mathbf{b} + \frac{\mu}{m\Omega} \mathbf{b} \times \nabla B + \frac{1}{\Omega} \mathbf{b} \times \left[\left(\frac{1}{\Omega} \mathbf{a} \times \mathbf{b} \right)^{\cdot} + (u\mathbf{b})^{\cdot} \right] \tag{1-49}$$

In order to iterate once again and preserve accuracy, one would have to restore the terms indicated by dots in eq. 1-17 and also evaluate $\boldsymbol{\rho}$ to the next order in ε.

The details of the gyration can be fixed, e.g., by choosing

$$\mathbf{e}_2 = \dot{\mathbf{R}}_\perp / |\dot{\mathbf{R}}_\perp| \tag{1-50}$$

in which event

$$\mathbf{e}_1 = \mathbf{e}_2 \times \mathbf{e}_3 = \mathbf{e}_2 \times \mathbf{b} \tag{1-51}$$

The derivation of the equation governing the evolution in time of the slowly varying phase function $\phi(t)$ can be found by returning to eq. 1-18 and viewing it as an inhomogeneous equation for the second approximation. That is, one writes eq. 1-16 in the form

$$[\boldsymbol{\rho}_\theta + (\boldsymbol{\Omega}/v) \times \boldsymbol{\rho}]_\theta = \mathbf{f} \tag{1-52}$$

where $v^2\mathbf{f}$ is given by the right-hand side of eq. 1-18 with the terms indicated by dots deleted, and $\boldsymbol{\rho}$ given by eq. 1-26. It is readily seen that $\mathbf{f}$ involves only the first and second harmonics of $2\pi\theta$, namely,

$$\mathbf{f} = \sum_{n=1}^{2} (\mathbf{f}^{(n)} e^{2\pi i n\theta} + \mathbf{f}^{(n)*} e^{-2\pi i n\theta}) \tag{1-53}$$

In order that the solution of the homogeneous equation associated with eq. 1-52 reproduce eq. 1-26, we must as before select $v = 2\pi/\Omega$. If then, as before, we resolve eq. 1-52 in a Cartesian coordinate system defined by the orthornormal vectors $\mathbf{e}_1$, $\mathbf{e}_2$, $\mathbf{e}_3 = \mathbf{b}$, on adding i times the two-component to the one-component of eq. 1-52, we obtain

$$(\rho_1 + i\rho_2)_{\theta\theta} + 2\pi i(\rho_1 + i\rho_2)_\theta = f_1 + if_2 \tag{1-54}$$

We require that $\boldsymbol{\rho}$ be represented by eq. 1-16, namely, that it be periodic in θ with period one and have no part constant in θ. That is, if we write

$$\rho_1 + i\rho_2 = \sum_{n=-\infty}^{\infty} c_n e^{2\pi n i\theta} \tag{1-55}$$

$$f_1 + if_2 = \sum_{n=-2}^{2} d_n e^{2\pi n i\theta} \tag{1-56}$$

where c_0 and d_0 are both zero, then the insertion of these expressions in eq. 1-54 and the equating of the coefficients of like Fourier factors $e^{2\pi n i\theta}$ yields

$$4\pi^2 n(n+1)c_n = d_n \tag{1-57}$$

Clearly, when $n \neq -1$, one has

$$c_n = -d_n/4\pi^2 n(n+1) \tag{1-58}$$

and the c_n vanish for $n = \pm 3, \pm 4, \ldots$. In order that a solution exist for $n = -1$, one must have $d_{-1} = 0$, or equivalently as follows from eq. 1-56 on multiplication by $e^{2\pi i\theta}$ and integration over one period in θ,

$$\mathbf{d}_1 = \int_0^1 d\theta(f_1 + if_2)e^{2\pi i\theta} = 0 \tag{1-59}$$

Equation 1-59 is equivalent to two real conditions resulting from taking the real and imaginary parts. These are effective equations for $\dot{\phi}$ and $\dot{\rho}$. On judicious combination they yield eq. 1-41. We shall not develop them in detail.

If one is not interested in analyzing the details of the gyration, it suffices to consider the equations governing the guiding center $\mathbf{R}(t)$: eq. 1-47 which gives the time rate of change of the component of the guiding center velocity along the magnetic field at the location of the guiding center and eq. 1-49 which gives the velocity of the guiding center perpendicular to the magnetic field at the location of the guiding center. The advantages of these equations over eq. 1-1 are twofold: first, they exhibit no fast gyrations on the scale of the gyration frequency; second, they constitute a fourth-order system of ordinary differential equations as opposed to eq. 1-1 which is a sixth-order system. These features are useful both for purposes of numerical calculation, and also for analytic work and qualitative analysis.

Higher approximations can be found by iterating the results just found, but in general the results are so complicated that the virtues of the reduced description are lost.

It has been shown (6) that these lowest significant order results represent the leading term in an asymptotic expansion of the trajectory of the particle in powers of the small parameter ε of eq. 1-11. That is, if one writes the partial sum,

$$\mathbf{S}_N(t) = \mathbf{r}_0(t) + \varepsilon\mathbf{r}_1(t) + \varepsilon^2\mathbf{r}_2(t) + \cdots \varepsilon^N r_N(t)$$

then, for any fixed time t,

$$\lim_{\varepsilon\to 0} \frac{|\mathbf{r}(t) - \mathbf{S}_N(t)|}{\varepsilon^N} = 0$$

This is distinct from what would prevail were the procedure convergent, namely,

$$\lim_{N\to\infty} |\mathbf{r}(t) - \mathbf{S}_N(r)| = 0$$

II. The Second or Longitudinal Adiabatic Invariant

A further reduction of the preceding guiding-center description can be made when the motion along the lines of force is quasiperiodic and much more rapid than the motion associated with the drift. The demonstration is assisted by writing the magnetic field in terms of two scalar fields $\alpha(\mathbf{r},t)$ and $\beta(\mathbf{r},t)$ via

$$\mathbf{B} = (\nabla\alpha) \times (\nabla\beta) \tag{2-1}$$

which clearly satisfies $\nabla \cdot \mathbf{B} = 0$. To show that eq. 2-1 is possible, recall that one can define lines of force by the equation

$$d\mathbf{r} \times \mathbf{B}(\mathbf{r},t) = 0$$

Let S be some surface nowhere tangent to the lines of force. In this surface choose a family of lines. The set of all lines of force through one of the lines of this family defines a surface. Let $\alpha(r,t) = \text{const}$ be the equation of such "magnetic surfaces." Now choose a second family of lines in S nowhere tangent to the first, and in a parallel manner associate with them a family of magnetic surfaces $\gamma(\mathbf{r},t) = \text{const}$. By construction

$$\mathbf{B} \cdot \nabla\alpha = 0 \qquad \mathbf{B} \cdot \nabla\gamma = 0 \tag{2-2}$$

and as follows directly from the above, since $\nabla\alpha$, $\nabla\gamma$, and $(\nabla\alpha) \times (\nabla\gamma)$ are noncoplanar on writing $\mathbf{B} = (\nabla\alpha) \times (\nabla\gamma)/\lambda + (\nabla\alpha)\mu + (\nabla\gamma)\nu$,

$$(\nabla\alpha) \times (\nabla\gamma) = \lambda\mathbf{B}$$

If one takes the divergence of the above equation and uses $\nabla \cdot \mathbf{B} = 0$,

$$\mathbf{B} \cdot \nabla\lambda = 0$$

and λ must be a function of α and γ. We shall now introduce a new variable $\beta(\alpha,\gamma)$. Clearly, $\mathbf{B} \cdot \nabla\beta = 0$. If we view γ as a function of α and β, and denote partial derivatives by the subscripts,

$$\begin{aligned}(\nabla\alpha) \times (\nabla\gamma) &= (\nabla\alpha) \times [\gamma_\alpha \nabla\alpha + \gamma_\beta \nabla\beta] \\ &= \gamma_\beta(\nabla\alpha) \times \nabla\beta\end{aligned}$$

We choose $\gamma_\beta = \lambda$. This yields the desired result

$$\mathbf{B} = (\nabla\alpha) \times (\nabla\beta) = \nabla \times (\alpha\nabla\beta) \tag{2-3}$$

Therefore, the intersection of any two surfaces $\alpha = \text{const}$ and $\beta = \text{const}$ is a line of force, and one can interpret the associated pair of values α, β as the coordinates of the line of force. Even though the pattern of lines of force may change in time, we shall identify that line labeled by a given pair α, β as the same line of force. The functions α and β need not be single-valued. See Figures 1, 2, and 3.

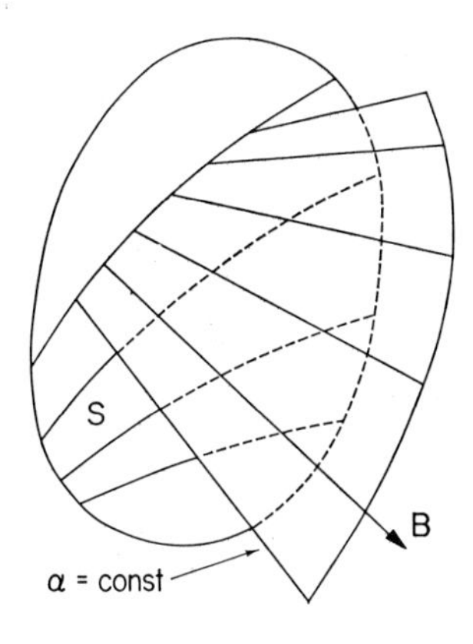

Fig. 1. Diagram illustrating the construction of surfaces. $\alpha = \text{const}$.

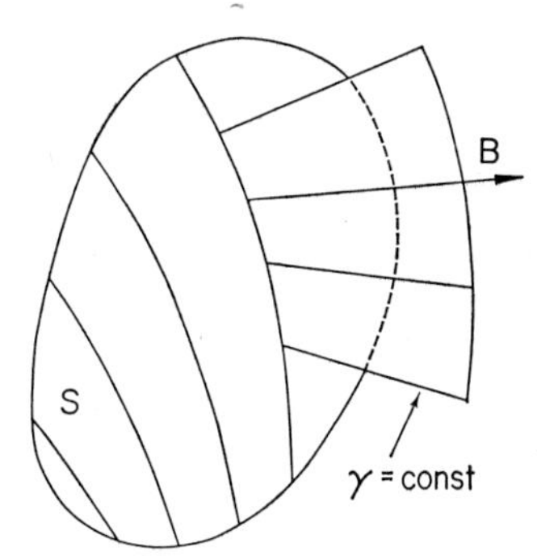

Fig. 2. Diagram illustrating the construction of surfaces. $\gamma = \text{const}$.

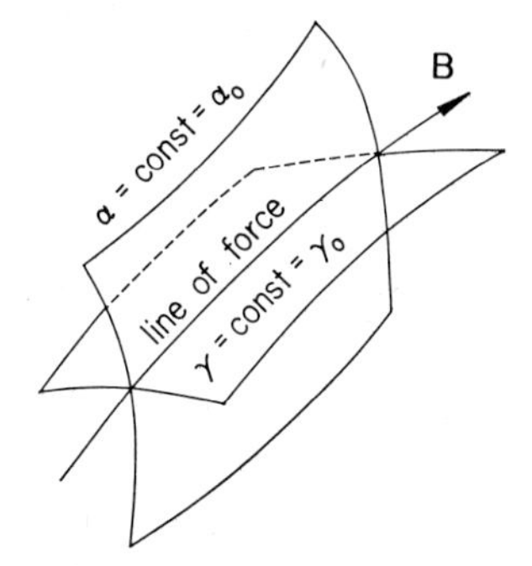

Fig. 3. Diagram illustrating the use of α, γ coordinates to label a line of force.

Now one can write the Maxwell equation

$$\begin{aligned} 0 &= \nabla \times \mathbf{E} + \frac{1}{c}\frac{\partial \mathbf{B}}{\partial t} \\ &= \nabla \times \left[\mathbf{E} + \frac{1}{c}\frac{\partial}{\partial t}(\alpha \nabla \beta)\right] \end{aligned}$$

whence

$$\mathbf{E} = -\nabla\phi - \frac{1}{c}\frac{\partial \alpha}{\partial t}\nabla\beta - \frac{1}{c}\alpha\nabla\frac{\partial \beta}{\partial t} \tag{2-4}$$

and

$$\begin{aligned} \mathbf{b}\cdot\mathbf{E} &= -\mathbf{b}\cdot\nabla\phi - \frac{1}{c}\alpha\mathbf{b}\cdot\nabla\frac{\partial\beta}{\partial t} \\ &= -\mathbf{b}\cdot\nabla\left(\phi + \frac{\alpha}{c}\frac{\partial\beta}{\partial t}\right) \end{aligned} \tag{2-5}$$

since both $\mathbf{b} \cdot \nabla\alpha = 0$ and $\mathbf{b} \cdot \nabla\beta = 0$. Thus, if one defines the potential

$$V(\mathbf{r},t) = q\phi + \frac{q\alpha}{c}\frac{\partial\beta}{\partial t} + \mu B + mG \tag{2-6}$$

the equation of motion for the parallel velocity, assuming that $m\mathbf{a} \times \mathbf{b}$ and $\mu\mathbf{b} \times \nabla B$ are of the same order of magnitude can be written to the lowest consistent order

$$m\dot{u} = -\mathbf{b} \cdot \nabla V + mu\mathbf{b} \cdot (\nabla\mathbf{b}) \cdot \dot{\mathbf{R}}_\perp \tag{2-7}$$

since to this order we make the parallel assumption that

$$\dot{\mathbf{b}} = \partial\mathbf{b}/\partial t + (u\mathbf{b} + \dot{\mathbf{R}}_\perp) \cdot \nabla\mathbf{b} \sim u\mathbf{b} \cdot \nabla\mathbf{b}$$

The associated expression for $\dot{\mathbf{R}}_\perp$ can be expressed as

$$\dot{\mathbf{R}}_\perp = \frac{\mathbf{b}}{m\Omega} \times \left[\nabla V + \frac{q}{c}\left(\frac{\partial\alpha}{\partial t}\nabla\beta - \frac{\partial\beta}{\partial t}\nabla\alpha\right) + mu^2\mathbf{b} \cdot \nabla\mathbf{b}\right] \tag{2-8}$$

Observe that, in the expression for $m\dot{u}$, the term involving $\mathbf{R}_\perp$ is ostensibly small compared with $-\mathbf{b} \cdot \nabla V$. We assume, moreover, that V_t is small in a sense that we shall make precise later.

Let s be the arc length along a line of force and suppose that V vs. s has the character of a potential well, as indicated schematically in Figure 4. When

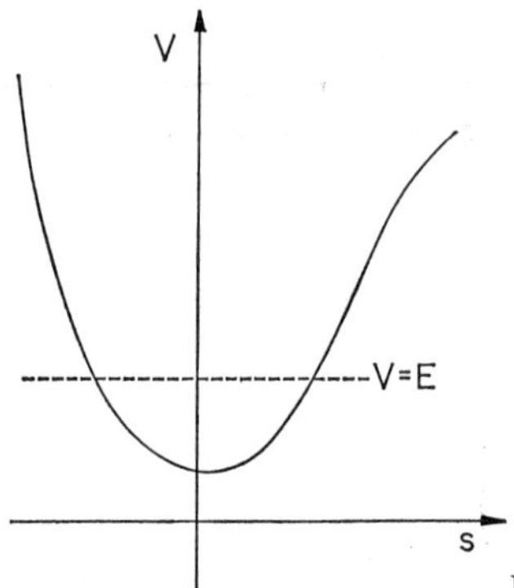

Fig. 4. A typical effective potential energy curve **V** vs. s.

both $\dot{\mathbf{R}}_\perp$ and V_t are zero, since

$$u = \dot{s}$$

there is a first integral of the equation of motion,

$$\tfrac{1}{2}m\dot{s}^2 + V = \text{const} = E$$

If one solves this for $\dot{s}$, it is easy to show that

$$t = \int^{s} ds\{2[E - V]/m\}^{-1/2}$$

Clearly, the motion is periodic with a period

$$\tau(E) = \oint ds\ \{2[E - V]/m\}^{-1/2}$$

The orbit of the particle in the s, $\dot{s}$ phase plane is the closed curve $E = \text{const}$. See Figure 5.

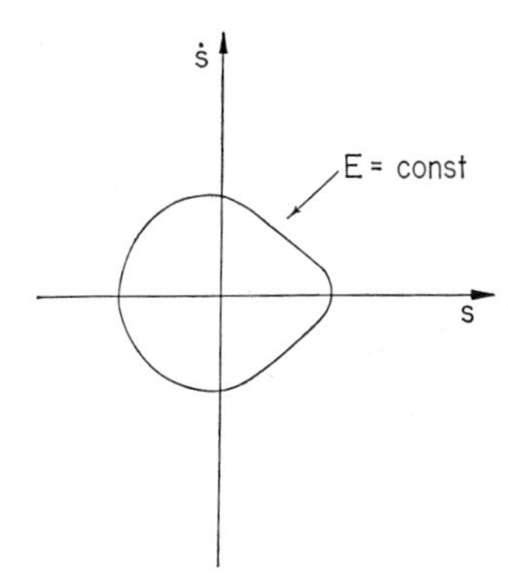

Fig. 5. A representative $\dot{s}$, s phase plane diagram for the case of constant E.

When $\dot{\mathbf{R}}$ and V_t do not vanish, the energy E will be a function of t. Suppose, however, that we extend the definition of the period $\tau(E)$ by means of the integral above to this case and assume that

$$\varepsilon = \tau|\partial \ln E/\partial t| + \dot{R}_{\perp}(2E/m)^{-1/2} \ll 1 \tag{2-9}$$

It seems plausible in this circumstance that the motion should be almost periodic. Let us assume so and seek a solution of the equation of motion via the introduction of an auxiliary variable $\theta(t)$ such that θ accounts for the rapid oscillation of period of the order τ, and any explicit dependence on t is associated with the slow time variation. That is, we write*

$$s = s(\theta,t) \tag{2-10}$$

whence if we define $\nu(t) \equiv \dot{\theta}$

$$\dot{s} = \nu s_{\theta} + s_t \tag{2-11}$$

* The symbols θ and ν are distinct from the quantities so labeled in the Introduction. We use the same symbols to illustrate the parallelism of the development.

where the subscripts denote partial derivatives. The equation of motion reads

$$v^2 s_{\theta\theta} + V_s = -2v s_{\theta t} - \dot{v}_t s_\theta - (v s_\theta + s_t)\mathbf{b} \cdot (\nabla \dot{\mathbf{R}}_\perp) \cdot \mathbf{b} \tag{2-12}$$

where we have used the fact that $\mathbf{b} \cdot \dot{\mathbf{R}} = 0$ to write

$$(\nabla \mathbf{b}) \cdot \dot{\mathbf{R}}_\perp = \nabla(\mathbf{b} \cdot \dot{\mathbf{R}}_\perp) - (\nabla \dot{\mathbf{R}}_\perp) \cdot \mathbf{b} = -(\nabla \dot{\mathbf{R}}_\perp) \cdot \mathbf{b}$$

The terms on the left-hand side above are presumably larger by a factor $1/\varepsilon$ than those on the right-hand side. Thus to the lowest order we require that the left-hand side above vanish. This requirement on multiplication by s_θ leads to

$$\left(\frac{m}{2} v^2 s_\theta^2 + V\right)_\theta = 0$$

whence on integration

$$\frac{m}{2} v^2 s_\theta^2 + V = E(t) \tag{2-13}$$

The constant of integration $E(t)$ is as yet unknown as a function of t. When one solves for s_θ from the above a further integration is possible, namely

$$\frac{\theta}{v} = \int^s ds\{2[E(t) - V(s,\alpha,\beta,t)]/m\}^{-1/2} \tag{2-14}$$

In the integrand we have indicated explicitly that the potential V depends on the point s on the line of force labeled by α and β, and by the time t. We have not indicated explicitly that it also depends on μ.

Let us pick $v = v(E,\alpha,\beta,t)$, so that θ is an angle variable; i.e., when s goes through one period of its motion for fixed α,β,t, we require that θ change by unity. Therefore,

$$\frac{1}{v} = \oint ds\{2[E - V]/m\}^{-1/2} \equiv \tau(E,\alpha,\beta,t) \tag{2-15}$$

and

$$\theta(t) = \int dt\, v \tag{2-16}$$

In order to determine $E(t)$, we revert to the equation of motion 2-13 and note that, if we retain terms to the next order in ε beyond that part which led to eq. 2-13, we find that

$$v^2 s_{\theta\theta} + V_s + 2v s_{\theta t} + v_t s_\theta + v s_\theta \mathbf{b} \cdot (\nabla \dot{\mathbf{R}}_\perp) \cdot \mathbf{b} = 0$$

If we multiply this equation by s_θ, the result can be written

$$(\tfrac{1}{2} v^2 s_\theta^2 + V)_\theta + (v s_\theta^2)_\theta + v s_\theta^2 \mathbf{b} \cdot (\nabla \dot{\mathbf{R}}_\perp) \cdot \mathbf{b} = 0$$

If we integrate this equation with respect to θ from zero to one, and recall that $s(\theta,t)$ is presumably periodic in θ with period one, we obtain

$$\frac{\partial}{\partial t}\int_0^1 d\theta v s_\theta^2 + \int_0^1 d\theta \mathbf{b}\cdot(\nabla\dot{\mathbf{R}}_\perp)\cdot\mathbf{b} v s_\theta^2 = 0 \tag{2-17}$$

Let us in the above equation use s as the variable of integration and recognize that to the lowest significant order we may use eq. 2-13 to express s_θ in terms of E and V. The equation then reads on multiplication by m:

$$\frac{\partial}{\partial t}\oint ds\,\{2m[E(t) - V(s,\alpha,\beta,t]\}^{1/2} + \oint ds\,\mathbf{b}\cdot(\nabla\dot{\mathbf{R}}_\perp)\cdot\mathbf{b}\{2m[E(t) - V(s,\alpha,\beta,t)]\}^{1/2} = 0 \tag{2-18}$$

Note that $\partial/\partial t$ acting on the first integral above means a time derivative holding the line of force fixed. We shall now show that $ds\mathbf{b}\cdot(\nabla\dot{\mathbf{R}}_\perp)\cdot\mathbf{b}$ is just the time rate of change of the element of arc length ds due to the velocity $\dot{\mathbf{R}}_\perp$. See Figure 6.

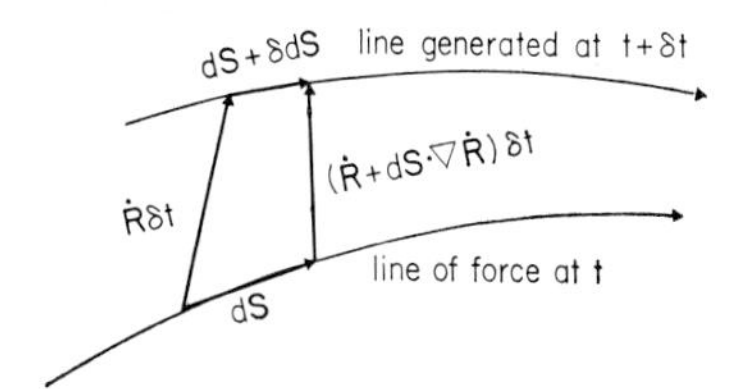

Fig. 6. Schematic diagram illustrating the calculation of the time rate of change of arc length along a line of force due to $\dot{\mathbf{R}}_\perp$.

Let us consider a vector

$$d\mathbf{s} = ds\mathbf{b}$$

In an infinitesimal time δt the end of $d\mathbf{s}$, as indicated in Figure 6, is carried a distance $\dot{\mathbf{R}}_\perp(\mathbf{R},t)\,\delta t$ by the guiding-center motion. The tip of $d\mathbf{s}$ is carried into

$$\dot{\mathbf{R}}_\perp(\mathbf{R} + d\mathbf{s},t)\,\delta t = [\dot{\mathbf{R}}_\perp(\mathbf{R},t) + d\mathbf{s}\cdot\nabla\dot{\mathbf{R}}_\perp(\mathbf{R},t) + \cdots]\,\delta t$$

The net change in $d\mathbf{s}$ is to lowest order

$$\delta d\mathbf{s} = d\mathbf{s}\cdot(\nabla\dot{\mathbf{R}}_\perp)\,\delta t$$

whence the square of the element of arc length is carried into

$$\begin{aligned}(d\mathbf{s} + \delta d\mathbf{s})^2 &= (d\mathbf{s})^2 + 2d\mathbf{s} \cdot (\delta d\mathbf{s}) + \cdots \\ &= (ds)^2 + 2ds\mathbf{b} \cdot [ds\mathbf{b} \cdot (\nabla \dot{\mathbf{R}}_\perp)\, \delta t] + \cdots \\ &= (ds)^2\, [1 + 2\mathbf{b} \cdot (\nabla \dot{\mathbf{R}}_\perp) \cdot \mathbf{b}\, \delta t + \cdots]\end{aligned}$$

Thus

$$ds + \delta ds = ds\, [1 + \mathbf{b} \cdot (\nabla \dot{\mathbf{R}}_\perp) \cdot \mathbf{b}\, \delta t + \cdots]$$

and in the limit $\delta t \to 0$,

$$\frac{\delta ds}{\delta t} = \mathbf{b} \cdot (\nabla \dot{\mathbf{R}}_\perp) \cdot \mathbf{b}\, ds$$

Equation 2-18 is then to be interpreted as a time derivative of the integral following the guiding-center motion, and

$$J = \oint ds\{2m[E(r) - V(s,\alpha,\beta,t)]\}^{1/2} \tag{2-19}$$

is an approximate constant of the motion. For a given value J and known potential V this expression is an implicit equation for E. The constant J is conventionally termed the second or longitudinal adiabatic invariant.

To recapitulate then, the motion along the line is determined by eq. 2-14 with E given by eq. 2-19. The motion perpendicular to the line is then given by $\dot{\mathbf{R}}_\perp$ (see eq. 2-8), where we may replace mu^2 by $2(E - V)$. To find the trajectory associated with $\dot{\mathbf{R}}_\perp$ requires only the solution of a second-order system of ordinary differential equations.

It is interesting to note that, if the technique of this section is applied to the equation,

$$\ddot{x} + \omega(t)^2 x = 0$$

corresponding to

$$V = \tfrac{1}{2}\omega^2 x^2$$

and

$$(\dot{\omega}) \ll \omega^2$$

then it yields the well-known, lowest order WKB results.

III. The Third or Flux Invariant

When the fields involved in $\dot{\mathbf{R}}_\perp$ are changing sufficiently slowly, a notion that will be made more precise later, a further reduction in the description is possible. To demonstrate this, it is convenient to write equations for $\dot{\alpha}$ and $\dot{\beta}$, instead of dealing with $\dot{\mathbf{R}}$. To this end we view $\dot{\mathbf{R}}_\perp$ as a function of s,α,β, and t and write

$$\dot{\mathbf{R}} = \dot{s}\mathbf{R}_s + \dot{\alpha}\mathbf{R}_\alpha + \dot{\beta}\mathbf{R}_\beta + \mathbf{R}_t \tag{3-1}$$

where the subscripts indicate partial derivatives. Moreover, by the chain rule for differentiation, if $\mathscr{I}$ denotes the unit dyadic,

$$\nabla \mathbf{R} = \mathscr{I} = \nabla s \mathbf{R}_s + \nabla \alpha \mathbf{R}_\alpha + \nabla \beta \mathbf{R}_\beta \tag{3-2}$$

whence on taking the dot product on the left-hand side with $\mathbf{b}$, one has

$$\mathbf{b} = (\mathbf{b} \cdot \nabla s)\mathbf{R}_s = \mathbf{R}_s \tag{3-3}$$

since $\mathbf{b} \cdot \nabla s = s_s = 1$. If one takes the dot product of eq. 3-2 on the left-hand side with $\mathbf{b} \times \mathbf{R}_\alpha$ and $\mathbf{b} \times \mathbf{R}_\beta$,

$$\mathbf{b} \times \mathbf{R}_\alpha = \nabla \beta \mathbf{R}_\beta \cdot \mathbf{b} \times R_\alpha \tag{3-4}$$

$$\mathbf{b} \times \mathbf{R}_\beta = \nabla \alpha \mathbf{R}_\alpha \cdot \mathbf{b} \times \mathbf{R}_\beta \tag{3-5}$$

The cross product of these two equations yields

$$\begin{aligned} -\nabla \alpha \times \nabla \beta (\mathbf{b} \cdot \mathbf{R}_\alpha \times \mathbf{R}_\beta)^2 &= (\mathbf{b} \times \mathbf{R}_\beta) \times (\mathbf{b} \times \mathbf{R}_\alpha) \\ &= \mathbf{b}\mathbf{R}_\alpha \cdot \mathbf{b} \times \mathbf{R}_\beta \end{aligned}$$

But, since $\nabla \alpha \times \nabla \beta = \mathbf{B} = B\mathbf{b}$, one has

$$\mathbf{b} \cdot \mathbf{R}_\alpha \times \mathbf{R}_\beta = 1/B \tag{3-6}$$

Now the dot product of eq. 3-1 with $\mathbf{R}_s \times \mathbf{R}_\beta = \mathbf{b} \times \mathbf{R}_\beta$ yields

$$\dot{\alpha} \mathbf{R}_\alpha \cdot \mathbf{b} \times \mathbf{R}_\beta = \dot{\mathbf{R}} \cdot \mathbf{b} \times \mathbf{R}_\beta - \mathbf{R}_t \cdot \mathbf{b} \times \mathbf{R}_\beta$$

or on using eqs. 2-8, 3-5, and 3-6,

$$\begin{aligned} \dot{\alpha}/B = & -\mathbf{b} \times \mathbf{R}_\beta \cdot (\mathbf{b}/m\Omega) \\ & \times [\nabla V + (q/c)(a_t \nabla \beta - \beta_t \nabla \alpha) + 2(E - V)\mathbf{b}_s] + \mathbf{R}_t \cdot \nabla a/B \end{aligned}$$

Since, by the chain rule for differentiation, $\mathbf{R}_t \cdot \nabla \alpha = \alpha_t$, $\mathbf{R}_\beta \cdot \nabla V = V_\beta$, $\mathbf{b} \cdot \nabla V = \mathbf{R}_s \cdot \nabla V = V_s$, $\mathbf{R}_\beta \cdot \nabla \beta = \beta_\beta = 1$, $\mathbf{R}_\beta \cdot \nabla \alpha = \alpha_\beta = 0$, while $\mathbf{b} \cdot \nabla \alpha = 0$, $\mathbf{b} \cdot \nabla \beta = 0$, and since $\mathbf{b}$ is a unit vector $\mathbf{b} \cdot \mathbf{b}_s = 0$, the above reduces to

$$\begin{aligned} \dot{\alpha} &= \alpha_t (\mathbf{R}_\beta - \mathbf{b}\mathbf{b} \cdot \mathbf{R}_\beta) \cdot (c/q)[\nabla V + (q/c)(\alpha_t \nabla \beta - \beta_t \nabla \alpha) + 2(E - V)b_s] \\ &= \alpha_t - (c/q)[\mathbf{R}_\beta \cdot \nabla V + (q/c)\alpha_t \mathbf{R}_\beta \cdot \nabla \beta - (q/c)\beta_t \mathbf{R}_\beta \cdot \nabla \alpha \\ & \qquad + 2(E - V)\mathbf{R}_\beta \cdot \mathbf{b}_s - \mathbf{b} \cdot \nabla V \, \mathbf{b} \cdot \mathbf{R}_s] \\ &= -(c/q)[V_\beta - V_s \mathbf{b} \cdot \mathbf{R}_s + 2(E - V)\mathbf{b}_s \cdot \mathbf{R}_\beta] \end{aligned}$$

But, on recognizing that $\mathbf{R}_{\beta s} = (\mathbf{R}_s)_\beta = \mathbf{b}_\beta$, and $\mathbf{b} \cdot \mathbf{b}_\beta = 0$, this can be written

$$\dot{\alpha} = -(c/q)\{V_\beta + [2(E - V)/m]^{1/2}([2m(E - V)]^{1/2} \, \mathbf{b} \cdot \mathbf{R}_\beta)_s\}$$

Finally, if we introduce the angle variable θ in place of s, since

$$ds = [2(E - V)/m]^{1/2}\,\tau\, d\theta,$$
$$\dot{\alpha} = -(c/q)[V_\beta + \{\tau^{-1}[2m(E - V)]^{1/2}\,\mathbf{b}\cdot\mathbf{R}_\beta\}_\theta]$$

If one integrates this expression over one period in θ,

$$\int_0^1 d\theta\dot{\alpha} = -(c/q)\int_0^1 d\theta V_\beta$$
$$= -(c/q)\tau^{-1}\oint ds\{2(E - V)/m\}^{-1/2}V_\beta \tag{3-7}$$

The right-hand side of eq. 3-7 can be related to the energy E, as defined implicitly by eq. 2-19, and considered to be a function of α,β,t, and of course the constants of the motion J and μ. If we take the partial derivative of eq. 2-19 with respect to β, we find that since J is an independent parameter,

$$0 = \oint ds\{2(E - V)/m\}^{-1/2}[E_\beta - V_\beta]$$

or on using eq. 2-15,

$$\oint ds\{2(E - V)/m\}^{-1/2}V_\beta = \tau E_\beta$$

Thus, if we interpret $\int_0^1 d\theta\dot{\alpha}$ as the time derivative of the average value of α associated with a particle over a period τ, we can cast eq. 3-7 in the form

$$\dot{\alpha} = -(c/q)E_\beta \tag{3-8}$$

In similar fashion we can show that

$$\dot{\beta} = (c/q)E_\alpha \tag{3-9}$$

where it is to be emphasized that α and β are the coordinates of the mean line of force on which the particle is gyrating and oscillating.

The equations of motion for α and β are in Hamiltonian form with $E(\alpha,\beta,t)$ playing the role of a time-dependent Hamiltonian. When $E_t = 0, E$ is a constant of the motion, and the orbit in the α,β phase plane is the curve $E = \text{const}$. Suppose that this orbit is a closed curve, as shown schematically in Figure 7. Then the motion is periodic with period

$$T = (q/c)\oint d\beta/E_\alpha = (q/c)\oint d\alpha/E_\beta \tag{3-10}$$

Suppose that

$$T|\partial \ln E/\partial t| \ll 1$$

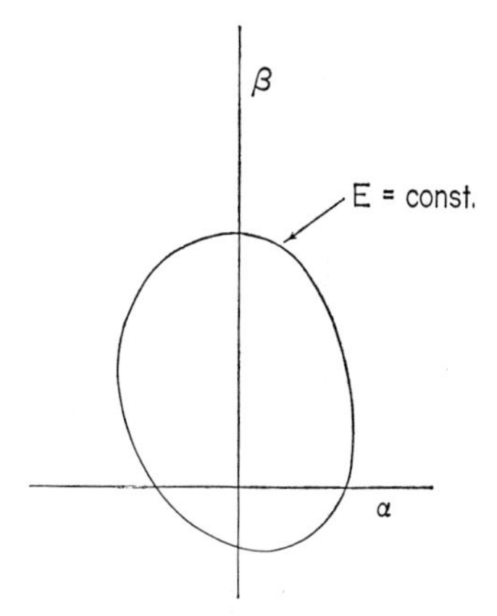

Fig. 7. Schematic diagram of constant flux surface illustrating a particle trajectory therein.

We anticipate as in the former cases that the motion will be almost periodic; we introduce an auxiliary variable $\chi(t)$ and write

$$\alpha = \alpha(\chi,t) \qquad \beta = \beta(\chi,t) \tag{3-11}$$

If we define $\omega(t) = \dot{\chi}$, we can write the equations of motion as

$$\omega\alpha_\chi + \alpha_t = -(c/q)E_\beta \tag{3-12}$$

$$\omega\beta_\chi + \beta_t = (c/q)E_\alpha \tag{3-13}$$

To the lowest order we delete the ostensibly small term α_t and β_t and note that then

$$\begin{aligned}(c/q)E_\chi &= (c/q)[\alpha_\chi E_\alpha + \beta_\chi E_\beta] \\ &= \alpha_\chi\, \omega\beta_\chi + \beta_\chi(-\omega\alpha_\chi) \\ &= 0\end{aligned}$$

Thus to this order

$$E = H(t) \tag{3-14}$$

where the constant of integration $H(t)$ is as yet undermined. Let us choose $\omega = 1/T$, where T is defined by eq. 3-10 but with the integrals extended over the closed curve $E = H$. This makes χ an angle variable, and one can formally integrate the approximate equations of motion, eqs. 3-12 and 3-13 with α_t and β_t deleted, to obtain

$$\chi T = (q/c)\int^\alpha d\alpha/E_\beta[\alpha,\beta(\alpha,t),t] = -(q/c)\int^\beta d\beta/E_\alpha[\alpha(\beta,t),\beta,t]$$

where $\beta(\alpha,t)$ is determined from $E(\alpha,\beta,t) = H(t)$, etc.

In order to determine $H(t)$, we note that, without approximation of the equations of motion,

$$\begin{aligned}(c/q)E_\chi &= \alpha_\chi(c/q)E_\alpha + \beta_\chi(c/q)E_\beta \\ &= \alpha_\chi(\omega\beta_\chi + \beta_t) - \beta_\chi(\omega\alpha_\chi + \alpha_t) \\ &= \alpha_\chi\beta_t - \beta_\chi\alpha_t \\ &= -(\alpha\beta_\chi)_t + (\alpha\beta_t)_\chi\end{aligned}$$

Thus, if we integrate this result with respect to χ from zero to unity, we obtain

$$\left(\int_0^1 d\chi\, \alpha\beta_\chi\right)_t = 0$$

and to the lowest significant order

$$\psi = \oint d\beta\, \alpha \tag{3-15}$$

is an approximate constant of the motion, where the integral is extended over the closed curve $E = H$.

We shall now show that ψ is a magnetic flux. To demonstrate this, we note that the flux crossing any surface in x,y,z space is, on using Stokes theorem and $\mathbf{B} = (\nabla\alpha) \times (\nabla\beta) = \nabla \times (\alpha\nabla\beta)$,

$$\begin{aligned}\int d^2\mathbf{r} \cdot \mathbf{B} &= \int d^2\mathbf{r} \cdot \nabla \times (\alpha\nabla\beta) \\ &= \int d\mathbf{r} \cdot \alpha\nabla\beta \\ &= \int d\beta\, \alpha\end{aligned} \tag{3-16}$$

The line integral above is extended over any closed curve resulting from slicing the magnetic surface defined in x,y,z space by the equation $E(\alpha,\beta,t) = H(t)$, as shown schematically in Figure 8.

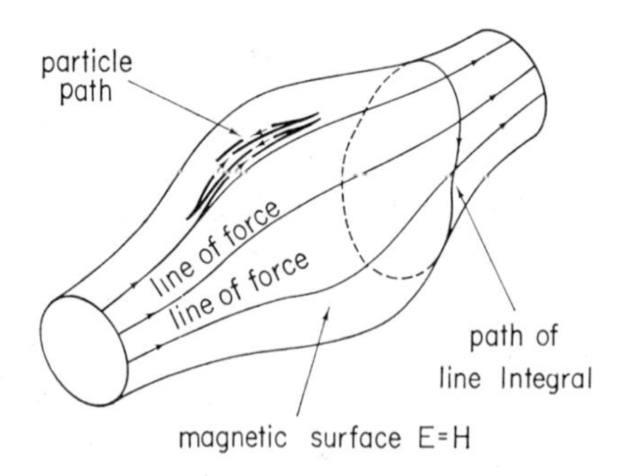

Fig. 8. Schematic diagram indicating particle path and line integral path in magnetic surface $E = H$.

Clearly, ψ is independent of the choice of line as long as it is topologically equivalent to that shown above. Equation 3-15 is then to be viewed as determining $H(t)$ implicitly, given ψ. The approximate constant ψ is conventionally termed the third adiabatic invariant, or alternatively the flux invariant.

Acknowledgment

The author is indebted to Professor E. B. Hooper, Jr., and Peter Cato for their careful reading of the manuscript. The work was supported in part by Atomic Energy Commission contract AT 31-1-3943.

References

1. H. Alfvén, *Cosmical Electrodynamics*, Oxford University Press, New York, 1950.
2. M. D. Kruskal, *J. Math. Phys.*, **3**, 806 (1962).
3. T. G. Northrup, *The Adiabatic Motion of Charged Particles*, Interscience, New York, 1963.
4. B. Lehnert, *Dynamics of Charged Particles*, North Holland, Amsterdam, 1964, pp. 26 ff.
5. J. Berkowitz and C. Gardner, *Communs. Pure and Appl. Math.*, **12**, 501 (1959).

Author Index

Numbers in parentheses are reference numbers and show that an author's work is referred to although his name is not mentioned in the text. Numbers in *italics* indicate the pages on which the full references appear.

Subject Index

Advances in Plasma Physics

Cumulative Index, Volumes 1-4

Author Index

Subject Index